# CHANGING

## *On the Biodynamics of Moral Courage in Turning Points of Love and Wisdom*

*By*

**Ira Rosenberg**

ISBN: 978-0-578-01184-4

BIOFEEDBACK
INSTITUTE   OF   MENDOCINO
ALBION BIOME PRESS

irarosen@mcn.org
www.irarosenberg.com

*In memory of my father, George Rosenberg*

.

# TABLE OF CONTENTS

# LOVE, WISDOM AND AGGRESSION

# 1.INSIDE HUMAN NATURE

## 1

The story we tell ourselves about ourselves is always one of opportunities taken or missed. We don't dwell on how we posed or strutted or thought or felt, but on what we did or didn't do. And we generally agree on the things that matter most. Inwardly, in the actual texture of our lives as lived, we struggle with the same priorities, because we share common physiological functions that cause us to experience, in every historical and cultural setting, the same passions and virtues. I hope to show that more than any other of our endeavors, love and wisdom, whether we succeed or fail in them, whether our cultures honor or suppress them, definitively shape our sense of self-worth and tune our enthusiasm for life.

Love or the search for it is the animating force in social life. Wisdom is the fruit of solitude. No solitude, no wisdom. No sociability, no love.

I think of love and wisdom as the vehicles that carry moral sentiments into the world. They're not moral sentiments themselves; they're biological imperatives, primordial rhythms that *predate* ethics. They even predate choice. The moral sentiments they carry can vary enormously in content, strength and resiliency depending on individual temperaments and historical needs.

**2**

Love is an Eros centered process that oscillates on alternating legs of approach and separation. Wisdom is a Logos centered process that moves by successive inward withdrawals and outward returns. Through love, we find each other. In wisdom, we discover ourselves. Love starts with a greeting. Wisdom begins with a departure. While love finds its peak experiences in close communion, wisdom gets its powers when a person descends alone from the surface of life to the depths of inwardness to make a discovery there and return with it. Alchemists called the process "rectification through going inward."

Both processes are dynamic engines of change, not states of being. They belong to the *vita activa* as *unfoldings* in time. Only in retrospect do they enter the *vita contemplativa* as truths to ponder.

**3**

Why do love and wisdom matter so much? What makes them so memorable to us? They matter because they employ our evolved physiology in ways that generate rhythmic resonances across many body systems. We *experience* their biological rightness with an augmentation of energy that, in its rhythmic consanguities, transmits a feeling of *meaningfulness*.

The strength and wholeheartedness with which we enter our turning points in love and wisdom is a measure of their rhythmic integrity. And rhythmic integrity can be measured in specific frequency bands that we will uncover in the course of our investigation (see # **7, 14, 19, 20**).

**4**

Love and wisdom are real only as 'round-trip' events. They move on legs to turning points. They intensify and reverse. The leg of approach leads to separation. Inward withdrawal provokes outward return.

| The Social Rhythm of Approach and Separation | The Solitary Rhythm of Withdrawal and Return |

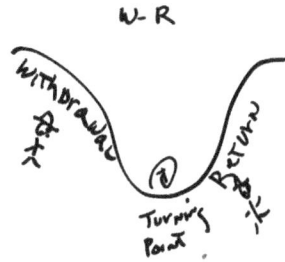

The phrases "perennial wisdom" and "eternal love" suggest that we climb the virtues like ladders to states of high constancy. But we don't. Love advances through reversals. Wisdom travels through stages marked by turning points. No single stage can stand for the whole. The worthiness resides in the completely unfolding pattern. The approaches, separations, withdrawals, and returns get their meaning because they join into dramas of engagement and discovery. The dramas, though they disclose themselves as events in time, can be illustrated graphically. I have sketched out the main patterns in still pictures. You can put them into motion in your mind.

Both love and wisdom are vulnerable to perturbation at the links between their mini-cycles, as illustrated to the left, and even more so, at the major turning in the pattern, the turning of turnings, where the temporal shape of the whole emerges from the congeries of its parts.

**Mini Turnings, vulnerable moments and Turning of Turnings**

### 5

Sex, belonging, touch, and affiliation all move in the rhythms of approach and separation built out of briefer units tied to frequencies in nature. When new life is conceived, always in closest approach, it develops with its own rhythmic course in a process full of frequency dependent stages leading to a specific gestation period followed by birth.

Some species only come together to mate. Others express approach/separation as they sleep and rise, feed and rest, travel and settle, attack and defend together. Sometimes the approach and separation occur in great flocking or herding populations. In existing human hunter-gatherer bands, group members separate to forage and hunt and then come together in the home camp to share food, maintain family life, socialize and sleep. Since the base-camp and the feeding range are often seasonal or even more temporary, the oscillations are tied primarily to the group members themselves, not to the place the group occupies; they approach and separation from and to each other; the particulars of the places can change but the good order of the troop persists.

Sam Keen caught the sense of the polar rhythms of approach and separation well when he wrote in *The Passionate Life*, "Loving is a continual dance between bonding and returning to our boundaries, coming together and going apart."[1] Distance and closeness enter not extrinsically but at the heart of the rhythm of nature itself.

**6**

Where the stimuli for approach and separation come mainly from outer sensation and involve sensory following and motor responses, the dominant sensations for withdrawal/return come from inside the body out of social view. Wisdom's venue is rumination, daydream, dream life, recollection, cognition, reflection, imagination, meditation and contemplation.

The journey of withdrawal and return manifests as a repeating series of inward-and-outward movements, pulsations between the surface of life and the depths, a pattern that weaves on until, after many dips and surfacings extending over many months, like fractals set in time, a larger inward-outward pattern forms.

Though we have internal sensory experiences regularly, in some situations, and at certain stages of life, they dominate the screen of consciousness, obscuring or even obliterating external sensory streams. This throws us into long inward journeys. Well conducted, they take us to turning points in our own depths. From them, by gradual shifts to outward sensory interests, we make our way back to the consensual world. The inner adventure changes us. Sometimes we are renewed by it. In happy circumstances, the world welcomes us back.

**7**

People vary greatly in the number of mini-cycles that come together to constitute their love lives or wisdom quests. The duration of the individual cycles and the overall pattern, as well as its susceptibility to perturbation depends on individual differences in constitution, upbringing and education. And chance plays a part.

Despite our individual differences, the component rhythms of love and wisdom do not play out at all possible tempos in everyone. They cluster within certain common frequency bands.

These frequency bands reflect the rhythms of nature to which we were tuned by our early evolution.[2] That is to say, by long processes of selection, organisms have come to mirror the rhythmicity of the earth environment. Its planetary movements, tectonic rumblings, magnetic moments, flowing water, cyclonic winds and radiant solar energies pulse through us.

## 8

We live in far-from equilibrium conditions. Tipping points occur in life. But not all tipping points are turning points. A tipping point becomes a turning point only when the sensitive nonlinear bio-oscillating systems of which we are composed respond to perturbations with rhythmic restoration in love or wisdom. We can speculate that this happens because:

1) Our thoughts and actions and all their cerebral, neural, humoral and skeleto-muscular components are designed to move in specific frequency bands with phase relationships to other oscillators;
2) That these unfold as events in a field of nature that is itself resonant,
3) That reality is structured in such a way as to generate resonances and dissonances that tune human nature (and all living natures) to the environment and to each other, and that
4) Love and wisdom are the highest expressions of these dynamics because they command the adaptive peaks of our most consequential solitary and social undertakings.

## 9
### *Perturbed Rhythms*

With their many contrasting frequencies, the oscillators in love and wisdom can clash in complex ways. We do not tick with the precision of quartz watches. External influences can draw our bio-oscillators away from their resident frequencies. Their phase relationships to other oscillators in the body can shift. Frequencies can pull on each other. In the interpenetrating fields of multiple vibrations that constitute inner and outer life, all kinds of dissonant interactions can disrupt body functions.

The complex phase relationships between the 24-hour day and the 25-hour drift of the circadian clock or the 27 and 29 day solar and lunar months, themselves may produce beat frequencies between body systems. The persistence of these out-of-phase harmonic relationships over huge spans of evolutionary time suggests that organisms have made use of them to adjust or reset their biological clocks.

However, there are limits to our resilience. Beyond a certain range of frequencies, beats and dissonances may derange the body, disrupt communications and degrade the instructions they bring from one body system to another. Derhythmization can lead to illness, ecological disasters or extinctions.

**10**

Glass and Mackay illustrate simple perturbations in an experiment that analyzes

"The effects of a brief electrical shock delivered to an aggregate of spontaneously beating cells derived from the ventricles of an embryonic chick heart. In response to a brief electrical stimulus, there is a resetting of the phase of subsequent action potentials, but the original cycle time is reestablished within several beats... Poincare called such oscillations *stable limit cycles*... If, on the other hand, a small perturbation induces a change in the dynamics so that the original dynamics are not reestablished, then the steady state or limit cycle is *unstable*.... Any value of a parameter at which the number and/or stability of steady states and cycles change is called a *bifurcation point* and the system is said to undergo a *bifurcation*... Since small changes in parameters lead to qualitatively different dynamics at bifurcation points, systems are not structurally stable at bifurcation points."[3]

Arthur Winfree, a pioneering biophysicist and chronobiologist, makes the case that small perturbations to rhythms, when precisely timed, can have big effects. In his studies on cardiac arrhythmias, especially ventricular fibrillation, he observed, "a small local depolarization can trigger transition to fibrillation...If the stimulus is too big or too small or wrongly timed, it won't work. There is an atrial vulnerable phase and, later, a ventricular vulnerable phase."[4]

An oscillator's vulnerability depends in part on its phase at the moment the perturbing force hits. They are more vulnerable at some times and less at others.

Lacey and Lacey, whose work Jerome Kagan draws on in his studies of childhood temperament, researched the timing of perturbations in the heartbeat cycle. Hit when they are most vulnerable, biological rhythms are forced to change. They change through rhythmic resetting or rhythmic annihilation. As Winfree demonstrated, it takes only a small, isolated stimulus to produce a resetting that changes the whole system.

## 11

Biophysicists call the rhythm-resetting moments, when they come, bifurcation points. They have been studied across an enormous range of physical, chemical, biological, social and even economic systems. Paul Davies describes them clearly: "At the bifurcation point," he writes, "the inescapable fluctuations, which in ordinary equilibrium thermodynamics are automatically suppressed, instead become amplified to macroscopic proportions, and drive the system into its new phase which then becomes stabilized."[5] We will encounter these fluctuations on all levels, from cells to civilizations. They work their way through human life in every venue. Possibly every biological rhythm by the time we get to see it is already perturbed, reset, become a compound rhythm, a carrier of near and distant echoes, timbres and modulations. You rarely encounter a timbre-less tone in nature or a pure sine wave in music.

**Hopi Journey Myth Shows Bifurcation Point**

## 12

I tentatively conclude from this that our vulnerability to perturbation, because it is itself subject to natural selection, has evolved to give us tools for restoring or replacing rhythms under varying assaults from inside and outside. Since rhythmic organization is the foundation for many life processes (and for the human meaning bestowing experiences we have already touched on) a selective process to restore rhythm to disturbed body processes, probably through epigenetic signals to clock genes, would have increased our fitness. This could explain why rhythm and phase resetting capabilities are found everywhere in nature.

We are creatures built of multiplexed, nonlinear bio-oscillating systems partly coupled to the surrounding medium, partly resistant to it and partly rhythmically dominating it. This is to say that when we are perturbed we are adapted to respond to it. We try to resist disruption. We make efforts to restore rhythmic functioning. We will fight off or get away from the perturbing forces. Our physiology is designed to return to homeostasis. A robust response to perturbation allows for adjustment, repair and renewal. The plasticity in the nervous system, the duplications of pathways, the alternate routes in the brain, the cascading interactions of biochemical processes on regulatory genes all cope with perturbation.

## 13

The most crucial messages sent and received in the rhythmic fields of life, from its most primordial expressions on, have to do with love and wisdom, the principal focus of the book. In primitive organisms these take the form of rudimentary approach/separation and withdrawal/return oscillations. The biochemical signals transmitted by these same dynamics tell us whether we are alone or not, coherent or not, safe or not. The perturbations to those rhythms, when they hit, raise alarms. All species have evolved means for restoring rhythmic functioning. The physiology in them regulates the temporal aspects of homeostasis in living organisms. To understand love and wisdom, then, even inwardly, as experienced, we have to see how we fit in primal nature generally.

## 14
### *Planetary Nature*

Look at it ecologically. That's where it starts. We live in a vibrating world whose waves and rhythms surround and course through us. The length of the day, the seasonal cycles of light and dark, variations in solar radiance, geomagnetic, electrostatic, hydrodynamic and many other frequencies are in our makeup. The month is in the menstrual cycle, the day is in the sleep/waking cycle. Even the simplest cyanobacteria have circadian clock genes.[6] All organisms without exception live by rhythm.[7]

Arthur Winfree described the coupling of environmental rhythms to the rhythms of life this way:

"We began with daily time organization in a piece of rotating machinery none of whose parts (mountains, oceans) has an intrinsic daily rhythm; but coupled together and pulling one another along in sequence, they constitute a clock. We passed through wide-ranging studies of chemical life-forms evolved on the surface of that clock, finding many chemical oscillators."[8]

Many of them move in frequencies associated with the hydrological cycle: ocean currents, tidal movements, temperature change, barometric pressure, evaporation, weather and climate.

**Global Ecological Rhythms**          **Global Water Rhythms**

Rhythmic influences are as fundamental to life as are the atoms of carbon, oxygen, nitrogen and hydrogen that give it mass. However, they are not, strictly considered, material *entities*. They are processes, patterns of movement that ride on matter. Our gait is rhythmical, our heart and breath interact rhythmically. Respiration inside the cells is rhythmical. Catabolism and anabolism, sodium/potassium transport through the cell membrane, nerve conduction, and the acid/base oscillations in blood chemistry all have oscillatory characteristics, reflecting prebiotic solar and hydrological rhythms (See # **24-26**.)

When we reach out to say or do something our words and gestures are pulsatile and move in specific frequency ranges. Every spoken word that comes out of us we deliver on an exhalation; the heart registers every arousal, and all musical tempos pulse at heart rates from largos of 40 BPM to prestos of 240 BPM. Melodic phrases follow breath lengths. All the rhythmic patterns of music conform to locomotion, gestures and dexterity. Every larger pattern of behavior that builds from these components in every animal species, from mating to feeding, fighting and healing, plays out rhythmically, with wave fronts interfering and entraining each other, amplifying, reflecting, diffracting, all in pulsatile movements that we can quantify by frequency and amplitude and locate in phase relationships to other oscillators in nature.

It is not the case that waves are secondary to the sea, or the heartbeat to the heart, or the spoken word to the vocal cords or music to the instruments that play it.

If you make this wrong assumption, you commit what Alfred North Whitehead called the *fallacy of misplaced concreteness*.

## 15

So far our electron microscopes have only let us view cellular functions in frozen stillness on the smallest scale, where they are breathlessly beautiful. We can even see molecules. As soon as new techniques in scanning microscopy let us do real-time live cell imaging on the molecular scale, the music will become apparent, unavoidable really. It will strike us when we peer into the cell and see its microfilaments and microtubules moving, growing, waving, pruning back, vibrating like cell harps. Scientists will want to figure out the temporal relationships in those and many other cellular processes. From that moment, the balance between the material substrate and the oscillations it contains, between the instrument and the music it makes, will become more apparent. We will understand life then as an event in time, resonant on every level. Moreover, discerning the musical qualities inherent in change — for music is the art *par excellence* for representing change through

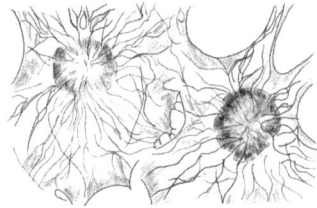

**Cell Harp Microfilaments**

time – we will develop tools to analyze biological processes in harmonic and octave relationships.

## 16

We can do this using a mathematics to the base two, rather than base ten, to quantify change, because it more faithfully represents the bifurcations, branching patterns, period doublings and halvings inherent in living nature. Using computerized musical analysis programs, we will develop methods that quantify nonlinear bio-oscillators clashing in complex ways. We will register out of tune notes, notes that are not quite noise, as beat frequencies. The influences of waveforms on each other, conceived biophysically, will link the non-living and the living world in a new way, joining physics with biology, biology with behavior, and behavior with culture.

The *I Ching* uses powers of two to express change. In the trigrams and hexagrams, there are only two symbols: the solid line and the dashed line. The solid line is about to bifurcate, the dashed line has already bifurcated and is in its resting condition, ready to come together. We can represent every curve on a waveform unfolding in time by trigrams built on powers of two. The notation makes perfect sense.

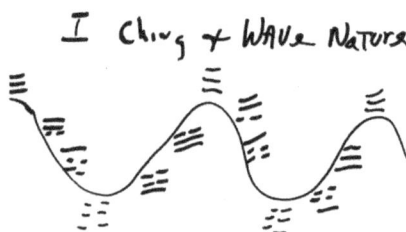

## 17

I'm arguing that to understand love and wisdom, you must recognize that integral nature, with us in it, is built on a thru-line of rhythm. Bio-oscillating systems interact in an ecology of rhythms. Organisms engage with each other by transferring their frequency and waveform characteristics. Winfree explained that "despite the variety of mechanisms that underlies rhythmic timing and its control in living organisms, several modest generalizations stand out. For example, any spontaneously rhythmic mechanism ('oscillator') can be expected to lock on to the nearby period of a strong enough rhythmic influence... This ability of most nonlinear oscillators to entrain or synchronize rests on their time-dependent sensitivity: exposed to some standard disturbance beginning at different times in the cycle, there will be different phase shifts inflicted."[9]

Rhythmic entrainment, resonance, dissonance, frequency pulling, constructive and destructive interference, and other kinds of modulation, link species together. In the simplest couplings, two rhythms fall into phase with each other and draw other close frequencies in. Frequency matching must have entered into evolutionary selection from the beginning. Species living in resonant relationships with each other, it follows, move, ship materials, transmit energy and share information in common frequency ranges.[10]

Interactions that work between pairs of oscillators also work in larger arrays, stimulating biochemical reactions between as well as within individuals, including those of different species.

Browsing animals follow the ripening grasses, plankton float and sink to the changing temperature gradients of the sun-warmed waters. Predators follow prey animals whose feeding patterns follow the vegetation. These rhythmic patterns repeat year after year. If the grass is late, the browsing herds are late. If the climate changes, the rhythms reset. At the mouth of the Navarro River near where I live, for instance, the sea birds gather to feed on the salmon fry as they swim down river at the end of the summer. But the fish can't get out to sea until the sand bar that blocks the river mouth breaks down with the high tides and the first fall rains. The species are rhythmically entrained through the weather and seasons.

## 18

Christian De Duve described the basic information and energy transfers that go on in these interactions. One or the other has to apply:

o One oscillator may provide or remove ingredients required by another, creating positive or negative feedback loops.

o They may intrude on one another's space of operation, take up room, and block activity, as in receptor site blocking.

o They can add or subtract energy. Temperature increases generally speed reactions.

o One oscillator may agitate the medium in which another oscillator functions. Wind on water influences the patterns of surface planktonic life.

o They may have their own clock cycles that open or close gates, adding or blocking materials or energy needed by another oscillator.

The same energy, information and material transfers undergird love and wisdom. They come into play as emergent properties of the evolution of the primordial rhythms themselves. In the rest of the chapter I will outline an approach to the life sciences that I call evolutionary chronobiology. If you only want to hear about love and wisdom skip ahead to chapter two. But if you stick with me now you will get the grounding you need to understand them better, most importantly as they function in your own life.

## 19

Presently only the day-length 'circadian rhythms' have been studied for their evolutionary links to environmental time cues. Molecular biologists have found them in clock genes that undergird the sleep/waking cycle, body temperature regulation, blood pressure and many other functions across phyla. Our very sense of ourselves is linked to the 24-hour day by internal clocks located in the super-chiasmic nucleus of the anterior hypothalamus. Certainly, the alternation between waking and sleep is the most telling existential expression of our rhythmicity. The contrasts between dream and waking life have long been subjects for philosophical inquiry.

However, circadian clocks are far from the only frequency driven processes in human physiology. Biologists have identified ultradian and infradian rhythms, biological oscillations moving faster and slower than a day respectively, working in animal life. The human menstrual cycle is an obvious example of a slower rhythm. Animal heart and breath rates move in the faster ultradian frequency bands. There are ultradian clock genes in the Saccharomysces cerevisiae yeast that orchestrate a 40-minute respiratory process that prepares the yeast for cell division.[11] C. elegans, biologists' favorite worm, has a 45-second defecation cycle controlled by an ultradian clock whose gene has been located. The fastest rhythms set the beat for biological processes on the smallest scale. They pervade our cellular chemistry and in some way must underlie our capacities for consciousness and through it for love and wisdom.

Biologists have calculated the rates and rhythms of many ultradian oscillators. They have studied the effects of perturbation on them. Nevertheless, the evolutionary sources of the ultradian rhythms and their perturbations, and the evolved responses we make to those perturbations, have rarely been sought.[12] Researchers have paid very little attention to the larger ecological, behavioral and moral consequences of actually living in a resonant world.

## 20

Of the ultradian frequency clusters, I consider the following the most important in our study of human virtues and passions.

o   The 90-minute rest/activation cycle, coordinated in the pontine reticular activating system in the brain, identified by Nathan Kleitman in the sixties, a cycle that runs continuously through the day and shapes the sleep stages and enters into REM sleep and dream life and so touches on the resources of human imagination. The same period is found in the somite deposition in vertebrate embryology.[13]

o   The 20-minute cell division cycle conserved from prokaryotes and carried into more complex eukaryotic cell cycles as the M phase in mitosis. Rossi and Lippincott see this as a trigger or a latency period for the basic rest/activation cycle. According to their literature search, "A 20 minute building up period of mediating

17

factors (maturation promoting factor and H1 kinase) is required to trigger the 90-120 minute process of genetic replication in the complete 24 hour circadian cell cycle in eukaryotes."[14] Murray, Solomon and Kirschner, quoted in Rossi and Lippincott tie the biology to behavior this way: "The neuroendocrine system is now well recognized as having prominent ultradian and circadian components related to a variety of psychobiological behaviors associated with mental and physical activity, nutrition, metabolism and reproduction. There are experimentally verifiable 20-minute couplings between peaks of associated hormones that are released in approximately 90-120 minute ultradian rhythms: luteinizing hormones peaks lead prolactin and testosterone peaks by 10-20 minutes (Veldhuis et al. 1987. Veldhuis and Johnson, 1988, Veldhuis, this volume); Glucose leads insulin by 15-20 minutes (van Cauter et al, 1989); cortisol leads B-endorphin by 20-30 minutes (Iranmanesh et al, 1989). These associations that extend from the molecular-genetic generation of these hormones at the cellular level to their expression at the neuroendocrinal level and their interaction with the mind-brain processes of memory, learning and behavior described below can hardly be accidental."

o The interaction of heart and breath, rhythmically driven by the respiratory sinus arrhythmia (RSA) organized in the medulla of the brain. Though it shows a good deal of cycle to cycle variability, heart rate variability is important to our study because it feeds and drives emotional states, paces relaxation, pushes anxiety, and contributes enormously to a moment to moment sense of ourselves. It entrains other body rhythms related to blood flow, blood pressure, fluid shifts, vasomotor and kidney functions. A power spectrum analysis of heart rate variability shows the presence of all of these systems. Jerome Kagan treats these conditioned cardiac cycle patterns as formative factors in a child's temperament.

o Calcium oscillations conserved across many phyla carry coded signals across cell membranes and bear instructions that start or stop or change the rate of re-actions. Spontaneous calcium oscillations have been discovered in the astrocytes of human brain cells. As-trocytes respond to chemical, electrical and mechani-cal stimuli with transient increases in intracellular calcium concentration ($[Ca2+](i)$).[15] Parri et al write that "astrocytes in situ display intrinsic $[Ca2+](i)$ oscil-lations that are not driven by neuronal activity. These spontaneous astrocytic oscillations can propagate as waves to neighboring astrocytes and trigger slowly de-caying NMDA receptor-mediated inward currents in neurons located along the wave path. These findings show that astrocytes in situ can act as a primary source for generating neuronal activity in the mammalian cen-tral nervous system."

These rate-determining factors play a role in syn-aptic neurotransmitter release and reuptake and in the balances between the neurotransmitters themselves, and these balances have distinct emotional and behav-ioral consequences.

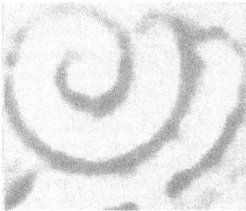

Calcium Wave
Microphotograph
from Jaffe

Fast calcium oscillations are found across all kingdoms of life. Jaffe and Creton, 1998, have been collecting and analyzing calcium wave traffic at a variety of frequencies.

Jaffe (1994)[16] provides a good review of the possible rates and rhythms divided into four frequency bands. These closely fit the ultradian frequency bands we have been describing.

o   Rhythm based motility. Organisms do not stay still. Their gliding, flying, twitching, swimming, walking and galloping gaits function rhythmically. The muscle contractions and relaxations behind them move rhyth-mically. Motility has been studied in organisms of many phyla. The rhythms of flagellated and ciliated locomotion in single celled organisms have been found in our own human bodies. The villi in our intestines and the cilia in our airways beat in tempos that are conserved across species. The molecular biology of actin-myosin is shared everywhere. So too with the biochemistry of the "movement molecules" actin, dynein, kinesin and others. The 9:2 rotors of flagellae and the coordinated movements of cilia of single-celled protista continue to function in the physiology of multicellular organisms. The 9:2 flagellar motor drives our sperm cells.

o  Many 8-12 Hz rhythms appear in human physiology, and may link together such divergent functions as theta brain waves, flagellar movements and the rhythm of muscle tone related to essential tremor. These frequencies recur all through the tree of life. We observe them in embryonic development. Rudolpho LLinas discuss these frequencies in primitive sea animal embryos, and notes "that intrinsic tremor in the musculature itself leads, via electrotonic coupling, to rhythmic, oscillatory movement, thus allowing water to flow through the gills and oxygen exchange with the external world and, so vital to life, throughout the egg sack. This form of motricity is "myogenic," for it represents movement born purely from the intrinsic properties of the muscle cells."[17]

Molecular biologists understand these intrinsic properties to involve the ratcheting movements in the actin and myosin motor molecules found even in primitive single celled organisms.

o  Micromotion can sometimes entrain rhythms of the near environment. They are transmitted to the movement of air, to slight temperature changes, electrical potentials, to the wafting of pheromones and other chemotactic communications. These tremulous dance-like movements, small as they are, produce big changes when they are timed right. They move into resonant relationships with other body oscillators and to the rhythms in the near environment, including the micromotion of other persons. There is some evidence that mother/infant bonding works this way. With current digital video motion analysis technology, we will very likely view these rhythmic interactions in all kinds of social exchanges.

**21**

In their separate frequency bands, the rhythms of living nature carry on six major biological functions:

1) They carry materials inside single cells along the endoplasmic reticulum, the Golgi complex and by transport through endosomes and exosomes.
2) They move substances rhythmically in larger body systems (digestive peristaltic rhythms follow this pattern of operation).
3) They drive metabolism, producing and transmitting energy in the mitochondria of cells.
4) They communicate information directly in the nervous system and from one organism to another through rhythm driven sensorimotor pathways, as in the mating dances of animals.
5) They provide motility to the organism. The actin-myosin molecular reactions in single muscle fibers move rhythmically. Muscle tone is a frequency driven phenomenon.
6) They serve as carrier waves on which other information is transmitted through complex modulations of the carrier waveform.

**22**
*Carrier waves*

In principle, the physiology of information exchange can be analyzed by power spectral analysis of carrier waves. We're always encountering and deciphering carrier waves in our lived experience. The voice is a carrier wave. The rhythmicity, the pulses and beats generated by vocal cords, the writhing hairs in the cochlea of the ear that receive these compression waves, the rhythms in words, timbres, gestures, touches, the changes in all senses, resonating at cell membranes, require senders *and* receivers. Some of these communications, sent and received as carrier waves, are modulated by AM and FM and other transforming technology embedded in our physiology. In real life, every verbal and gestural communication has frequency, amplitude and phase characteristics that can be quantified.

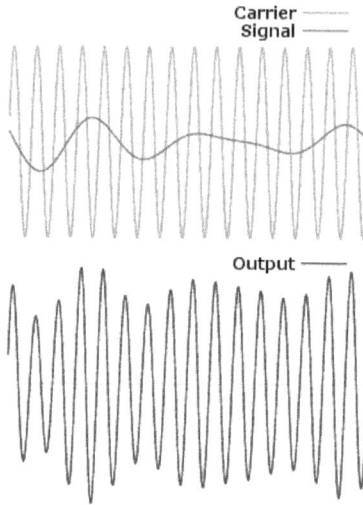

Carrier
Signal

Output

Radio transmissions depend on carrier waves. The carrier wave, produced in the radio station, contains no information until it is modulated by another information-bearing wave. A piece of equipment called a modulator adds information to the carrier wave, usually voice or music, and turns it into a complex wave train. In AM radio transmission, the carrier wave is modulated by amplitude, as shown above. Its waves get taller or shorter while the frequency remains the same. The changes in amplitude encode information on the carrier wave. FM radio works by frequency modulation of a carrier tone of constant amplitude. Your radio receiver is a demodulator. It extracts the signal from the carrier wave. You hear the music. Nature uses many other carrier techniques besides amplitude and frequency modulation.

They resonate in living cells in the molecular traffic across plasma membranes. Here biologically useful information is carried on very fast ultradian oscillators as modulations on their waveforms.[18] Brain waves are information carriers. However, to be used as carrier waves, the neurons have to establish resonant relationships by entraining to each other.

In his studies of brain physiology, Rudolpho Llinas considers these rhythmic carriers basic to higher neural functioning.

"Studies indicate that 40hz coherent neuronal activity large enough to be detected from the scalp is generated during cognitive tasks… What does it mean? We are confronted with a system that addresses the external world not as a slumbering machine to be awoken by the entry of sensory information, but rather as a continuously humming brain, This active brain is willing to internalize and incorporate into its intimate activity an image of the external world, but always within the context of its own existence and its own intrinsic electrical activity."[19]

The humming brain with its neural synchronies in the 40 Hz gamma range spanning wide areas of the cortex may be implicated in consciousness. The "binding problem" has been under active study by Gyorgy Buszaski, Francis Crick and others. Timothy J. Walter in *REM Illumination*[20] applies it to memory consolidation. What researchers say of coherence in the brain applies to signal generation across living nature generally. Every organism hums at characteristic frequencies in the bands we have identified at least in some of its physiological subsystems. Biosystems interact through harmonic and dissonant processes. Some rest on simple number ratios. Glass and Mackay in *Clocks and Chaos* explain that

"periodic stimulation of spontaneously oscillating physiological rhythms has powerful effects on the intrinsic rhythm. As the frequency and amplitude of the periodic stimulus are varied, a variety of different coupling patterns are set up between the stimulus and the spontaneous oscillator. In some situations the spontaneous oscillator is entrained or phase locked to the forcing stimulus so that for each N cycles of the stimulus there are M cycles of the spontaneous rhythm, and the spontaneous oscillation occurs at a fixed phase (or phases) of the periodic stimulus (N:M phase locking. )"[21]

The entrained ultradian oscillators commonly do double duty. They carry substances while they transmit the encoded instructions for their use.

## 23

Why these rates and not others? Where do they come from? What ties to nature do or did they express? Why are they so widely conserved? And crucial for our exploration: what do they tell us about love and wisdom and the other basic virtues? To make sense of the biological influences on human nature, we will have to speculate on their evolutionary origins.

Lynn Margulies suggests that many of the functions in eukaryotic cells (cells with nuclei and organelles) entered the stream of life when bacterial symbionts invaded or were ingested and were eventually co-opted by their host cells and lost their independence. They became the mitochondria, ribosomes, mictrotubules, and other organelles of primitive single celled life. "The descendants of the bacteria that swam in primeval seas breathing oxygen three billion years ago," she wrote, "exist now in our bodies as mitochondria. At one time, the ancient bacteria had combined with other microorganisms. They took up residence inside, providing waste disposal and oxygen-derived energy in return for food and shelter."[22] Earlier still, there were horizontal gene transfers between microorganisms. Carl Woese has done research on the rapid evolution this caused in the archea. We can guess that clock genes and rate-determining gene sequences were likely to have been among the transferred material. Later, Darwinian evolution carried them from species to species.

Margulies speculated that neuronal functioning itself might have been structured originally around the behaviors of imported bacterial spirochetes. She wrote:

"I continually play with an idea, the origin of thought and consciousness is cellular, owing its beginnings to the first courtship between unlikely bacterial bedfellows who became ancestors to our mind-brains... The microbes are not just metaphors; their remnants inhabit our brain, their needs and habits, histories and health status help determine our behavior."[23]

By knowing the lineages of the conserved frequency bands, from horizontal gene transfers to bacterial endosymbiosis, – and by uncovering the primal functions of the ur-rhythms carried by them – we can learn more about ourselves. We can learn how we love and think.

But a crucial question remains: where did the first organisms get their rhythms?

Answer: from the sun, water and atmospheric electricity.

**24**
*Old Sol*

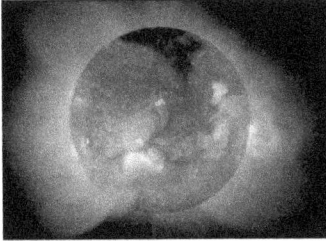

The day/night rhythm has been conserved in almost every plant and animal species. Why? Because every day the energies that control food supply and reproductive possibilities are parsed out by dark and light. However, the light of the day in the day/night cycle also opens a window through which faster solar frequencies can act on living matter.

Guy Murchie in his interesting survey on physics and cosmology, *The Music of the Spheres*, pointed out that in Greek mythology the seven strings of Orpheus' lyre were said to be tuned by the sun.

On the fast end of the solar spectrum UV light, visible color and infrared radiation penetrate the magnetosphere and reach the earth. These oscillate in the electromagnetic spectrum billions of times a second, with frequencies ranging from $3x10(12)$ for deep infrared through $4.3x10^{14}$ to $7.5x10^{14}$ for visible light to $3x10^{17}$ for ultraviolet. At even higher frequencies, x-rays and gamma rays bombard us. The faster the oscillation the more energy it carries. UV radiation is a known mutagen. Over long periods, the fast solar spectrum must have influenced species evolution

**Solar acoustic vibrations**

These fast solar radiances carry slower rhythms to the earth. The fusion furnace in the sun's core rumbles and shakes, causing the surface of the sun to ring like a bell as its diameter expands and contracts rhythmically. Heliosesimologists observing the solar pulse from the South Pole put the fundamental period at 160 minutes, with a harmonic series moving by octaves to 80, 40, 20, 10 and 5 minute periods. The frequencies generated by the periodic release of solar radiation show up prominently in the cell cycle.

From the beginning of life, and particularly *at* the beginning, and probably as a condition of the beginning, particularly in the first, tentative, submicroscopic bondings of organic molecules, organisms had to adapt themselves to the rhythms in the solar radiance, especially those transmitted as thermal gradients in water on an ocean-girdled earth. Biochemical reaction rates change with heat, visible light, UV and other radiative energies that come from the sun. The biochemicals set going by solar radiation in some way must form to the beat of these rhythms, vibrating to the seven strings of Orpheus' solar lyre.

Ilya Prigogine, the Nobel laureate whose work on dissipative systems helped create the field of chaos and complexity, identified the sun as the source of biological rhythmicity. With his collaborator Isabelle Stengers, he wrote, "… life seems to express in a specific way the very conditions in which our biosphere is embedded, incorporating the nonlinearities of chemical reactions and the far-from-equilibrium conditions imposed on the biosphere by solar radiations."[24]Sunspot activity oscillates with an 11-year period and this affects the weather. A belt of increased rainfall in the northern hemisphere oscillates between 60-70N and 70-80N during the sunspot cycle. A similar pattern occurs in the southern hemisphere. Lightning strikes in England increase during sunspot maxima. The Little Ice Age in Europe (c1600-1750) correlated with the Maunder sunspot minima. Solar magnetic reversals follow a 22-year cycle. The equatorial regions of the sun rotate faster than the polar regions, causing electromagnetic patterns described as the fast solar tacholine oscillations. There are solar rhythms that come to earth at sound wave frequencies. The sun in its fusion furnace rings like a bell, its photosphere beating many times per second over a five-minute repeating period and these pulses reach the earth through the magnetosphere, permeable to visible light. Other solar frequencies cause electrical disturbances. These ionospheric discharges paint the aurora borealis on the sky.

The sun delivers this whole array of energies to earth in nested rhythms. The 11-year sunspot window, the 29-day solar rotation window, the window of the day, and the 5-160 minute acoustic solar pulse play within each other's waveforms. The energies that come to earth from the sun are deflected by the magnetosphere. What gets through to the surface of the planet varies with changes in the strength and shape of the magnetosphere itself. Together the faster solar influences that get through produce thermal agitations, random Brownian movements in water and many other kinds of radiative influences on small bits of matter. The molecules of the chloroplasts of photosynthetic plants and the rhodopsin molecule in the visual purple of the eye are exquisitely sensitive to trains of photons.

Since biochemical reaction rates change with heat, visible light, UV and other electrical field energies, the reactions streaming in through the solar window must have fundamentally shaped the chemistry of life.

The role of these influences can be investigated by studying the solar spectrum and correlating its frequencies to those we have identified in life.

## 25
### *Sun on Water*

The sun's rays play on the great mirror of the world's ocean surface. The $H_2O$ molecules and their hydration complexes are its first receivers. The thermal oscillations of $H_2O$ in water influence the delivery times and reaction rates of biochemicals in solution.

The action of sunlight on water is crucial to life. One can see it impressing its daily rhythms on the movements of plankton. It warms and cools them rhythmically, draws them up and down into darkness and light, induces them to float or sink, to photosynthesize or respire, to get active or rest, even to align themselves along the shining wave crests where the solar radiance is strongest (or away from it when it is too intense). Surely the direct influences of sun on water would have acted on emergent life. These radiative and thermal rhythms set up agitations and movements that would have influenced molecular combinations in the microscopic realms where life originated in the sea.

Motility itself may have found its rhythmic pulses in the frequencies generated at the water/solar interface. The first living things that learned to swim by their own power must have had to adjust themselves to patterns of fluid motion through water. By evolutionary selection, the rhythms of water must have been carried into their swimming motions. The whipping tails, the screwlike turning of single celled organisms and the biochemical motors that drove them must all reflect the frequency spectra of sunlight on water. Daily warming and cooling, windowing the fast ultradian solar radiance rhythms running in the five to 160 minute range, as mediated by the fluid characteristics of water, must have gotten into the tissue of life early on. The smaller the proto-cells from which the first life emerged, the larger the influence of the radiation hitting it. The rhodopsin molecule in the visual purple of the eye can register the presence of single photons. So too with the influence of hydrogen bonding in water molecules on nucleotides and amino acids. If creatures swam in oil, they would have different motion frequencies.

On the reasonable assumption that life started in the waters, and biological organisms *are* made mostly of water, and seeing that cellular motility has adapted itself to aqueous rhythms, watery wave motions <u>inside</u> cells very likely play a part in intra-cellular processes. In the fluid medium of the cytosol, materials move in vesicles or as free molecules. They gather and combine or

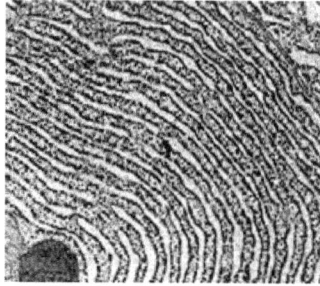

**Rough Endoplasmic Reticulum**

disperse around docking sites in the organelles of cells. Charged molecules with electromotive potentials move in wavelike patterns through the cytosol. Protoplasmic streaming shows the movement of currents in the cytosol, often channeled along microtubules and microfilaments, the tiny cytoskeletal and cytokinetic elements visible under the electron microscope that are themselves responsive to watery currents, temperature and radiation.

Further, the wave motion in the cytosol itself travels from cell to cell through gap junctions in their membranes. Typical diffraction patterns may form around each opening. Does the cell water in the cytosol serve as an information transfer medium? Do the radiative energies of the sun, passing through these nearly transparent cells, slightly bent by the wave patterns in the cytosol, inscribe a holographic design on the apical face of the opposing membrane?

**Standing acoustic waves in water. Hans Jenny photo**

We can conduct experiments to explore these possibilities by showing first that water itself can pick up and transmit frequencies capable of carrying detailed information. By applying sound to a water drop, we can show that water picks up auditory frequencies and dances to them in symmetrical standing wave designs.[25] If we

28

add salts or other basic chemicals of life to the water, the wave traffic, though perhaps changed in frequency and amplitude, persists.

Microscopic examination of living cells will show that the cytosol does carry waves in a variety of frequencies and that these have AM and FM characteristics. Further, we know that these vibrations touch the anchoring points of microtubules. I conjecture, then, that the microtubule and microfilament networks, especially as they ramify along the inner surface of the cell membrane and as their single fibers stretch across the cell, may work as water harps, sensing the movements of the cytosol and transducing them at the cell or organelle membrane end of the strand. Further research might show that thermal influences generate rhythms in the cytosol. The microfilaments could pick up these rhythms in the wave motions hitting the cell harps and transmit them to organelles. It follows that the docking sites along the microtubule/microfilament mesh to which signaling substances attach would be occupied and vacated in ways reflecting the tempo of the waves pushing on them.

## 26

For hundreds of millions of years before life appeared amino acids and protein chains must have been organizing themselves in water under the influence of sunlight. Moving through seawater, developing hydrophilic and hydrophobic sites, these molecules took on specific chemical bonding characteristics that influenced their folding patterns.

Surely these rhythmic radiative and thermal sources must have created agitations and

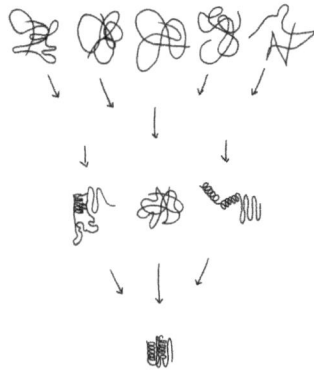

**Protein Folding Patterns**

movements among water molecules and solutes that influenced molecular combinations in the submicroscopic realms where life originated in the ambiance of these very small comings and goings.

No wonder the most primitive organisms take on the oscillatory characteristics of water.

It further follows that with life temperatures limited to a tiny slice of the temperature spectrum the frequencies of the biochemistry of life could not proceed at all possible reaction rates. In addition, on this small scale, the mass of molecules and the frequency characteristics of light would have distributed the rhythms of life into discrete spectra. All the more so when you consider that water as a basic ingredient of life itself can take only certain thermal agitations at normal atmospheric pressures before being bound in an ice lattice or boiling away into steam. In addition, while they were moving, joining, rejecting each other, evolving their autocatalytic properties, the solutes were swimming in water according to their own ionic attractions and repulsions and their stereochemical properties. Moreover, the wind, sun, the earth's rotation and the moon's pull drove them too, together imposing their frequencies, flow forms and agitations on the molecules of life.

The thermal agitation that starts chemical reactions going in living protoplasm may have randomness in it, but the responses of the cytosol to those agitations must be limited by the nature of water and by the chemical constituents dissolved in it. Since the major biological energy transactions involve phosphorylation, the rhythmic characteristics of phosphorous in water probably helped shape the original frequencies of mitochondrial chemistry and they are still latent in it. In *Investigations,* Stuart Kauffman reasons that in this reaction a pyrophosphate molecule carries the added energy. This PP molecule is cleaved to form P+P and the energy released pushes forward the autocatalytic reactions permitting life. However, the energy dissipates quickly and Kauffman acknowledges that unless we "add energy to resynthesis PP from P+P" the reactions break down. "To do so," he continues,

"I invoke an additional source of free energy in the form of an electron e, which absorbs a photon, hv [hv is the formula for Planck's constant (h) times the frequency of the photon (v)]; is driven endergonically to an excited state e+, and falls back endergonically to its low energy state, e, in a reaction that is coupled to the synthesis of PP from P+P."[26]

Here is Kauffman's problem: where do the photons come from? They must exist in nature. If they find their way into life through sunlight on water, as we have argued, they will drive the autocatalytic reactions not randomly at odd moments but in wave fronts or packets with specific frequency and amplitude characteristics. These then find their way into life by generating resonances in the oscillatory media of the molecules.

Water itself exists in various configurations in the presence of other elements, and bulk water takes on different characteristics according to the ions dissolved in it. Geometric shells of $H_2O$ form around solutes. This structured water remains distinguishable from the bulk water around it. So in principle we should be able to detect the rhythmic interplay between bulk water, the hydration structures it builds around other molecules in the water, the water mediated redox reactions and the molecular foldings flowing through it.

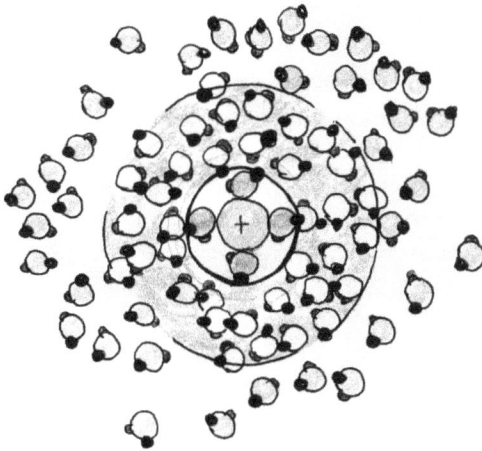

**Hydration Structures Build Around Solute**

Cell membranes themselves are organized by their responsiveness to water; the cell membranes from protozoa to man are constructed of a lipid bilayer with its hydrophobic areas sandwiched between two hydrophilic surfaces punctuated by numerous membrane proteins forming valves and channels, through which all nourishment and information must enter and leave the cell, usually borne by water. You can find a good description of these processes in Christian De Duve's marvelous study, *A Guided Tour of the Living Cell.*[27]

The hydrogen bonds in the water and the ions and charged particles of the matter dissolved in it in turn interact with photons from the sun—and these interactions when life was first forming on the surface of the waters must have influenced the folding of polypeptide chains. Proteins still show their watery influences. They form hydrophilic and hydrophobic regions and turn toward and away from water. They twist into helices that look like tiny whirlpools; they have chirality and turn clockwise or counterclockwise. Perhaps they reflect the Coriolis forces induced in them by the turning earth. The bonding characteristics of water shape its kinetics on small and large scales. Hydrogen bonds are continuously broken and reformed as water rolls and turns.

In the rates of these tiny spinning vortices, we may find the evolutionary sources for ultradian rhythms that shape biological processes. Arthur Winfree found these whirlpools everywhere in living nature.

"We found it... in tissues made of clocks—and then found the clocks themselves dispensable: in excitable tissue the singularity became the rotating pivot of a spiral wave. We saw it not only in excitable tissue but in nonliving chemical media as well, no longer an abstraction about timing relations but now a visible rotating source of waves. And there we saw the first singularity in its fullest development, as a tornado-like filament arching through three dimensional space to close in a ring.

"Each kind of organizing center is woven in its own distinctive way from fibers of phase singularity, as though from the funnels of chemical tornadoes, organizing centers are little chemical engines made of rotating parts...except that the rotating parts are only patterns of chemical activity, like ghosts in the material substrate."[28]

Carl Woese, the microbial geneticist who helped develop the paradigmatic theories of horizontal gene transfers between primitive cells, recently wrote "It is becoming increasingly clear that to understand living systems in any deep sense, we must come to see them not as machines, but as stable, complex dynamic organizations." Freeman Dyson explained in a recent talk "Woese likens organisms to eddies in a turbulent stream that reappear no matter how often they are disturbed." He invokes the image of the whirlpool. Its rhythmicity is basic to life, carrying and sustaining the information for origin, development and behavior in its spin.[29]

Ivanitsky, Krinsky and Mornev, Russian scientists writing near the end of the Soviet era, showed how oscillating systems generate vortices in all living tissue. Speaking of slime molds they state,

"Here again revolving reverberator vortices are the fastest of all local sources of autowaves, because autowave sources have identical properties in all active media, and all other sources are, therefore, suppressed... This is an example of how nature uses reverberators for building up a structure in extreme conditions."[30]

## 27

Vortices spin with fractal independence of scale all through the biosphere and beyond in solar and galactic media. The vortices in near contact with each other influence each other and are subject to influence on all scales.

Vortices in nature do not spin in perfect circles or cylinders. They are bent spirals. They move from one place to another. Matter is pulled up and through them. A hurricane is a vortex that moves in the fluid medium of the atmosphere along a track influenced by barometric pressure differences, prevailing winds and currents, water and air temperature and whatever else produces turbulence within the medium, including other vortices. These spiral movements have been permanent factors for change on earth. They pull on each other to produce the frequencies we have found in the tissue of life on all scales.

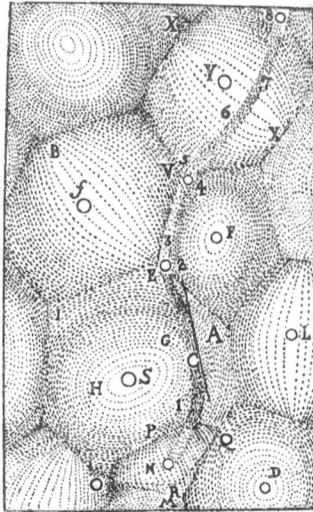

**DesCartes Notion of the Vortex**

## 28
### *Myths of Whirlpools*

**Newgrange spiral
tomb carving**

The spiral wave, the whirlpool, and the vortex are symbols deeply set in the human imagination. Their figures were incised on rocks and painted on pottery thousands of years ago. Northern European myths associate whirlpools with transformation, with communications between realities, between the living and the dead, with a rupture of levels, with the turbulence of the turning point moments in our own lives. When we confront major changes, we are sucked into a whirlpool, and perhaps jetted out the other end. A new myth with similar content appears in the science fiction of galactic black holes as in Stephen Baxter's *Ring*.

Historian of science Giorgio de Santilliana with Hertha Von Dechend trace some of these myths back to prehistoric sources. They write,

"The engulfing whirlpool belongs to the stock-in- trade of ancient fable. It appears in the Odyssey as Charybdis in the straits of Messina – and again, in other cultures, in the Indian Ocean and in the Pacific. It is found there too, curiously enough, with the over-hanging fig tree to whose boughs the hero can cling as the ship goes down... But the persistence of detail rules out free invention. Such stories have belonged to the cosmographical literature since antiquity. Medieval writers, and after them Athanasius Kirchner, located the *gurges mirabilis,* the wondrous eddy somewhere off the coast of Norway,[31] or of Great Britain.... For the Norse the whirl-pool came into being from the unhinging of the Grotte Mill [the Sampo]...No localization is indicated here, whereas the Finns point to directions which are less vague than they sound. Their statement that the Sampo has three roots – one in heaven, one in the earth, the third in the water eddy – has definite meaning, as will be shown."[32]

Triskele, the symbol of ancient Sicily, found in Neolithic Sicilian artifacts, shows three feet bent in apparent circular motion emerging from a sun center.

Sicilian triskele image

From early on the sun was an object of awe and worship as the energy source that enlivens the whole process of evolution on earth. The Triskele, I fancy, behaves like a whirlpool caused by the sun's heat on the waters. The Triskele, in this im-agery, would represent a whirlpool with both cosmic and oceanic elements. But why three legs? Three is a recurrent theme in vortex imagery. Odysseus dies in a whirlpool. His ship goes around three times.

Tides, currents, thermohaline oscillators, the great globe span-ning subsurface streams, all of them periodic, animate the oceans of the world and generate eddies and whirlpools in water and air. A power spectral analysis of the kinetics of moving water at the sur-face of the sea might very well show traces of all the main fre-quency bands of life. We would find the circadian in the daily warming and cooling at the surface layers, the infradian in the monthly pull of the moon on the tides, and the ultradian rhythms in the fast components of the solar radiation interacting with water molecules and hydration structures, and the movement of wind waves with their swells pulsing like respirations. Even the water in space, observable in water masers in the Milky Way, shows a re-lated power spectral distribution. Studies of the rhythms in water masers would bring a cosmic dimension to the origins of life not yet explored.

But water waves and chemical vortices seem to be sluggish sources for the fastest oscillations in living tissues. Could the electrical pulses in nerve cells have an independent source in nature? Were there environmental pulsations moving in these frequency ranges?

## 29
### *Electrostatic Rhythms*

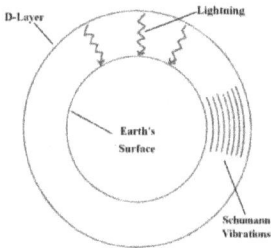

Schumann resonances
at 7 Hz

One source could be the lightning discharges resonating in the cavity between the earth's surface and the ionosphere. One can reasonably conjecture that these events provided the energy for reactions between organic molecules that produced the earliest amino acids. The atmosphere becomes a sounding tube for electrical discharges moving at the speed of light. They circle the earth more than seven times a second, resonating at specific frequencies. W. O. Schumann in fact detected these frequencies in the ionosphere in 1952. With a fundamental oscillation at 7.8hz and an overtone series at 14,20,26, 33, 39 and 45hz, the so-called Schumann waves match many physiological fast rhythms, most notably in the brain, in nerve traffic, gap junction traffic, and the even more basic calcium oscillations.

Prebiotic chemists consider the early earth atmosphere to have been much more electrically active than it is now. Solar flares entering the ionosphere unprotected by the ozone layer added electrostatic energy to it. Those were turbulent times. Earth chemists speculate that hellish core temperatures

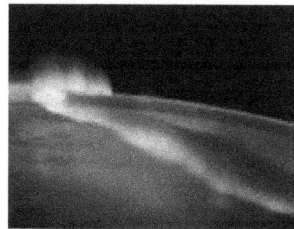

Aurora seen from space

and meteorite impacts repeatedly boiled off the oceans. With solar and lightning-generated radiations playing between dense vapor clouds saturated with aerosols ejected from volcanic eruptions, one can plausibly suppose that partial protein chains could have been formed from small organic molecules in reactions catalyzed by electrostatic atmospheric discharges from both lightning strikes and auroral phenomena. Experiments conducted by Harold Urey and Stanley Miller in 1953 confirmed this. Putting methane, ammonia and water in a glass bottle and exposing it to electrical sparks, they produced almost all of the amino acids found in life in a matter of days. But at what rate were the sparks pulsed? It might be worthwhile repeating their experiments using frequencies matching the Schumann wave spectra

The stormy eons on the prebiotic earth that lasted for scores of millions of years, provided time enough for primitive life to evolve and for the electrostatic rhythms to enter bacterial cells and from them, by endosymbiosis, to become part of us. Though fanciful, this frequency driven model for the origin for life deserves consideration. It puts the radiative frequencies and the organic molecules in close contact. Though the Schumann wave in its current amplitude seems too weak to influence living organisms (the earth's magnetic field is 5000 times more powerful than it,) evolution is conservative and the echoes of the early electrostatic discharges high in the atmosphere may still resonate in living tissue.

Itzhak Bentov, using an ultra-sensitive ballistometer, found the Schumann wave frequency in the descending aorta driven by the ejection of blood from the ventricles. He traced its resonations elsewhere in the body, carried through the bony skeleton in the body's micromotion. According to him, the Schumann frequency even jiggled the brain. Bentov connected this with sensations felt as the Kundalini rising in the coccyx and speculated (half humorously perhaps) that since "the resonant frequency of the earth—ionosphere cavity—is about 7.5 cycles per second and that the micromotion of the body is about 6.8 to 7.5 Hz. This suggests a tuned resonant system."[33] From this observation, he took a huge leap, speculating that this wave might actually couple with the electrostatic field of the planet and act as a carrier wave sending human messages around the planet every 1/7th of a second. George Leonard in *The Silent Pulse* carried this notion even farther, supposing "the Schumann waves are sometimes 'hooked into' the pulsing of human brains, connecting them at a distance."[34] These *jeux d'esprit* explore the notion that radiative influences shaped life on earth and continue to interact with it. Though inspired guesses have led to new understandings before, we will put these playful thoughts aside for now.

## 30

I think we've made the argument well enough for our present purposes. Circadian, ultradian and infradian rhythms <u>do</u> enter cellular processes through evolutionary selection. At least in low-level cellular and organ functions they do. To get from here to love and wisdom and the other meaning-bestowing rhythms in human life is harder. We have bigger steps to take. We have to discover how high up the same rhythms go. Do they penetrate behavior? Do we love as well as make love in rhythm? Do the fruits of wisdom deliver themselves on rhythmic trains of association? Do the oscillatory characteristics of behavior influence our values via our sense of timing? Do our attitudes, motivations, passions and possibilities congeal from rhythmic units? Do rhythms drive history? Our review of the rhythms of nature taken up by life, though providing us with important information, does not seem to help us answer the most important question – how the *meaning of life* could in any way depend on the interactions of bio-oscillators. Or that love and wisdom are oscillatory processes to which we resonate. Or how they could have become so by evolving from primitive environmental precursors. But when you look at life in the frequency domain, when you take on that perspective, when you envision the whole thing as a temporal unfolding, when you see all of "Nature naturing" (Spinoza's phrase,) when you understand the continuity of rhythms, a pathway to love and wisdom set deep in the tissue of life will open up to you.

I will now show you how two primordial rhythms, the very ur-rhythms of animate nature, lead the way to love and wisdom.

## 31
### *Ur-Rhythm one: Aggregation and dispersal*

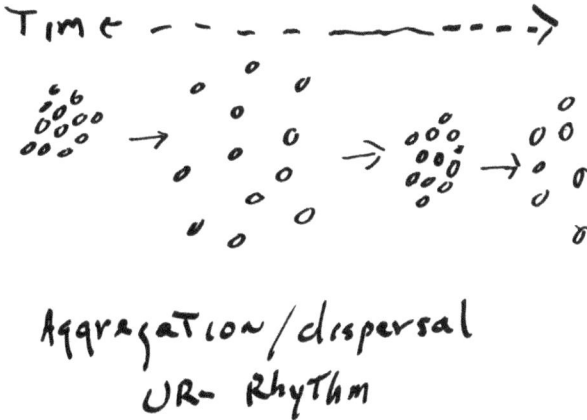

The oldest dance is the pavanne of molecular aggregation and dispersal. It predates life. It operates in the dynamics of ionic attraction and repulsion. The aversions and attractions of charged particles express it, as do the clumping together and pulling apart of polymer chains, and on higher levels in the hydrophilic and hydrophobic foldings of proteins in aqueous environments.

Polarization at all cell membranes depends on concentration gradients of sodium, calcium and potassium ions that gather and disperse through channels and pumps. These molecular movements are essential to cell viability. Enzymes in turn regulate the rates of reactions in the cell, and they aggregate and disperse from their receptor sites. These enzymes themselves are created by the aggregation of amino acids along ribosomes.[35]

## 32
### *Ur-rhythm two: Expansion and contraction*

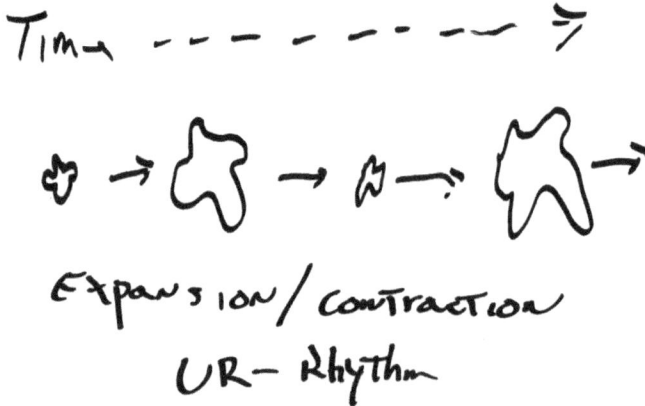

Time ⎯ - - - ⎯ - - ⎯ - ⎯ ⌁

Expansion / Contraction
UR- Rhythm

As soon as bioactive proto-cells exist, aggregation and dispersal produces expansion and contraction. You can observe it in the pulsing rhythms of whole organisms. Their movements, outward from a center and back toward it, depend on the presence of a bounding membrane that isolates the internal world of the organism from the external environment. Even the simplest bacterial cells have bounding membranes. Christian De Duve considered encapsulation essential for the creation of life.

In expanding and contracting, cells necessarily pull on their microtubular and microfilament skeletal structure, the harp inside them, which changes the tuning of the harp. The strings play an active part in cellular growth and development. They appear and disappear, lengthen and shorten. The cell creates and disassembles them as needed. They form the mitotic spindles in the rhythmic dance of cell division. The actin-myosin molecules seated in the anchor points of the microfilaments have been conserved for billions of years.

Animal cells expand and contract by passing substances through the cell membrane, and this membrane structure, the lipid bilayer, is widely conserved across every phylum. The membrane is itself alive. Its trans-membrane pumps and channels open and close as certain molecular signals glom onto them. The cell *senses.* It has chemical senses that separate self from other.

The same two-part movement of expansion/contraction and swelling/shrinking underlies motility. Healthy skeletal muscle tissue maintains muscle tone by oscillatory processes organized from the rhythmic inflow and outflow of calcium ions at 5-10 HZ from the sarcoplasmic reticulum.[36] The calcium ion signals are themselves basic bio-oscillators. The muscle movements in turn produce behavioral rhythms.

When an amoeba moves away from an acid droplet, for example, it shrinks away by sending its protoplasm inward in one direction and then outward in another creating a temporary pseudopod to move it to safety. The amoeba must be able to tell inside from outside and make distinctions between good and bad outer environments. It does this through chemical senses that work by binding environmental molecules to sites along the cell membrane. Chemical sensing is a simple process conserved through life. We have it in our sense of smell. An aggregation and dispersal of molecules on membrane binding sites shrinks and expands cell organelles. In amoeba, it produces gliding motility.

You see expansion/contraction everywhere inside cells. It is there in intracellular transport vesicles moving to and from docking sites. They upload and offload their cargoes either outside the cell or in other compartments inside the cell. The release of neurotransmitter substances into synapses depends on the filling, transport and emptying of vesicles, and the diffusion of neurotransmitters across membranes itself requires an aggregation and dispersion correlated with the binding and reuptake of molecules. Through these expansion-contraction events, nerve cells secrete neurotransmitters, secretory cells send out substances.

Primordial expansion and contraction is bi-phasic. There is no filling without emptying, no dilation without compression. The extremes are connected; they are part of one polar process, not antagonisms or oppositions but continuities.

## 33

The two outcome rhythms, aggregation-dispersal and expansion-contraction, interact to produce higher level behaviors. In our study of human nature, the most important high level behaviors are social approach-separation and internal withdrawal-return. The evolution of approach-separation and withdrawal-return from aggregation-dispersal and expansion-contraction respectively, returns us to the main theme of the book: the meaning-bestowing rhythms of love and wisdom. I'll trace out some possible routes we took to get here. They start almost at the dawn of life.

Lynn Margulies speculates that the first social movements of approach-separation occurred in archaic bacterial mats.

"In some cases, like swarms of cyst-forming myxobacteria (for example, Chondromyces, Myxococcus/), the component genomes sense each other and fuse, forming a larger structure—no membranes are breached. In others, as when the akinetes of a cynobacterium float away, the genomic systems disperse."[37]

In primitive eukaryotic cells, the two ur-rhythms combine to produce quasi-social rhythms. They probably reflect the bacterial endosymbiosis from which they originated.

Approach-separation rhythms become functionally social in the clustering phases of single celled colonial organisms. Typically, they move into agglomerations of 8, 16 or 32 members in their social stages. Through the microscope, you can see them vibrating together. Their social vibrations are themselves rhythmical and move at tremor rhythm frequencies.

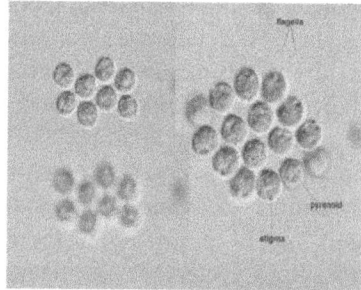

**GONIUM COLONIES**

Wherever single celled life merges into multicellular life, we find approach-separation working. You can observe it in colonial sponges, in coral formations, and in slime molds that cluster and disperse in different stages of growth, signaled by intracellular chemical changes that pulse rhythmically as signals sent out into the surrounding medium. The slime mold has been well researched. It shows many of the rhythmic components later carried into the complex social lives of multicelled organisms. The ameboid cells that become the slime mold live as independent individuals. However, when they are starving, they emit periodic waves of cyclic AMP. Jeremy Campbell describes it this way:

"In the metamorphosis of slime molds, periodic waves of cyclic AMP are a medium of communication. A cell acts as a 'center of attraction' by putting out pulses of the chemical and other cells start to converge on the center in a rhythmic motion… This cell mass is called a slug, and before it settles down to produce spores, it moves along the ground by means of rhythmic contraction."[38]

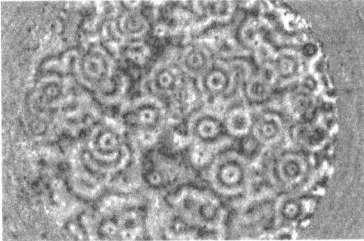

**SPIRAL WAVES IN SLIME MOLD AGGREGATION**

RNA molecules approaching and separating from docking spaces along the ribosomes in each single-celled organism themselves drive the rhythmic waves of cyclic AMP. You can see it unfolding in the spiral patterns. Nearby cells receive the cAMP message and begin to move toward the source, at the same time emitting cyclic AMP signals of their own. They form into a spiral wave with a central focus. Then the individual cells join into a wormlike creature that moves along with synchronous contractions. Rhythms compound upon rhythms. The slime mold next migrates to a new site better supplied with resources. There it builds a stalk, produces fruiting bodies that release thousands of spores, completing its reproductive cycle. In all of these stages, we can detect the basic features of the rhythmicity in life, all of them quantifiable: signal rhythms (cAMP), rhythmical movement (spiral waves), synchronization (migration), cooperation (building the spore tower) and dispersion (casting of the spores.) Winfree comments on the slime mold oscillations:

"We witness here, perhaps, a living fossil replaying events that were common during evolution from unicellular to multicellular organisms two billion years ago. This process has been familiar to biologists for many decades, but its frequent organization by rotating spiral waves was documented for the first time only in 1965, in the laboratory of Gunther Ferisch in Germany."[39] Now that we have located the urrhythms at the foundations of life, we can get back to love and wisdom to investigate whether our human virtues and passions also express themselves in primordial frequency-driven units of behavior.

## *End Notes*

[1] Sam Keen. *The Passionate Life.* p. 62

[2] From a heart beat to an ice age, to a galactic revolution, the frequency of an oscillation no matter how slow or fast does not in itself alter its dynamics. That is to say, these cycles are neither brief nor long in themselves. There is no quick or slow except in relation to another oscillation. Human beings choose which to use as frames of reference. The time-keeping oscillations always belong in some intimate way to the observer's life experience. So to you and I, a second is pretty fast and a year is pretty slow.

We can directly perceive the biological oscillations only when they fall into the range of our senses and our sensitivity to time. We easily observe the systole and diastole of the heart, the phases of the breath, the sleep/waking cycle. We infer the slower rhythms of the seasons and years by their partial effects on our bodies. But to a mayfly, though born in May, the cycle of the year is nonexistent, and we are mayflies to other oscillations that may influence us.

[3] *From Clocks to Chaos.* p.22-26

[4] *When Time Breaks Down.* p. 298

[5] Paul Davies, *The Cosmic Blueprint.* Touchstone. 1988. p. 89

[6] Kondo T., Ishamura M. "The Circadian Clock of cyanobacteria". Bioessays 2000. Jan; 22 (1)

[7] Reppert, S. M., and Weaver, D.R (2002) Coordination of circadian timing in mammals. Nature. 418, 935-941. Scibler, U., and Sassone-Corsi, p. (2002) A web of circadian pacemakers. Cell 111, 919-922

[8] Arthur Winfree. *When Time Breaks Down.* Princeton University Press. 1987. P. 252

[9] Winfree. *When Time Breaks Down.* P. 12

[10] Glass. 1996. "Synchronization and rhythmic processes in physiology". *Nature.* 410 (6825) p 277-284

[11] Klevecz et al. Proceedings National Science Academy 101 1200-1205

[12] John D. Palmer has done research on the 6-hour tidal rhythm impressed on inter-tidal creatures exposed to successive drying and wetting cycles. He has shown that these tidal rhythms persist in organisms taken far from their feeding grounds and kept in constant light in another time zone. But even with this independent six hour rhythm before his eyes he sees nature running on a single circadian clock. Accordingly, he calls his recent book *The Living Clock*, in the singular. For him there's one circadian timepiece, like the clock on the wall. It has de-

signed into it various escapement mechanisms, like the second, minute and hour hands on a spring wound wristwatch. All rhythms are shaped by these escapement mechanisms. He speculates that "when the escapement mechanism is finally deciphered, a Nobel Prize should be the reward for the discoverer(s)."[12] Again he uses the definite article, as if there was one gear train distributing the various units of the one 24 hour clock. Why not several escapements attached to several clocks, each originally coupled to a natural rhythm that got woven into the evolutionary process? Mightn't these independent clocks establish relatively stable phase relations to each other? Even though core temperature, salt balance, urinary electrolytes, melatonin and other physiological functions are considered to follow a circadian rhythm, don't they reach their peaks and troughs at different hours of the day or night?[12] Perhaps inside the day they have their own ultradian ups and downs regulated by their own clock genes. These peaks and valleys and their timed relationships to each other must serve a biological purpose. Biological clocks moving in a variety of frequencies, controlled by genes that start and stop the enzymes in their separate reactions, could adjust to each other without a master clock controlling them all. Reaction products from successive rhythmic processes could become the starting ingredients for the next reaction in a biochemical cascade. These possibilities have been explored in an interesting recent research article, (with excellent bibliography,) by Tauber, Last, Olive and Kyriacou, who tie biological rhythms to clock genes and trace their divergences over evolutionary time.[12]

[13] Pourqui, 2003 – geneve
[14] Murray, Solomon and Kirschner, quoted in Rossi and Lippincott.
[15] Neuroscience. Parri, H.R., Gould, T.M. and Crunelli, V. (2001) Spontaneous astrocytic Ca2+ oscillations in situ drive NMDA receptor-mediated neuronal excitation Nature Neurosci. 4, 803-12
[16] Jaffe. 1994
[17] Rudolpho Llinas. *I of the Vortex.* MIT press 2001 p.203
[18] Schuster et al "Modeling of simple and complex calcium oscillations." *European Journal of Biochemistry* 269, 2002.
[19] Llinas. *Vortex.* p. 124
[20] Buzsaki, G. "The Gamma Buzz: Gluing by Oscillations in the Waking Brain," in *Rhythms of the Bain.* Oxford U. Press. 2006. Timothy J. Walter. *REM Illumination.* Lotus Magnus. 2007.

[21] Glass and Mackey. *From Clocks to Chaos, The Rhythms of Life*. Princeton U 1988 p. 119

[22] Margulis and Sagan. Microcosmos. UC Press. 1986. p.31

[23] Sagan and Margulis. Speculation on Speculation. In *Slanted Truths*. Springer Verlag. 1997. p. 119

[24] Prigogine and Stengers. *Order out of Chaos*. Bantam. 1984 p. 14

[25] Image from Hans Jenny. Cymatics. Macromedia. 2001.

[26] Kauffman. *Investigations*.ibid p. 65

[27] *Student Edition*. Scientific American Books. 1984, as updated by recent texts in cell biology.

[28] ibid p.246

[29] Woese. "A new biology for a new century." Microbiology and Microbiology Reviews. vol 68. p.173. Dyson. New Scientist. Feb 11-17, 2006. p. 36-39.

[30] In *Cybernetics of Living Matter*. Mit Publishers. Moscow. 1987 p. 73

[31] Landsat image taken east off Greenland, corresponding to Hamlet Mill site.

[32] Giorgio de Santilliana with Hertha Von Dechend. *Hamlet's Mill, An Essay Investigating the Origins of Human Knowledge and its Transmission Through Myth*. Godine. 1977 p.204-5

[33] Itzhak Bentov. *Stalking the Wild Pendulum*. Bantam. 1977 p. 53

[34] George Leonard. *The Pulse of Life*. Dutton. 1978. p.72

[35] *Molecular Cell Biology*. Lodish, et al. Chap3

[36] Becker, Deamer. *The World of the Cell*. p.593

[37] (Margulis and Sagan. reprinted in *Slanted Truths*. Springer Verlag 1997. p.65

[38] Jeremy Campbell. *Winston Churchill's Afternoon Nap*. Simon and Schuster. 1986. p. 221

[39] When Time Breaks Down. p. 175-6

# 2. REAL LOVE IS MUTABLE

## 34

Behavioral rhythms play out in the annual, seasonal, daily, and ultradian frequency bands I listed in the last chapter.

The strongest, most dependable environmental rhythms engage all the members of a species living within a geographical range in coordinated behaviors. The sleep/waking cycle is the clearest example of this. Its stages and turning points link the daily and seasonal rounds connected with the sun and with the revolution of the earth around it to basic human activities. Over long periods of time, even the precession of the equinoxes and the advance and recession of ice ages drive patterns of aggregation and dispersal that motivate species to migrate in search of habitat.

Though we can cite many examples of the operations of the ur-rhythms in every organ and animal species, the important point is not to multiply instances but to establish a sound conceptual scheme showing how aggregation and dispersal, physiological expansion and contraction, social approach and separation and cognitive withdrawal and return interact to produce complex social and solitary behaviors. With this scheme in hand, we will be able to understand how organisms build ecologies of rhythm, often across species lines, and how these ecologies prepare the way for emergent love and wisdom.

## 35

In its earliest expressions in primitive animal life, the social and solitary rhythms very likely functioned reciprocally. That is to say, when one was active the other was dormant. Otherwise, the inward tendencies would have blocked social expressions and the social drives would have inhibited reflective learning. But this reciprocity is only one kind of interaction among many. In our actual experience, one rhythm never utterly vanquishes the other. The social and solitary behaviors move through life together. Neither has priority. Depending on conditions, they develop complex phase relationships sensitive to stimulation from cycles of nature, from other organisms, from internal drives, etc. The interacting rhythms either adjust themselves to each other or break each other down.

In primate, hominid and human life, certain signals encourage approach and bonding, others bring on enmities, competition, aggressive tendencies and pain. Along our line of development, nightfall, body warmth, infant dependency, food storage, dangers, sexual pleasure strengthen the aggregation impulse and its component parts. On the other hand, daylight, intelligence, cunning, curiosity and self-seeking support dispersal and produce social separation.

## 36

Picture yourself viewing the life of a city from a space satellite. You will see the great daily rhythms, and within them briefer rhythms. Alternating approaches and separations will reach a peak in the day and subside at night. To model these patterns, we would have each individual's patterns follow straightforward "ant-like" rules. From this, enormously complex patterns of dispersal and aggregation would develop within which approach-separation and withdrawal-return was contained in a circadian pulse.

Imagine we have been videoing city life with high-resolution cameras and we have a futuristic instrument that lets us see in the infrared and even take real-time MRI's. And suppose we have fast computers that can crunch the numbers. What will we see? We will recognize, first, that even on the smallest scales social and solitary behaviors are composed of simpler rhythmic components down to the molecular level. We will observe the social rhythms combining and taking on their larger approach-separation patterns. With our advanced monitoring equipment, we would witness the rhythms within and between cells. We would see cellular expansion, contraction, molecular aggregation, and dispersal. Viewed in brain imaging studies, we would see them producing outcome behaviors built on approach-separation and withdrawal-return patterns.

**37**

In both patterns we would observe fractal processes. That is to say, each larger behavioral pattern would be built from briefer rhythmic units. The social and solitary rhythms, represented *as changes over time*, would weave together threads of briefer approach/separation and withdrawal/return rhythms.

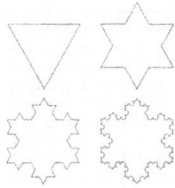

Figure 9.7 The first four stages in the construction of the Koch snowflake.

**Fractal Iterations of Koch Snowflake**

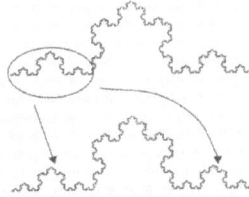

Figure 9.8 Magnification of the Koch snowflake.

**Imagine them Unfolding Over Time**

The fractal quality is more than a metaphor. The granularity is real. All living systems are made up of oscillating fragments, which look like wholes when seen from one perspective but look like fragments when seen from another.

The fractal qualities of rhythmic life are adaptive. Approach-and-separation and withdrawal-and-return, because nature compounds them out of many mini turnings, can adjust themselves to perturbation in many ways and still keep oscillating. These fractals unfold in time. To perceive their rhythmic qualities we have to see nature in movement: Not the full-grown tree or the finished seashell, but nature in process, *doing its thing*, nature naturing. Even the mountains heave. Everything moves. Everything unfolds in temporal patterns shot through with rhythms marked by reversal moments by which we establish their frequencies.

**38**

The notion that when events reach an extreme they turn around is basic in the West. Heraclitus emphasized it. Plutarch paraphrases him as saying: "One cannot grasp any mortal substance in a stable condition, but it scatters and then again gathers; it forms and dissolves, and approaches and departs." (Heraclitus Fragment LI.) John Burnet, a scholar of pre-Socratic philosophy, explains that in Heraclitan philosophy "'The strife of opposites' is really an 'attunement'. From this it follows that wisdom is not knowledge of many things, but the perception of the underling unity of the warring opposites."[1] "This conceptual scheme we are seeking here."

Polar rhythms play in the world on every scale. The myths of the heroes all show polar dynamics. They deal with reversal of fortune in fundamental ways. Death and rebirth—the symbol of the fullest kind of reversal—is at the core of shamanic and heroic transformation. According to Joseph Campbell, myths show that things move in a cosmic round "out of which they rise, which supports and fills them during the period of their manifestation, and back into which they must ultimately dissolve."[2]

The extremes are neither unnatural, regrettable nor in contention with each other. They do not engage in a tug of war over the fate of events. They are the dynamics by which the events are formed. The extremes belong to each other. They're part of a whole. They're mutually inducing.

From this polar movement human life takes on its essential characteristics. The extremes play parts in a coherent sequential design. They occur only at points of tension in rhythmic processes. Taoists locate this dynamic in the deepest tissue of nature. Lao Tzu writes of the Tao: "Being great, it is further described as receding, receding, it is described as far away, being far away it is described as turning back."[3]

*The Book of Balance and Harmony* says:

"Waxing is the beginning of waning: waning is the end of waxing. Waxing is the massing of energy; waning is the dissolution of matter. Growth and development is called waxing; returning to the root, submitting to destiny is called waning."

I will now show how our quintessential human behaviors follow these patterns.

### 39

Take approach-separation. Observe young lovers. They build their love from many brief approaches and separations. He calls, she says yes, they meet, they share, they part; she calls him, he calls her, they meet again. And so it proceeds. Or finally ends when one person's separation stage moves too far out of the other's range to sustain ties. Sometimes they both mutually retreat.

A single lovemaking incident itself contains many caresses, many approaches and separations, and within these caresses, many heartbeats happen, and in each of these there is an oscillation, a pulse and interval, an arousal and relaxation. Moreover, many arousals and relaxations are enclosed in yet larger oscillatory patterns as the lovers come to know each other. A love relationship construes itself, then, from thousands of small rhythmic gestures, many of them founded on approach-and-separation, shown in facial expressions, eye-contact, certainty and doubt, noddings of assent, demurrals, reciprocated caresses, crises of confidence, restorations of trust, interruptions, perturbations, oppositions. The mechanisms for these interactions – following responses, imitation, introjection, identification, are all mediated through sensation carried on rhythms of approach-and-separation. The pattern is not uniquely human. All social animals build their relationships from alternating movements of approach and separation. To acknowledge the biological substratum does not degrade or disparage human love. It shows its ties to nature and points to the evolutionary thru-line that all life carries on – the ur-rhythms of dispersal/aggregation and expansion/contraction. In other words, when we quantify the frequency-related behaviors conserved all through the tree of life, we find the building blocks culminating in love.

## 40

In mammals, the feeding/satiation rhythm evolves into a basic rhythm for approach and separation. Nursing at the breast is rhythmical. And some say food is the source of love.

The birth trauma itself is a separation, followed by a close reunion on the mother's breast. Infant feeding arrangements have natural contours based on the body's hunger and satiation rhythms. Even the way the nipple is given and taken away shapes it. The transactional elements in the contact of an infant's lips with a mother's breast are conveyed rhythmically. The two move together: there's an underlying dance, a rhythm; the mouth moves towards the stimulation of the breast; the rate at which the breast moves away determines the rate at which the mouth moves towards it, dampened by the softness of the breast. The following response in this setting builds upon physiological entrainment – we swing and sway to each other's movements. By responding to or leading the mother, and by making eye contact, the infant begins to acquire a sense-based kind of knowing that by their rhythmic underpinnings probably fall into patterns of duple and triple meter musically. And rhythmic contiguities open the way to deeper connections that develop into dances of entrainment. By entrainment I mean the sequences in which the parent's movements and the child's function as signals to each other, eliciting responses that elicit other responses in a train of rhythms that takes on something of the qualities of a pantomime of approaches and separations, analyzable on a time-series grid as a resonance phenomenon. Facial recognition between infant and mother goes back and forth.

In the social setting, normal tremor, at 6-10 Hz, may function as a fractal unit of approach and separation, and serve as a carrier for body language, gesticulation, emotional empathy, imitation and communication. The sexual embrace may transfer the tremor rhythms from one lover to another. Sexual encounters are fundamentally rhythmical. Even the lovers' heartbeats and brainwaves may be synchronized. The mutually reinforcing giving and receiving of pleasure is rhythm-driven. Riding to the peak, at the climax, in the moments leading to orgasm in sexual love right before the rhythm itself breaks down in its moment of fulfillment, lovers send many signals. A profound exchange bearing on bonding and commitment takes place in many bodily systems.

Social tremors working beneath the threshold of consciousness (they are accessible to human consciousness with sensory awareness training) establish routes along which individuals pass packets of information to each other. They travel first through touch, but normal social tremor can be seen and heard too. Perhaps these tremors of intent even surface in political contexts. They may become dominant when larger cultural rhythms break down.

## 41
### *Romeo and Juliet*

Shakespeare built the tragic drama of Romeo and Juliet entirely on turning points of approach and separation. The lovers meet at the masked ball and have to separate quickly because their families are feuding. However, Romeo approaches under Juliet's balcony in the night. Until morning breaks, they explore their intimacy, then separate, resolved to marry.

The next day they are secretly married in friar Laurence's cell. But the marriage is not consummated. They have to separate. The rhythm of love is here perturbed at its most vulnerable moment.

In their time apart, the world intrudes again. Romeo slays Tybalt and has to run away. Before leaving, he comes to Juliet in her bed chamber. They spend the night together. They make love. That's their closest approach. At dawn, they part. Romeo flees to Mantua, their furthest separation. Only their sense-memories remain. Their hopes attach to them. But further perturbations to rhythm thwart their plans. They end up dead in each other's arms, twin suicides. Final separation.

## 42

All through life, the cycles of approach and separation play out. They are there when a youth leaves home. They are resonant in every erotic desire. They play out over many cycles in the course of a marriage. And each signal for approach or separation is carried in sensation.

People learn early that they cannot touch someone from far away, but can see them and call to them; that closer in you can touch them and smell them and feel their heat. Certain experiences attach to sight, certain ones to hearing, to touch, to smell and to taste in distinct combinations. Approach brings data from certain ranges of sensation in a certain order, separation others.

In human social life, tone of voice, the modulation of the amplitude and timbre of the voice, carries meaning independent of and sometimes in contradiction to the words used. And many non-verbal body movements and gestures have social content. We receive and transmit them like attitudinal, spinal, semaphore signals that cover specific distances and communicate messages encouraging approach or separation appropriate to those distances.

These sensory messages are always transactional. They happen between persons. Someone is either coming or going. Two or more people are always engaging, refusing, or otherwise provoking each other through movement. The rhythmicity is in the "we" not the "I." And the engagement or refusal is actual, tangible, and physiological. These boundaries and crossings become measures of value because individuals prefer and seek out and feel safe in different sensory distance combinations.[4] From this process, starting in infancy, a primal sense of place and belonging grows.[5]

The social rhythms we receive through the senses always transfer information at a distance—in fact over *specific* distances depending on the mix of senses. They move in sequence from sight to hearing to touch to smell to taste. Moreover, we are sensitive to violations of this order. In *The Hidden Dimension*, Edward Hall mapped four basic distances underlying the approach-separation dance in all social interactions. He described them as the public, social, personal and intimate distances. The boundaries between them he called their structure points.

"By using one's self as a control and recording changing patterns of sensory input," he wrote, "it is possible to identify the structure points in the distance-seeking system. In effect, one identifies, one by one, the isolates making up the sets that constitute the intimate, personal, social and public spaces...For example," he continues, "the presence or absence of the sensation of the warmth from the body of another person marks the line between intimate and non-intimate space... By using one's self as a control and recording changing patterns of sensory input it is possible to identify the structure points in the distance-seeking system." [6]

## 43
### *Love Songs*

Love songs and poems clearly show the rhythmic movements through the four sensory distances. They detail the kinds of information and the feelings that occur in each range. Many of the standard tunes in the American songbook rely on this structure. The overriding tendency in the early comings and goings is for the lovers to get closer and closer. Though they have their separations, the mini cycles favor approach; they combine to form a big approach process.

Somewhere in the very midst of the compounded approaches, however, the balance between the approach and separation begins to shift; the tide turns, and the lovers recede from each other. Sometimes the new approach is weak or never comes.

*Our little dream castle with every dream gone,*
*Is lonely and silent,*
*The shades are all drawn.*
*And my heart is heavy as I gaze upon*
*A cottage for sale.*

The rhythms have been caught charmingly well in Shake-speare's *Venus and Adonis*, beginning with sight and ending with smell:

*Had I no eyes, but ears, my ears would love*
*That inward beauty and invisible*
*Or, were I deaf, thy outward parts would move*
*Each part in me that were but sensible:*
*Say, that the sense of feeling were bereft me,*
*And that I could not hear, nor see, nor touch*
*Though neither eyes nor ears, to hear nor see,*
*Yet should I be in love by touching thee.*

*And nothing but the very smell were left me,*
*Yet would my love to thee be still as much:*
*For from the still'tory of thy face excelling*
*Comes breath perfum'd, that breedeth love by smelling.*
*But, O, what banquet wert thou to the taste,*
*Being nurse and feeder of the other four!*
*Would they not wish the feast would ever last,*
*And bid Suspicion double-lock the door?*
*Lest Jealousy, that sour unwelcome guest,*
*Should, by his stealing in, disturb the feast.*

## 44

Every kind of loving relationship resides in these distances. Nevertheless, each distance favors one kind of love over another and pauses there, orients itself to that distance and finds its central values there. The virtue of compatibility, for instance, has different content and meaning in sexual love than social love because it finds its center in a different sensory circle, social love in sight and hearing, sex in touch, smell and taste.

The medieval troubadours chronicled the procession through the sensory distances (often approach to someone else's wife) with particular clarity. Each stage of love they tag with certain sensory expectations that make their lives meaningful or empty. Take *The Romance of the Rose,* (circa 1225) where the Rose symbolizes the beloved. The troubadour starts with sight and closes the distances:

*Amongst them all*
*My rapturous eyes on one did fall,*
*Whose perfect loveliness outvied*
*All those beside it. Then I spied*
*With joy its lovely petals...*

*And then is moved closer by smell that awakens a desire for touch.*
*When I caught its odor, I was wholly fraught*
*With strong desire that now I might*
*Snatch for my own that sweet delight.*[7]

As the intimacy rushes in towards its culmination, its rhythm is checked. The poet of the Rose stops in his tracks. Thorns and thistles around the rose "wound the profane hand" and block the touch.

## 45
### *The Song of Songs*

Of all love songs, the *Song of Songs* most profoundly portrays the movement through the sequence of sensory distances, with crucial meanings attached to each distance, showing the kinds of gains and losses involved and their inevitability. You can treat it as a mystical manual of instruction in approach and separation. The *Song of Songs* catches the content of all love songs.

From its first verse, it shows this search for union through deep sensory approach.

*Let him kiss me with the kisses of his mouth—For thy love is better than wine. Thine ointments have a goodly fragrance...*

It affirms the truth that a series of approaches and separations is the human condition for love, not only the soul's love for God, but of man and woman, full fleshed and vibrant with erotic desire. All have change at their core.

*Thou hast ravished my heart, my sister, my bride;*
*Thou hast ravished my heart with one of thy eyes,*
*With one bead of thy necklace.*
*How fair is thy love, my sister, my bride!*
*How much better is thy love than wine!*
*And the smell of thine ointments than all manner of spices!*
*Thy lips, O my bride, drop honey—*
*Honey and milk are under thy tongue...*

One wants to keep the union but it passes. The *Song of Songs* says that our wish for constancy is an illusion.

*I opened to my beloved; but my beloved had withdrawn himself, and was gone: my soul failed when he spake: I sought him but I could not find him; I called him, but he gave me no answer.*

(Song of Songs 5:6)

For biological reasons, the union of the lovers is tremulous and temporary and cannot last. In their heart of hearts, the lovers know that the world is wild and full of accidents and that reality, because of our mortal condition, has inconstancy in its core, and heartbreaks come and when they hit us in our phase vulnerable moments they change us.

Solomon understood that lovers, whether of each other or of God can't hold steady against the rhythmicity or the hazards of life, which is the tidal pull of the creation itself. Every love has its tensions and reversals, its little deaths and its transports of passion and agonies of despair. This suggests that the eternal union in love, though sought after, can never be achieved. As Adin Steinsaltz, the translator of the Babylonian Talmud into English, commented, "The Jewish approach to life considers the man who has stopped going – he who has a feeling of completion, of peace, of a great light from above that has brought him to rest – to be someone who has lost his way."[8] Knowing this and accepting it and still having the heart to seek love is the poem's homiletic fulcrum.

*I will rise now, and go about the city,*
*In the streets and in the broad ways,*
*I will seek him whom my soul loveth.*

**46**
### *The Four Kinds of Love*

Different kinds of love develop from the turnings at the extremes of approach and separation at different social distances. Following classical traditions, Rollo May distinguishes four kinds of love: "One is sex, or what we call lust, libido. The second is Eros, the drive of love to procreate or create—the urge, as the Greeks put it, toward higher forms of being and relationship. A third is Philia, or friendship, brotherly love. The fourth is Agape or Caritas as the Latin calls it, the love which is devoted to the welfare of the other, the prototype of which is the love of God for man."[9]

We must add to this that each of the four kinds of love moves with the polar dynamics of approach and separation. In each kind of love, the frequencies of approach-separation play out in episodes as brief as a tremor, to a season of joy or woe, to a lifetime of parental or conjugal commitment. However, in each love the contents are different. Though all four social distances participate, each love uses sensation differently, and emphasizes different sense organizations, and each love differs in the way it focuses attention, in how the intention behind it shapes events – and on what we remember afterward of it. And each love has its own neural circuitry shaping how the body moves, how the autonomic and endocrine systems engage with consciousness, how we seek out or avoid others, and with which senses out front the love achieves its turning points.

### *Sex*

The dynamics are clearest and closest to the body in sexual love, where intercourse, conception and the birth of the child are the manifest outcome of the encounter. Sexual love requires merging during closest approach. The lovers exchange bodily secretions. They communicate energetic rhythms in both directions. And they exchange information: tenderness, passion, consideration, mastery, revulsion, etc.

Successful conception depends on more than biology. Broader kinds of sexual compatibility and receptivity come into play. They too ride upon the rhythms of approach and separation. In the intimate distances, using the close-up senses, we enjoy smell and taste compatibility. Bonding is tightened by it. All kinds of sexual attractions and repulsions go on that shift the chemical content of the vaginal and seminal fluids.

In all transactions, a dance goes on. The chemical senses vibrate with approach and separation rhythms of their own down to the molecular scale. Signal molecules seek and enter receptor sites in each other's body. They disperse from their sources. They aggregate in the somatic being of the other. In many interactions, their rhythms are entrained.

Maybe the lovers meet in the morning, spent the day together, and part at night. They share a meal, walk together, sing and dance. Approach and separation work in many frequency ranges simultaneously.[10]

# 47
## *Eros*

Where sexual love manifestly uses the approach and separation rhythms to transfer body secretions in the closest sensory circle, Eros is less wound up in physical intimacy. It lives and breathes in the province of emotion. Physical approach and separation is carried by emotional changes. Like sex, Eros uses all the senses to move through all four social distances, but the interest centers in Hall's personal distance range. Here what counts most is voice timbre, glow of the skin, close eye contact, glances met and averted, small touches full of feeling, caresses, holding hands, listening, and making conversation.

Eros seeks an exclusive and unique connection with a particular beloved. Because of this desired intimacy, we want to get to know the other, but a knowing based more on mimesis, empathy and emotion than on intellect. In Plato's Symposium, Diatoma the Priestess explains that the defining characteristic of Eros is the completion of the self in the other. Two incomplete half-souls come together into a whole soul. The philosopher Robert Nozick describes this sensibility very well in his essay "Love's Bond". He writes, "The intention in love is to form a *we* and to identify with it as an extended self, to identify one's fortunes in large part with its fortunes."[11]

Like sexual lovers, erotic lovers enter each other's being through rhythmic sharing and modulation and find marvels of attraction and mutuality there. When there is mutual engagement in Eros the approach and separation is not only desired, it carries meaning: the bond says I am wanted, I belong, and I have my place in another's heart. Lovers even believe they are thinking the same thoughts at the same time.[12]

But Denis de Rougement, a mid 20th century French Catholic thinker, argued that romantic love actually seeks opposition and repulsion because it cannot stand its own unremitting intensity. He argued that "the erotic process introduces into life an element foreign to the systole and diastole of sexual attraction – a desire that never relapses, that nothing can satisfy, that even rejects and flees the temptation to obtain its fulfillment in the world..."[13] Eros relieves itself from its own relentless intimacy by creating crises.

De Rougemont only experiences union and completion in God's unconditional love. It takes Divine fiat to deliver us from the maelstrom of human love. Only with its help can we hold on to an ideal family life, resistant to change, based on fidelity, balance, obedience and wholesome caring.

"A fidelity maintained in the Name of what does not change as we change will gradually disclose some of its mystery: *beyond tragedy another happiness waits*. A happiness resembling the old, but no longer belonging to the form of the world, for this new happiness transforms the world."[14]

## 48
### *Philia*

Philia, the camaraderie that draws group members together in shared beliefs, rides on different sensory pathways than sex or Eros (again with overlap) but it too is a kind of love, sometimes a battle-field love.[15]

Because Philia is inclusive of groups, its sensory range, in Hall's social space, takes up a room, a town square, a gathering place, a dining table. Its focus is on the shared social space where groups convene and people gesticulate and speak. Plato's Symposium, the rather wild dinner and wine-fest of philosophers, was bound together by Philia.

In Philia the approaches and separations are multiple and simultaneous. It takes a different kind of attending to keep up with it. However, Philia does not achieve the relaxation of tensions that Rollo May hoped for because the rhythms of approach and separation never stop driving it. It too has its turbulence and turning points. Factions form and dissolve, conflicting loyalties and enmities build up. Some people are accepted, others rejected. "In-groups" divide themselves from "out-groups." In every heart-to-heart revelation, the possibility of a slander arises. Philia, from within its camaraderie, can build a world of "us and thems."

Even within the "us", conflicting approach and separation rhythms form as coalitions come together and dissolve and group members jostle for places. Each kind of love has its own kind of hate, every attraction its revulsion.

## 49
### *Agape*

Agape is charitable love, unconditional acceptance, a love that gives without thought of receiving. In this sense, it is a love that transcends personality. Ruysbroek wrote, "When love has carried us above and beyond all things, above the light, into the Divine Dark, there we are wrought and transformed by the Eternal Word..."[16] Agape is the only love that can be fully realized in the cosmic expanses. But Agape can also express in the closest physical intimacy; Philia cannot. Christian saints washed leper's sores. In crossing all distances and excluding no persons, Agape, according to its adherents, transcends the contradictions in nature, and so gives the lover access to eternity.

But even the great saints suffered reverses after their beatific enlightenment experiences. St. Francis had repeated spiritual crises. St. John of the Cross, the great Spanish mystic, wrote of his own experience in "The Dark Night of the Soul":

"That which this anxious soul feels most deeply is the conviction that God has abandoned it, of which it has no doubt; that He has cast it away into darkness as an abominable thing... All this and more the soul feels now, for a terrible apprehension has come upon it that thus it will be with it forever."[17]

**50**

The dynamics of approach, turning, separation, and turning back, though they are common to all loves, manifest differently in each kind of love, they pursue different aims at different physical distances, and each lover in the pursuit of that aim, discovers his or her own distinct knowledge of the beloved. Sexual lovers know each other one way, erotic lovers another, friends another, divine lovers another. Moreover, lovers in congress, and groups in fellowship, in every kind of love, find the significance, the pulling power, the hold of their bond in the belief (and experience) that they belong to something or to each other and that when they are moving in the field of love they are getting closer or further from it. Together the knowing, the being known, and the physiological resonances that carry it, which is the music of love, create its lived reality.

All lovers in every kind of social interaction sustain some kind of physiological congress. Sense receptors mediate the connection, though the communications array themselves differently among the senses. Each person has his or her own way of prioritizing the movements between approach and separation and may conceive of them distinctively, according to his sensory equipment and the learned responses he has acquired at the different distances over a lifetime. In each kind of loving congress, crucial exchanges of information happen, some volitional, some based on the following response and imitation that ordinarily occur below the threshold of consciousness. Each kind of love carries information unique to it.

Still, after every intimacy distance comes. Every separation sets up a possible approach. Pablo Neruda wrote:

*Sorrow rises and falls, comes near with its deep spoons, and no one*
*can live without this endless motion; without it there would be no*
*birth, no roof, no fence. It happens: we have to account for it.*[18]

It happens, but the rising and falling makes for an enormous problem. Neither blissful union nor terrible separation last. Furthermore, we repel as well as attract each other in all four kinds of love. The antipathies of the soul are as real and immediate and rhythmically expressive as its desires – as sexual arousal and performance. Hate is close to love. Emotional rejection and sexual satiation both reverberate on the legs of approach and separation. At every distance, both possibilities come up.

## 51
### *Closest Approach*

Where closest approach raises certain transformational possibilities, furthest separation deals out others. In the moment of union, closest approach, changes us through a primary exchange of materials, flavors, touches, caresses, but also with resonances, information and energy carried up from the deeper workings of the body.

At closest approach, where all the body rhythms, desires, and emotional and spiritual hopes have mingled, giving and receiving are joined and perhaps made indistinguishable, as Pablo Neruda says,

*like a double drum in the forest, pounding*
*against the thick wall of wet leaves.*[19]

Only those who have been close enough for rhythmic entrainment can separate with truly lasting after-effects. Those who have endured the greatest intimacy are those who through self-forgetting in those moments have been most profoundly changed by the other.

In closest intimacy, in some real sense the being of the other crosses over and is shared. Rhythm based physiological entrainments carry profound information about the tonalities of life of the other. Language, touch, tremor, thermal changes and aromatic biomolecules cross between lovers. They carry messages by frequency and amplitude modulation decoded in the primeval being of the other whether in desire or aversion, love or hate.

The contents of the giving and receiving may travel differently in each kind of love but the silent language of transmission and reception, the giving that is receiving and the receiving that is giving, is always there. As Robert Graves wrote:

*After when they disintwine*
*You from me and yours from mine,*
*Neither can be certain who*
*Was that I whose mine was you.*

*To the act again they go*
*More completely not to know.*

(The Thieves)

The giving that is receiving either lets us be changed by the other or, when the mutuality fails, prevents it. In all four kinds of love, the giving that is receiving and the receiving that is giving form the natural substratum of the ethics inherent in human nature. Kin and reciprocal altruism show the outer surface of it in ethological studies, but in human life love, not altruism, is its basis.

## 52
### *Furthest Separation*

By stages, separation takes lovers out of sensory range. Finally they reach a distance where the senses no longer meaningfully pick up the presence of the other. The beloved now only resides in memory, in sense memory. The impressions can be reviewed in the reverse order they were acquired until at the furthest distance the last thing remembered is the first encountered.

In furthest separation, we nurture the raw materials for transfiguration in loneliness. The seed of the next approach that is planted in the moment of union in visceral and intellectual knowledge, the scent of the other, the touch of the other, exchange of fluids, exchange of eye-contact, breaks into consciousness in furthest separation.[20]

In the farthest extremes of loneliness, at separation, longing burns the memory of the beloved into the heart. And the heart changes when memories and introjections burn themselves into it. That is what Solomon meant when he said to "Set me as a seal upon thine heart." In absence, we reconstruct the other in the body. Hart Crane's poem *Carrier Letter* says it well:

*My hands have not touched water since your hands, –*
*No; – nor my lips freed laughter since 'farewell'.*
*And with the day, distance again expands*
*Between us, voiceless as an uncoiled shell.*

*Yet, – much follows, much endures… Trust birds alone:*
*A dove's wings clung around my heart last night*
*With surging gentleness; and the blue stone*
*Set in the tryst-ring has but worn more bright.*

In all kinds of love, at farthest distance, imagery floods into consciousness and begins to shape our expectations for the next approach, whether with fear or hope.

## 53
### *Turn My Beloved: Turning Points in Love*

*Until the day breathe, and the shadows flee away,*
*Turn, my beloved, and be thou like a gazelle or a young hart*
*Upon the mountain of spices…*

"Turn, my beloved" is the operative phrase here, the defining act, the shape-giving event. The turning point moments matter most. In them, we find or lose each other. We celebrate or mourn, love or hate. Our weddings, divorces and funerals mark these passages in the larger social world. They express not only in conjugal love, but in every kind of love. A child's love for its parents, brother and sisterly love, camaraderie, lifelong friendships, marriages and affinity groups all show it. We define the larger contours, the 'story content' of love, by its turning moments.

However, turning always involves personal change. The way back is never a mirror image of the way here. Not constancy but mutability gives love its true character. The yearning for constancy comes as a response to its mutability.

The crucial test for love is living with change and being true to change. The pledge taken in intimacy, as if the intimacy would last forever, remains empty until we test it in the fires of loneliness. In the sonnets, Shakespeare frequently tries to mend the eroding effects of time on commitment. We soothe it mainly by denying it. In protesting, however, he must admit there is a breach that needs mending.

*Let me not to the marriage of true minds*
*admit impediments. Love is not love*
*Which alters when it alteration finds,*
*Or bends with the remover to remove.*[21]

Elizabethan poets usually celebrated their constancy when they were preparing to leave their lovers. Here's John Donne.

*Our two souls therefore, which are one,*
*Though I must go, endure not yet*
*A breach, but an expansion*
*Like gold to airy thinness beat.*[22]

What lovers lose in constancy, they stand to gain from the power of change itself, even when the change follows terrible losses. Freedom makes its presence known in these changes. Turning points bravely endured set us on new paths.

## 54

We are predisposed by nature and nurture to specialize in one kind of love, to prefer one kind to another. In each kind of love, we follow our proclivities. We have intense or mild turning points, strength in approach or separation. We are histrionic or flat, extroverted or introverted. We feel safe and comfortable in sex, Eros, Philia or Agape. Or on certain legs thereof.

Our ways of proceeding mutate. In our greatest turnings, one kind of love can turn into another. At the extremes of closest approach or furthest separation, near the reversal moments, sex can become Eros, or Philia Agape, and these can happen in all permutations.

The introjected components of the beloved, sealed in the heart in furthest separation, return to the world transformed. They usually go to the person who gave them, but not always with the same kind of love. The introjected presence of other lovers plays a part. Past loves shape us. The shadow or light from our mothers or fathers can darken a new relationship. All those one has ever loved may return to the world through each new intimacy. When one kind of love changes into another, and something of the states and traits of all our past loves mingle, love changes tracks.

The sensory boundaries that shape the character of each kind of love also function as gateways to other kinds of love. We go through the gateways at vulnerable moments. The vulnerable moments tend to come during turning dramas.

After almost any calamity, love can reestablish itself. It relies on the power of rhythm to restore balance when perturbations hit at vulnerable moments near turning points. In the resetting process, love can change tracks. Eros can become Philia, or Philia sexual love, in all kinds of combinations. These transformations happen because all four loves have turning points in them that ride on the common tides of approach and separation. Moreover, they do so in the same frequency bands, reflecting sleep/waking rest/activity, desire/satiation rhythms in the circadian and ultradian bands.

These common dynamics are what keep the four loves potentially interchangeable, so that after significant perturbations, we do sometimes come back to our lovers on fresh foundations, sometimes shifting to a different kind of love. Sometimes the beloved accepts the new configuration, but by no means always. Sometimes the new love has to seek out a different object.

### *End Notes*

[1] John Burnet. *Early Greek Philosophy.* Meridian. 1957. p.143
[2] Campbell. *The Hero With a Thousand Faces.* Meridian p-.257
[3] Tao te Ching Chapter 25
[4] "Not only are there introverts and extroverts, authoritarian and egalitarian, Apollonian and Dionysian types and all the other shades and grades of personalities," Hall tells us, "but each one of us has a number of learned situational personalities. The simplest form of the situational personality is that associated with responses to intimate, personal, social and public transactions. Some individuals never develop the public phase of their personalities and, therefore, cannot fill public spaces; they make very poor speakers or moderators. As many psychiatrists know, other people have trouble with the intimate and personal zones and cannot endure closeness to others." ibid p. 115
[5] different aims, different neural centers are stimulated and distance specific responses are generated along motor pathways. Together these produce approach and separation rhythms that are communicated between people with specific nuances of

meaning transmitted as modulations on the mini-cycles of the broader approach-separation rhythms. The different senses are set to receive data from specific ranges in each kind of social interaction and these distances unfold in a certain order reflecting movement closer to and farther away from somebody. While in a sensory range, interpersonal communication occurs in the frequency bands natural to the body at those distances, generally with sight first and furthest away, then hearing closer in and touch, smell and taste the closest. You react viscerally whenever your space is violated and someone "gets in your face."

Each distance conveys its information on a dominant carrier wave and these shape our rhythmic units of behavior, depending on its frequency and amplitude characteristics. The closer the distance, the deeper the intimacy and the slower the frequency. The visual world, the longest ranging human distance sense, picks up frequencies in the 400-700 megahertz range and then steps them down by biochemical reactions in the retina. Along with sight hearing dominates in the public and social distances, but hearing depends on pressure waves in the air vibrating at frequencies from 20 to 20,000hz. And these frequencies are carried through intact from the eardrum to vibrate the cilia in the cochlea where they are turned into nerve frequencies. Speech, the dominant sensory transmitter in the personal distance range, has an even narrower and slower frequency spectrum. Its information conveyed by oscillations in the larynx over stretched vocal cords, and humans only speak on an exhale, and since the exhalation of the breath is partly controlled by the rhythms of the autonomic nervous system it is stepped down to the general tonus of arousal. In the intimate distance, touch, smell and taste run on even slower carrier frequencies. If it is true that at each sensory distance, moving from the public to the intimate range, people extract the most meaningful information from a slower carrier, it would mean that while certain faster information cannot be conveyed by the dominant senses in intimate connections, certain slower communications, as in the undulations of sexual love, that carry their own special meanings, *can* be conveyed – or rather sublimated – in the faster and more distant sense arrays.

[6] Edward Hall. *The Hidden Dimension.* Anchor 1969. p.115
[7] The Romance of the Rose, (circa 1225)
[8] Adin Steinsaltz. *The Thirteen petalled Rose.* Basic Books. 1980. p. 132

[9] Rollo May. *Love and Will.* Delta Books. 1969

[10] Rollo May characterized it this way: "In sexual intercourse we directly and intimately experience this polar rhythm. The sex act is the most powerful enactment of relatedness imaginable, for it is the drama of approach and entrance and full union, then partial separation... It cannot be an accident of nature that in sex we thus enact the sacrament of intimacy and withdrawal, union and distance, separating ourselves and giving ourselves in full union again. For this eternally repeated participation in each other, the touch and withdrawal, is present even in the hesitant beginnings of acquaintanceship and is the essence of courtship in birds and animals as well as men and women. In the rhythm of participation in a union in a dual being and the eventual separation into individual autonomy are contained the two necessary poles of human existence itself..."

[11] Nozick. *The Examined Self.* Touchstone. 1989. p.78

[12] Rollo May contrasts Eros, as a continuing, deepening experience, with sex, which he sees as more strictly rhythm bound. "Whereas sex is a rhythm of stimulus and response," he explains, "Eros is a state of being. The pleasure in sex is described by Freud and others as the reduction of tension; in Eros, on the contrary, we wish not to be released from the excitement but rather to hang on to it, to bask in it, and even to increase it."[12] Experience contradicts this. There are dramatic reversals in Eros and all kinds of instabilities and conflicts. Love and Will p. 71.

[13] Denis de Rougement. *Love in the Western World.* Harper. 1956 [Originally published in 1940] p.62 "Our human passions are always connected with antagonistic passions, our love with hate, and our pleasures with our pains. Between joy and its external cause there is always some gap and some obstruction—society, sin, virtue, the body, the separate self. Hence arises the ardor of passion. And hence it is that the wish for complete union is indissolubly linked with a wish for the death that brings release."[13]

[14] p.323

[15] Rollo May locates the greatest degree of concord in Philia. Here a more resilient and durable love can flower and bring on a further rhythmic relaxation. To May, Philia stabilizes Eros. "We have said that sex is saved from self-destruction by Eros, and that this is the normal condition. But Eros cannot live without Philia, brotherly love and friendship. The tension of continuous attraction and continuous passion

would be unbearable if it lasted forever. Philia is the relaxation in the presence of the beloved which accepts the other's
being as being..."

[16] Ruysbroeck, *The Book of Truth*, chap ix.

[17] Ruysbroeck reported similarly: "Here all the storm, the fury,
the impatience of his love, grow cool: glowing summer turns
to autumn, all its riches are transformed into a great poverty.
And the man begins to complain because of his wretched-
ness: for where now are the ardours of love, the intimacy, the
gratitude, the joyful praise... How have all these failed him."

[18] Pablo Neruda, from Sonnet 55

[19] Neruda sonnet 79

[20] Neruda's sonnet to Mathilde shows this:

*I crave your mouth your voice, your hair.*
*Silent and starving, I prowl through the streets. Bread does*
*not nourish me, dawn disrupts me, all day I hunt for the liquid*
*measure of your steps.*
*I hunger for your sleek laugh,*
*your hands the color of a savage harvest,*
*hunger for the pale stones of your fingernails, I want to eat*
*your skin like a whole almond.*
*I want to eat the sunbeam flaring in your lovely body,*
*the sovereign nose of your arrogant face,*
*I want to eat the fleeting shade of your lashes,*
and I pace around hungry, sniffing the twilight, hunting for
you, for your hot heart, like a puma in the barrens of
Quitratue. Neruda sonnet 11

[21] *O, no! It is an ever fixed mark,*
*That looks on tempests and is never shaken;*
*It is the star to every wandering bark,*
Whose worth's unknown, although his height be taken.**Love's
not** Time's fool, though rosy lips and cheeks
*Within his bending sickle's compass come;*
*Love alters not with his brief hours and weeks,*
*But bears it out even to the edge of doom.*
*If this be error, and upon me proved,*
I never writ, nor no man ever loved.
The last couplet "If this be error" has a certain ring of irony to
it. In the very excessiveness of the claim "I never writ, nor no
man ever loved", Shakespeare questions the constancy if not
the love itself.

[22] John Donne. From A Valediction: Forbidding Mourning

# 3. WISDOM AND THE SOLITARY LIFE

**Going Down In and Coming Back Out**

55

Wisdom like love is rhythm based. Its pulses come from conserved expansion/contraction rhythms evolved by single-celled organisms.

What began as a movement between the nucleus of a cell and its parts becomes, in human life, a dialogue between the soul, the self and the world.

What began in the swelling and shrinking of independently living cells, by multiplication became vast number of cells moving in synchrony across cells and organs. Rudolpho Llinas suggests that we generate consciousness as an emergent property of neural synchronies.

"Neurons that display rhythmic oscillatory behavior may entrain to each other via action potentials. The resulting, far-reaching consequence of this is neuronal groups that oscillate in phase—that is, coherently, which supports simultaneity of activity. Consider the issue of coherence from the perspective of communication, for coherence is what communication rides on."[1]

## 56

Withdrawal and return generates its own distinct sensory grids. They do not build up from Hall's four social distances. Instead, the sensations have interoceptive and proprioceptive origins. They are gut sensations stimulated by humoral secretions and nerve traffic. They influence breath and heartbeat, skin flush or pallor, pheromone flows, arousal, relaxation and to some extent all moods and states of mind.

Fallings, risings, elations, depressions, aches and chills, little scintillations, gut rumblings, sensual thrills, tastes and revulsions, balance and imbalance are the sensory landscape of the withdrawal-return journey. The inward senses register pains and pleasures, warmth, coolness, pressure, visceral states orchestrated to the rhythms of breath, heartbeat, digestion and tremor. "The various permutations of which these organic changes are susceptible," William James said, "make it abstractly possible that no shade of emotion should be without a bodily reverberation as unique, when taken in its totality, as is the mental mood itself."[2]

From this palette of sensations, we derive our sense of inner security or insecurity, well-being, despondency, despair or joy. No animal, however primitive, does entirely without these inward sensations and their inward-outward, efferent-afferent connections. In human life, they form the basis for self-knowledge.[3]

Soren Kierkegaard, the nineteenth century Danish existential philosopher, observed that "The only reality to which an existing individual may have a relation that is more than cognitive is his own reality, the fact that he exists; this reality constitutes his absolute interest..."[4] From this place a person makes choices. Here ethics lives. Kierkegaard insisted, "The ethical, as being internal, cannot be observed by an outsider. It can be realized only by the individual subject, who alone can know what it is that moves within him."[5]

## 57

During the earlier phases of withdrawal, the senses turn away from the outer world. Over the course of numerous mini withdrawal-return events, withdrawal comes to dominate.

Like love, wisdom has a fractal structure. As with love, wisdom combines, perhaps down to some irreducible quick social turning, its mini-cycles of withdrawal and return into larger patterns of meaning. As in love, no stage of wisdom in the succession of stages is wisdom itself. No single realization, insight or epiphany constitutes wisdom, only the process as a whole; as with love, constancy is the illusion, change is the reality. We may seek the perennial wisdom, but we cannot find it. God's truth is inaccessible. Mystics who climb close fall away. In Jewish mysticism, God himself withdraws and returns, comes closer and gets further away depending on our readiness to turn toward Him. *Teshuvah*, the Jewish notion of repentance, depends on these turnings.

## 58

The full course of a wisdom quest can last for months or years until finally its small polar units arrange themselves into an overall pattern of withdrawal-return with a central turning point, a turning of turnings, as its shape-giver. In both love and wisdom, the big turning passes quickly compared to the arduous trail of tests and attempts that leads to and from it.

However, unlike love, we experience the force of wisdom in solitude. In solitude, the mini-withdrawals become progressively longer. They pull us further inside ourselves. The mini-returns become brief and superficial. Internal sensation draws us away from the surface, not always willingly. In pursuit of wisdom, we dig our way into our own depths wherever else we go. After a deep turning, another fractal journey takes us back to the social world.

## 59

The daily sleep cycle is a fundamental unit for withdrawal and return. Animal species widely distributed through the tree of life sleep and rise. Sleep divides human life into its narrative units. Waking is always a return, sleep a withdrawal. We remember our days and sequence them into stories. Sleep itself is divided into stages lasting 90 minutes or so, and these stages oscillate in patterns of withdrawal and return. Nathaniel Kleitman first identified these patterns.

Allen Hobson describes the sleep rhythm as a shift in sensory interest from "stimulated neuronal information" (i.e. from the outside world) – to "spontaneous neuronal information" (something the brain is doing internally) with dreams as outcomes. When we include dream life in the sleep/waking cycle, we have the basic ingredients from which human inwardness very likely developed.[6]

Beside sleep and dream, other briefer oscillations structure the inward/outward journey. Perhaps the rhythms of normal tremor, as with love, set the beat for the interchange between behavior and cognition. Perhaps it is the briefest unit that communicates between deeds and the effects of deeds on character. In those frequencies, life carries the transformational power of events back into physiology and from physiology into solitary awareness. Single-celled colonial gonium, perhaps the earliest social species, vibrate in the tremor frequency range. More than one great creator has shaken like a leaf in the moment of discovery when the "frisson", the thrill felt on the skin, accompanied the epiphany. We can consider the 300-millisecond P300 wave, the evoked response potential, as a candidate for the basic cognitive unit. That is the time it takes for the EEG to register the brain's response to a stimulus, and may be the fractal unit that first organizes cognitive processing into patterns of withdrawal-return. The amplitude of the P300 itself varies with circadian rhythms, and along with the rest/activation rhythms, and other rhythms we will treat later in the book, it may represent one of the "ticks" of the body's many living clocks.[7]

### 60

Neurobiological evidence suggests that, just as we bind together the separate data from the outside world into whole perceptions by means of cortex-spanning neural synchronies, in similar ways we bind the inner world and its internal sense data together. As sensory following responses, imitation, introjection and projection bind us to others, a parallel entrainment occurs in the world of solitude inside the bounding membrane of the organism. It links us to our own rhythms, our basic cognitive units and the nuances of our emotional lives. This pulse of coordination suggests that we construct our sense of self from a drama of withdrawal/return.

## 61
### *The Phenomenology of the Wisdom Journey: Withdrawal*

Because the way inward and the way back rarely run smoothly, we experience the wisdom journey as a succession of adventures and ordeals. The stages, because they come together from combined nonlinear biological oscillators, acquire great complexity. The world subjects them to perturbing forces. The stages differ in intensity, content and outcome. Though the root pattern of withdrawal and return holds, each quest has its own time course and branching moments. They take detours and hit dead ends. Every wisdom path, as lived, is unique.

Withdrawal-return patterns differ, first, by the circumstances that draw a person inward. Suffering, disappointment or illness, or loss in material, political or economic life can trigger a withdrawal. Relationship problems can force one away from the social world. Some individuals go inward to find or preserve sanity, to leave behind a consensus reality they find stifling. Such departures can happen in the midst of apparent success, when success comes up empty or seems absurd. Depending on a person's makeup, withdrawal can turn a man introspective, ruminative, concentrated, agitated, obsessed, and bewildered or depressed, each in various successions and combinations.

In deep withdrawal, the map of custom falls to pieces. The context by which we orient ourselves in the world breaks down. The habitual sequencing of personality parts shudders apart; the pieces shift or fall away. We become awkward. Our characteristic gestures and patterns of muscle use, our habitual expostulations, sighs, grimaces, seem empty. We jumble our competencies and emotions. Our ways of accessing the body falter. We realize that the "I" that departed on the inward voyage, whether with high hopes or dread, can no longer be counted on as the familiar "I". It has turned into a stranger whose motives are uncertain and whose worth is untested. Our plans no longer fit our circumstances. Our emotional stability is shaken. We can no longer rely on ourselves.

At some point, the retreat of the senses from the outside world to the sensing of the body may become troubling and threatening.

Periods of anguish and anxiety mark this condition. Unable to receive affirmation from outside for what we no longer are, and reluctant to ask the world to acknowledge what we have not yet become, we retreat into solitude. Life impels us to question our steadiest assumptions.

## 62

Hans Castorp's experiences in Thomas Mann's novel *The Magic Mountain*, show these stages of withdrawal and return with clinical clarity. In his first weeks at the Swiss tuberculosis sanitarium, Castorp feels tired, enervated, his senses dulled. Then, though he had nothing wrong with him when he arrived he gets faint and feverish, the thermometer confirms his fever.[8]

A similar sequence occurs in C. S. Lewis' description of his suffering after the death of his wife. In the first entry in his journal, later published as *A Grief Observed,* he writes, "No one ever told me that grief felt so like fear. I am not afraid but the sensation is like being afraid. The same fluttering in the stomach, the same restlessness, the yawning. I keep on swallowing."[9] Notice how he focuses on interoceptive sensation. Lewis is moving into darkness, into distances of removal. "At other times it feels like being mildly drunk or concussed. There is a sort of invisible blanket between the world and me."[10] The long period of indwelling sensation drained his interest from the outer world of the senses. "I see the rowan berries reddening and don't know for a moment why they, of all things, should be depressing. I hear a clock strike and some quality it always had before has gone out of the sound. What's wrong with the world to make it so flat, shabby, worn out looking? Then I remember."[11] Uninvited memories well up, shaping a world increasingly phantasmagorical.[12]

The absence of outward anchoring consensual validation in our wisdom quests amplifies the impact of the disruptive forces in the stages of withdrawal. "To some I am worse than an embarrassment," Lewis wrote, "I am a death's head. Whenever I meet a happily married couple I can feel them both thinking. 'One or the other of us must someday be as he is now.'"[13]

The world Lewis left behind had given him his map coordinates. But when sufficient emotional and physical pressure bore down on him for long enough, the consensual world slipped away.

Under those pressures, our image of how others see us recedes. They no longer remind us of who we are and where we fit into the world. People retreat from us, or we push them away, and the loss of amity and comity opens us up to chaotic shifts in our internal fields of sensation. Something in us is cut loose. Where we were once deferential, we become aggressive; where gregarious, now remote; where motivated, now lax. What was opened seems closed; what was closed seems open. What we once knew we do not know now, what we never noticed before plagues us. We live with shocking uncertainty.[14]

## 63

We get no comfort either in the extremes of loneliness or the extremes of engagement. Our very hold on ourselves thins out then. The threat of loss of self is characteristic of extreme situations and sometimes brings with it tremendous accessions of fear. In these moments human nature—the lived reality of it, I mean— throws us into disarray. We respond to pressures from inside and outside by decohering. Turbulence degrades the body's oscillators. A person under remorseless pressure can break down emotionally or get ill. The ordinary sense of wholeness retreats.

Most people experience the perturbations and the biochemical cascades as a kind of inner dissociation. The I-sense is denuded. Subpersonalities move in and out of frontality seemingly on their own volition. In this condition, we yearn to hold on to the self. We do not want to let go. But there comes a time when, unless we let go, we must abort the turning process toward which we have striven. The anxiety that comes with *holding on* to what is determined to slip away anyway can alienate a person from his or her real needs. To experience and endure transformation you must hang tough, even while in doubt, adhering to doubt. But you hang tough by letting go of the old self-system. Like Odysseus, you stay present in the storm even while the boat founders.

Whatever provokes it, and whether you go willingly or not, you experience withdrawal into the depths as a painful, turbulent and destabilizing time. To reject the surface of life and turn inward (and to be rejected for rejecting it) gravely tests the spirit.

Dante's Divine Comedy opens with the poet lost.

*In the midway of this our mortal life,*
*I found me in a gloomy wood, astray,*
*Gone from the path direct...*

Gilgamesh, the legendary king of Uruk in ancient Mesopotamia, travels alone through solitary perils following the death of Enkidu, his only friend and soul brother. In his story, we are told, "When he had gone one league the darkness became thick around him, for there was no light, he could see nothing ahead and nothing behind him. After two leagues the darkness was thick and there was no light, etc...." Then "In his bitterness he cried, 'How can I rest, how can I be at peace? Despair is in my heart?'"

In this earliest written record of a mythical journey, Gilgamesh travels to the ends of the earth seeking the secret to eternal life. Only this boon (which he intends to offer to the respected elders of his city first and then use himself,) will console him for the death of Enkidu. At the ends of the earth, he learns about a plant of immortality growing at the bottom of the sea. He dives for it and sinks into the unknown. The adventure of withdrawal reaches its nadir. "He tied heavy stones to his feet and they dragged him down to the water bed. There he saw the plant growing; although it pricked him, he took it in his hands; then he cut the heavy stones from his feet, and the sea carried him and threw him on the shore." The turning moment comes when he cuts the heavy stones from his feet and bobs up to the surface.

## 64
### *Phenomenology of the Wisdom Journey: return*

After the turning, a sensory shift away from interoception to exteroception occurs, and with it a redistribution of energy. One experiences a change from frustration to enthusiasm for action, often accompanied by a sudden experience of simplification and a sense of the heightened power of the moment. The autonomic nervous system shifts gears. Imagination surges. It discovers new motivations for action. It leaves behind its compensatory fantasies and ruminations.

In the return phase, we find ourselves saying yes where we used to say no and no where we used to say yes. Our sense of our own prospects changes. We talk to people differently. Time flows differently. Space opens up. Meaning is new. We revise our model of reality. And these changes bring with them not only fresh understandings but the chance to attain new deeds, and as a side-benefit, if one is attentive to it in ways I shall describe later in the book, we acquire a measure of voluntary control over autonomic functions previously inaccessible to consciousness. These new skills and powers change the way we distribute our energy on the long leg back.

The circumstances endured and surmounted on the inward journey serve us best when the turning point lines up with a readiness of the world, a readiness to receive us when we come back – a readiness to receive the gift we are offering. How wonderful when the world seems to want us and opens its doors! However, the world may refuse our gifts or use them wrongly. What then? We have to keep offering them wholeheartedly. In this our strength lies. When we travel into the depths with sufficient motivation and bravely endure its trials, the ordeal becomes transformative. It has ethical force. It provokes a creative crisis that expands our range of choices and enlarges our circle of freedom. But choosing something and doing it are different. In the turmoil of the turning, we must find the strength to return *in actuality*. We must take real steps in that direction. You take the first real step with a sealing gesture; it puts you on the road back to the world of others.

Without a return to the world commensurate with the pull of the inward journey, a person's growth is inhibited. The half journey makes one angry, bitter, or crazy. Only the full round-trip confers the benefits of wisdom. Every cultural tradition, from Gilgamish to Odysseus to Dante, celebrates the full round-trip journey as the source of renewal in life.

## 65

Joseph Campbell wrote the clearest description of the wisdom journey I know. *In The Hero With a Thousand Faces,* he divided the quest into three stages. "The first step, detachment or withdrawal, consists in a radical transfer of emphasis from the external to the internal world, macro- to microcosm, a retreat from the desperation of the waste land to the peace of the everlasting realm that is within."[15] The second is the "supreme ordeal" at the center, through which we accomplish transformation. Campbell tells us:

"When [the hero] arrives at the nadir of the mythological round, he undergoes a supreme ordeal and gains his reward."[16] He describes the central turning experience as a triumph. "The triumph may be represented as the hero's sexual union with the goddess-mother of the world (sacred marriage), his recognition of the father-creator (father atonement), his own divinization (apotheosis), or again—if the powers have remained unfriendly to him—his theft of the boon he came to gain (bride-theft, fire-theft); intrinsically it is an expansion of consciousness and therewith of being (illumination, transfiguration, freedom)."[17]

The third stage "is to return then to us, transfigured, and teach the lesson he has learned of life renewed." But the turning, interestingly (and importantly for our study,) is a thing of the moment. Gilgamesh kicks the weights off his feet. He floats up with the herb of immortality. The sea throws him on the shore.

The quickness of the starting deed usually seems insignificant in comparison to the quest as a whole. But this decisive stroke marks the turning moment. It is always emphasized. Intention alone will not do. Only a deed can cap the long interneuronal process paralleling the withdrawal. Only a *meaningful action* will bring together, by deep tissue and organ changes, the intention to act with a sufficient implementing gesture. Without the actual deed, we make empty choices, but without the consciousness of choice, we leave our deeds deprived.

Ethical authenticity, therefore, develops from our conscious presence in our own turning processes and results in the coalescence of intention, choice and action.

This brief outflow, however, only provides the first impetus for a long homeward journey full of perils.

The way back can take strange twists. Often the new self assembled in the depths, which is neither simple, single nor unbreakable, has to undergo great ordeals to survive with a sense of its present purpose intact. Moreover, one may not have the strength to do it. One may not be able to offer oneself repeatedly in the face of rejection.

As it turned out, Gilgamesh never did bring back the herb of immortality. On the way home, a serpent stole it. Gilgamesh reentered Uruk empty handed. All he could do was tell his story. He engraved it in stone. By doing so, he became the first to record the depth journey, and this turned out to be his real accomplishment.

**66**

Human beings acquired inwardness a long time ago. It happened when personal withdrawal and return became distinguishable from larger group dispersal/aggregation rhythms.

Jung, drawing from Levy-Bruhl, referred this condition as the breaking of the "participation mystique", a necessary precursor for individuation.

At some venue deep in the past, we were thrown upon ourselves in solitude. The daily pulse of group dispersal and aggregation fell into the background then. We knew we were alone. From that *felt* solitude, in those settings, we *remembered* rather than *encountered* others. We reassured ourselves that, though absent, they had not ceased to be. At first, the situation may not have been quite clear. As Julian Jaynes pointed out in *The Origin of Consciousness in the Breakdown of the Bicameral Mind,* we may have heard voices in our heads that we attributed to the presence of the gods. According to Jaynes, it took us scores of thousands of years to own these experiences by assimilating these auditory hallucinations to consciousness. [18]

It took aloneness to make us aware that we belonged. The others had not abandoned us. We may even have kept our composure by recalling the perceptions of the voices, faces, bodies and expressions of those closest to us: mother, siblings, father, group leaders, and companions. At some illuminating moment, we came to understand the *significance* of the recollected voices as something distinct from the persons to whom they belonged, as reminders not presences. Once we had that perspective, we very quickly started engaging in imaginary conversations with our mental companions. We do it still, rehearsing arguments and exploring our possibilities through a dialectical process in dialogue with imagined others.

Language must have become a basic processing unit in the withdrawal-return cycle, one of the narrative vehicles with which the adventure in solitude was conducted. The language buzzing inside us must have had its parallel in real-life exchanges. In conversation, we could lay down the markers of our values much more

81

clearly. We could throw a conceptual net over a moment, a day, or a lifetime. Just imagine the creative insight it took to devise words for time, space, belief, truth, choice, spirit, energy, soul, etc. – and how long ago it occurred and how permanent an endowment it has become.

As familiarity with the inward experience developed, some of our forbears must have been better at it than others. Some must have intentionally sought out places of solitude in order to explore inward consciousness, to think deeply, to model possibilities, to commune with the gods, or to become possessed by waking dreams and fall into trances. They became our first bards and shamans.

### 67

To acquire their professional powers, shamans needed periods of isolation. The American Indian vision quest illustrates this vividly, as does the even earlier tracery of the Australian aborigine dreamtime. Everywhere isolation of some sort becomes the road to spiritual attainment. The shamanic sensibilities revealed in the cave paintings of the high Paleolithic must have had a developmental history going back scores of thousands of years. A penchant for solitude must have been strong in human evolution.[19]

**Shamanic illumination.
Rock painting**

Shamanic wisdom is directly responsive to the rhythms of life. It seeks a resonant relationship with them. The Paleolithic shaman saw birth and death as the most powerful, mysterious manifestations of the rhythms of nature.

Birth was the first aggregation, death the last dispersal. The killed being, human or animal, once dispersed, could not aggregate to the group again. Through death, the dispersal door, the male door, all life passed into another realm, the underworld. The female, by contrast, was the gate to life.

The vagina and the grave were the doorways through which the greater aggregation and dispersal rhythm swept into the world. They were the apertures between our middle world and the upper and lower realms. Life aggregated through the sex door of the female and dispersed through the death door. A killed being, human or animal, was dispersed. S/He had a journey to take; s/he could not aggregate to the group again without being born again. The shaman's calling was to keep both approach-and-separation and withdrawal-and-return balanced on the greater turning wheel of dispersal and aggregation. The secret was to correlate these with the broader rhythms of nature, read from the cosmic wheel in the starry sky, the retreat and return of the light, the passing of the seasons, and the journey of the sun through the firmament. As Giorgio de Santillana speculated, these observations, conducted over centuries and millennia, encouraged the naming of constellations and the partitioning of the heavens into the twelve signs of the zodiac.

By a journey of withdrawal and return the shaman tried to keep the social order running smoothly. Shamanic wisdom saw all creatures sweeping back and forth through the birth and death doors, dispersing, aggregating, separating from, and approaching on the stage of the living world. The shaman sought the secret doors and went through them. They knew that the membranes separating life and death could sometimes, on certain occasions, soften and become permeable. With their esoteric skills they could travel between the worlds in dream, trance and hallucination. Siberian shamans, for example, in a dream or trance state, ascended a ladder pole representing the world axletree set between our world, the underworld and the over-world.

The shaman, moving in the fields of dispersal and aggregation, was the intercessor with the dead, the seeker of lost souls, human or animal. The lost soul could be caught between the membranes, sometimes held captive by other dispersed beings among the dead. The shaman could travel in the underworld and induce one who could not fully die to die, helping them find what they needed to complete their journey, or by helping one who could not be born to get born or one who was ill, whose soul was stuck between the worlds, to return and heal. The shaman might meet with an esteemed ancestor and persuade it to yield special wisdom needed by the group.[20]

The shamans made seasonal journeys. They timed their séances to coincide with the solstices, the appearance of migrating herds, or if need be to the rhythms of the crisis in a patient's illness. The shamanic séance itself was immersed in rhythm. Drums and rattles accompanied it.

The shaman succeeded by actually influencing a turning point in broader nature by means of a withdrawal-return turning point of his or her own. The shaman aligned his or her moving whirl, the vortex of ascension and descent, into the greater rhythm of dispersal and aggregation that formed the vortex of the cosmos itself.

## 68

The wisdom process, judged from the great age of shamanism, preceded dramatic changes in the tool kit by many thousands of years. The evidence indicates that the achievements of the High Paleolithic period came from essentially non-material inner and interpersonal advances, not from advanced tool technologies. By my reckoning, the shamanic techniques, and the inwardness they relied on, culminated the emergence process by which love and wisdom rose from embeddedness in dispersal/aggregation and entered the cultural world as independent properties of human nature.

We became human by virtue of the realizations we made in the turning points at the extremes of approach-separation and withdrawal-return, not by cognitive processing alone or "problem solving abilities", but by prototypical love and wisdom. Shamanism was the first cultural fruit of wisdom.

Our disposition for love, as we have seen, goes back farther still. We see its form in parent/offspring relations and pair bonding in other species. Love as we know it in the human experience, like wisdom, breaks free from the rhythmic hold of dispersal/aggregation. It becomes not only a personal attainment, but also a lauded virtue of life.[21]

In my story of human origins, then, the cultural explosion that gave us love and wisdom came in the high Paleolithic when ongoing trends in conscious approach and separation and withdrawal and return *discovered each other.* In the blinking of an eye, our species destiny became open-ended.

After that, change comes quickly. The cultural environment fills with art and invention. Calendrics, agriculture and husbandry and settled towns appear, then written script. An evolutionary instant later astonishing urban centers and governments appear, with bureaucracies, priesthoods, laws, literature, trade and war and all the provisions of a civilized life fed by memorable individuals, creators, inventors, leaders, tyrants, slaves and workers.

**69**

The historian Arnold Toynbee saw the individual withdrawal and return journey in this larger context. To him, the journeys were part of a challenge and response dynamic that fueled the development of world civilizations. He wrote, "the withdrawal makes it possible for the personality to realize powers within himself which might have remained dormant if he had not been released for the time being from his social toils and trammels." Each civilization he studied thrived to the extent it made successful responses to environmental, demographic, geopolitical and economic challenges during its 'Time of Troubles'. Not everyone joined in the process. It took the members of a creative minority, often independently of each other, to complete the journey. In his view, the return of the creative minority from withdrawal shaped the course of civilization.

Toynbee recognized that the pattern only worked as a whole; withdrawal needed a return.

"A transfiguration in solitude can have no purpose and perhaps even no meaning," he relates, "except as a prelude to the return of the transfigured personality into the social milieu out of which he had originally come...The return is the essence of the whole movement as well as its final cause."[22]

**70**
*The Four Kinds of Wisdom*

Through withdrawal and return we seek four distinct kinds of wisdom: creative accomplishments, self-knowledge, spiritual insight and physical healing.

1) **Creativity.** Creative breakthroughs in the arts and sciences fit the withdrawal-return pattern very well. Stages of preparation, gestation, insight, and implementation follow each other. Together they bring into being a work of art, an invention, a theory, a new paradigm. Koestler's bisociation, Getzel and Jackson's divergent and convergent thinking, Poincare's account of his own creative work all follow this pattern.

**71**

2) **Self Knowledge**. Personal growth, integration, self-realization, psychological insight, balance and maturity also move along the inward-outward route. Here personal growth comes in response to the breakdown of the personality in the depths. One's values and constraints, motivations and coping mechanisms can be pushed to their limits in horrifying situations, real or imagined. Under these pressures, the self decoheres and falls into chaos. The wisdom process draws on inherent neural somatic functions of disassembly and re-assembly. The body brings them to bear when it needs to right itself in trying situations. The crisis William James suffered in the course of his medical studies took this route. Notice how the strong interoceptive bias promoted the play of his imagination.

"There came upon me, without any warning, just as if it came out of the darkness, a horrible fear of my own existence. Simultaneously, there arose in my mind the image of an epileptic patient I had seen in the asylum, a black haired youth with greenish skin, entirely idiotic... That shape am I, I felt, potentially. Nothing I possess can defend me from that fate if the hour should strike for me as it struck for him... it was as if something hitherto solid in my breast gave way, and I became a quivering mass of fear. After this, the universe was changed for me altogether."

For Freud, psychological growth came when unconscious contents were brought to consciousness in what amounted to a withdrawal-return process supervised by a psychoanalyst. In Jung's psychology, the withdrawal and return process was marked by the appearance of a series of archetypal figures emerging from the depths of the collective unconscious and breaking into consciousness. Their integration enlarged the conscious self. All personal growth, seen dynamically, involves two-way communication between the periphery and the center, and this is homologous with and an extension from the biology of individual cellular functioning writ large in whole body systems.

**72**

3) **Religious Inspiration**. The road of withdrawal/return informs religious conversion experiences too. In withdrawal, we pass through a series of crises that take us beyond ourselves to what William James described in *The Varieties of the Religious Experience* as the experience of the "More". The spiritual seeker breaks ties with an old way of life, and then by an inward solitary exploration struggles to discover a new center. The *More* sweeps in to reveal a deeper meaning, a redemptive purpose to life, hitherto inaccessible. St. Augustine's *Confessions* contains the clearest, most wonderful account of this withdrawal-return process I know. He was full of backslidings, reversals and spiritual discoveries.

But the withdrawal-return rhythm, though it manifests in our experience of spiritual access, has a deeper aspect still. According to Jewish mystics, God's withdrawal created the world. That act, what Isaac Luria (Safed, 1534-72) called *Tsimtsum,* made room for the universe. Gershom Scholem described it as "one of the most amazing and far-reaching conceptions ever put forward in the whole history of Cabalism." He explains that "according to Luria, God was compelled to make room for the world by, as it were, abandoning a region within Himself, a kind of mystical primordial space from which He withdrew in order to return to it in the act of creation and revelation." And later that "every new act of emanation and manifestation is preceded by one of concentration and retraction." From this come the basic rhythm underlying the mystical *experience* of withdrawal/return.

**73**

4) **Healing.** Self-directed healing follows the wisdom path. Whenever we are deeply injured, either physically or psychologically, our healing process becomes a wisdom project if and only if we take it on with specific intentions involving a search for its causes and origins in self-examination. In successful cases, we either induce or ac-

company the cure with a turning point in the depths. Reports on spontaneous remissions show that ill persons regularly turn illnesses around through the power of intention, sometimes helped by imagery or prayer or by placebo medications. The "crisis in the illness" marks a turning point. In those moments, a person's conscious qualities can come into the mix of influences on the body with special force.

## 74
### *Breakdown in the Depths: Dark Night of the Soul*

Western spiritual seekers sometimes experience the transformation at the turning point of deepest withdrawal as a harrowing crisis of loss. St. John of the Cross called it the Dark Night of the Soul. He dreaded the emptiness and loss of self he experienced in it.

"This is one of the most bitter sufferings of this purgation. The soul is conscious of a profound emptiness in itself, a cruel destitution of the three kinds of goods, natural, temporal and spiritual, which are ordained for its comfort. It sees itself in the midst of the opposite evils, miserable imperfections, dryness and emptiness of the understanding, and abandonment of the spirit in darkness."[23]

One encounters moments of cruel destitution in every kind of wisdom quest. It unpins us. It disrupts our creative efforts, harrows us and fills us with doubt. The membrane between inwardness and outwardness grows porous. We experience ourselves as unstable, multifarious, and threatened. We break into factions. The parts of us go to war with each other.

"Now I found myself dividing into parts," Lawrence of Arabia writes of his transformation.

"The spent body toiled on doggedly and took no heed, quite rightly, for the divided selves said nothing I was not capable of thinking in cold blood...they were all my natives. Telesius, taught by some such experience, split up the soul. Had he gone on to the furthest limits of exhaustion, he would have seen his conceived regiment of thoughts and acts and feelings ranked around him as separate creatures, eyeing, like vultures, the passing in their midst of the common thing that gave them life."[24]

Accounts linking wisdom with disintegration and reassembly go as far back as we care to look. They surface in myths of physical dismemberment, as when Isis tried to reassemble the parts of Osiris' body but could not find his penis.

In the Gilgamesh tale, Enkidu dreams of dismemberment. He tells Gilgamesh:

"Last night I dreamed again, my friend. The heavens moaned and the earth replied; I stood alone before an awful being; his face was somber like the black bird of the storm. He fell upon me with the talons of and eagle and he held me fast, pinioned with his claw, until I smothered; then he transformed me so that my arms became wings covered with feathers."

## 75
### *Dismemberment Reversed*

**Wounded Shaman**

If the quest journey makes the self decompose, reassembly is wisdom's response. In one furnace, character splits apart. In another, it amalgamates. The physiology behind the experience of dismemberment and re-patterning comes from rhythmic breakdown and restoration. Under the stress of withdrawal, when we are most in turmoil, and are functioning in near chaos conditions, our body rhythms are particularly vulnerable to perturbations. All kinds of resettings, extinguishments, bifurcations and rhythmic reorderings occur then, arising from the dynamics of interacting nonlinear bio-oscillating systems. In the creative crises associated with turning points in wisdom, the modular components of our personalities force themselves into view.

## 76
### *Modular Personality*

If we attend closely to the experience of fragmentation and dismemberment, we will discern behind it clashing personality parts. We will feel them struggling for dominance. Each fights for its own style of access, contesting the use of organ systems, imposing its postures and physical attitudes on the body. The meek part of us wants to run; the bold part wants to fight.

The withdrawal-return process, as it functions in the depths, allows the physiological turbulence to break personality into its modular parts *adaptively*. It pulls us apart in order to let us become whole in a new way. In response to sustained shifts in the needs and conditions of life, it breaks the arrangement of part-personalities down and builds it up again differently. In the process, meaning, identity and life choices change.

Michael Gazzaniga, a pioneering neuroscientist who worked under Roger Sperry doing split brain surgeries for epileptic patients, describes modularity as follows: "Human brain research urges the view that our brains are organized in such a way that many mental systems coexist in what may be thought of as a confederation. "[25]

Moreover, they are not hardwired. They keep changing. They develop differently in each of us. They compete to command body energy. They struggle to occupy frontal consciousness.

## 77
### *The Passing Rules*

Each module commands a different sense of space and time and has different hopes, fears, apprehensions, regrets, involving differential access to episodic and semantic memory, hence different allowances and prohibitions and subtly different expectations for life. What we call identity, then, are those discrete neural organizations that have a history and a sense of future prospects that, burdened with hopes and fears, have at some stage of their development broken from their transient flowing absorption in the field and learned to think, worry and wonder for themselves, to want and avoid, forget and remember.

Each identity state, it follows from this, commandeers its own personal history.[26] It does this by establishing what I call passing rules between the part personalities that come together to form it. It shifts between certain modular complexes with relative ease under the selective pressure of changing fields. Others it prohibits. From this perspective, our character is an operational device compounded of incumbent and insurgent semi-independent "personalities" evolved to shift into and out of frontality as needs change by following a set of self-organizing passing rules.

The passing rules direct the neural pathways along which nerve traffic moves between personality modules. The passing rules serve as the stage directions by which our partial personalities, with their distinct vocabularies and values, make their entrances and exits in the frontal drama of life. They're the way the I-sense, the feeling of consistent identity, is passed from one leading character to another.

The passing rules establish pathways for dealing with change. The culture teaches us from earliest childhood when it is right to override one personality module in favor of another. The passing rules incorporate these structures. They tell us which part of us ought to be in charge in different circumstances. The *you* that drives the car is different from the *you* that loves your wife. The predatory business person in the day becomes the couch potato at night. Parents display a different face with their children than they show to friends or strangers. The occasions on which it is proper to switch from one module to the other are shaped by the culture, mapped by sensory distances and inscribed in the passing rules.

## 78

The modularization of the mind occurred along with language acquisition. The functional ordering of the mind into modules quickened as we acquired rudimentary use of language. We store our inventory of internal voices, built up from the voices of others, from our own speech heard through our own ears and internalized, and from the sounds of nature, in modules. Our access to them, as to other kinds of memory, is limited, oriented to present needs and state-dependent. The clinical evidence for language recovery following trauma, as for the retention of singing and swearing in aphasic patients, supports this point of view.

Language expression uses not only the meanings and connections between words, it animates the tongue and breath and vocal cords; it expresses its force and urgency in motor activity. We

communicate not primarily as an intellectual activity. You open
your mouth when you have something to say. Language use rides
on will and desire. You carry expressiveness as much in word
tempo, emphasis, rising or falling pitch, tremolo and other timbral
qualities as in vocabulary and word order.

A carefully conducted experiment would distinguish each
front-facing component of our modular personality system by its
vocal characteristics, its inflections, vocabulary and semantics, its
frequency of word use, and its facial gestures and habitual expostulations, sighs, grimaces, and tonalities. We would almost certainly
come upon subtle differences in skeletal-muscular bearing and carriage, autonomic underpinnings, hormonal and neuropeptide secretions, and even styles of breathing and heart and blood vessel
regulation.

## 79
### *The I-Sense*

We maintain our sense of ourselves by the smooth succession
of modular parts. It proceeds under the illusion of a constant "I".
The I-sense, organized and stabilized by the *passing rules,* floats
from module to module. The passing rules insure that the present is
always "my present", and that there is always a "me" in it to comprehend, respond, suffer, enjoy and take responsibility for it. In
ordinary circumstances on the long legs between major turning
points, the neural networks, being robust, swell, shrink and compete without generating global breakdowns in function. Certain
preferred companion personalities (almost like a social system of
temporary alliances in primate bands) cooperate. They pass the
lead back and forth without great competition. However, in great
turnings, the old alliances decohere, trusted bonds fail, and agreements end. During these times, we fight major battles between
modules in the electrochemical fields of the brain. The passing
rules come under increasing strain. They split into competing
fragments until they eventually re-form and stabilize in a new more
adaptive configuration.

The passing rules are not hard-wired in the brain. They form
part of its modifiable software. They are adaptive. They show our
neuroplasticity. Stable passing rules keep the self together on the
long legs following the turning points in love and wisdom. They

shape memory and motivations. They persist until the next big turning impends. Then they destabilize and break down again.

The passing rules organize my modularized memories into a selective *almost fictional* set of narrative devices that, despite its distortions and deletions, feels real to me in the flow of the moment. But while in my present "I-sense", I can only trace one set of memory strands. In reflective moments, I may recognize that my history and goals have many strands and that many paths wind back to different pasts. But today, at this moment, in the heat of action, because I can only countenance one me, the one I am now, I forget my other memory trains. Tomorrow I might toss out another net to catch the memories of another me, but when I did that, I would forget the me that is saying this now. On both occasions, a somewhat different "I" would be present but without special training, I could not tell the difference. However, one can learn to thread the labyrinth.

The wise person knows how fortunate we are to have this illusion of constant identity as a practical aid. It is an island of stability rising in a sea of ceaseless change. Without it, the waking life would seem as malleable and unpredictable as our dreams.

William James in one of his letters catches a glimpse of it when he speaks of the function of an "enveloping ego to make continuous the times and spaces not necessarily coincident of the partial egos."[27]

In practice, we sustain the illusion of identity by filtering out all kinds of borderline perceptions that might let us see out through the cracks in ourselves. Without the narrowing of attention, we would be overwhelmed by data.

When we narrow the focus, we do more than cut down the flow of data. We re-channel it to conceal contradictions from ourselves. Even toward these contradictions, the wise person is tenderly supportive. Nevertheless, s/he knows that the evolved illusion of the "I-sense" can become bloated beyond its own utility. When that happens, s/he feels herself seduced away from the doing of deeds. Instead, s/he wants to preen and fuss in the mirror. But the false face, though it makes for a pleasing mirror image, fastens us to false needs and foolish posturing.

Given how we come together to make a self, the wise person knows that it is sometimes serves our real needs to let the illusion of wholeness go. However, you have to know when to let go and when to hold on to it. To rise to the occasion one has to know when the occasion arises. Much can go wrong.

## 80
### *Three Root Dualisms*

The passing rules build a map of the psyche by organizing the three root dualisms into a system of basic relations. The three root dualisms. They are: Mind/body, self/world, and I/you. They give relational order to experience. "I am me, you are you", "this is my mind, this is my body;" "this is me, this is the world" give structural integrity to the flow of qualia. They interact to conjure up a continuous reality.

Spiritual teachers tell us these dualities are illusions, part of Maya, measure, separation. Underneath they say the world is whole, One. Duality may be an illusion but I am sure we cannot function without it. We may get transient moments of non-dual awareness, but they do not last. The three basic dualisms, as Ken Wilber analyzed them, do not point to a way out of illusion so much as describe our experience of the binding energy of our gestalts.

When the passing rules fail and the rhythmic interplay that makes the mind cohere falters, the I/you, mind/body, self/world break apart. They reveal themselves as illusions.

What really happens in our turning points crises in wisdom, then, is that the moral sentiments held in stable dualisms in the different modules of the personality – our current gut understanding of what I owe to myself, and what I owe to others – gets reconfigured. New resonant systems throw neural nets across the brain. The I/you, mind/body, self/world stabilize. But they're not what they were. Our core beliefs, we find, have shifted. In the aftermath of the wisdom turning, our moral intuitions seem new to us. Some are held over; some are revised, from earlier times; some after falling out of favor come back again.

Personality parts affiliate differently. Foreground cultural values recede, while background values surge forward. We reprioritize. Our motivations shift. However, we never mend the dualisms finally. Even in successful turnings, the resolution of dualities is temporary, an interim solution, good only until the next big turning comes, whose contents are unknowable.

**81**

In the turning points of life, the breakdown of the passing rules and the reorganization of memory that follows it, constitute the main changes. In the lead-up to the turning, turbulence loosens the organization of the parts of the identity. Even parts of me that the world has reaffirmed through consensual validation start to fade and fail. After the turning, the pieces come together in new configurations. I redeem lost meanings. What once made no sense reveals its deeper purpose. But at a cost: much that was prized is lost or suppressed.

The wisdom turning "at heart" serves as a reworking of the connections between the parts of the personality. At "mind", it offers a reworking of our cognitive take on reality. It shifts our hopes and fears. Our senses of time and space and even our access to our memories change. At "muscle", it reworks our understanding of what we can and cannot do.

Metaphorically, rough handling from life has separated the wholeness of our genius into parts, each secluded in a fragmentary personality center, split off from the others by the organizational needs of the psyche. Embedded in the fragments, state dependent voices cry out from behind their walls. The I-sense, disempowered, having now lost its glamour of enchantment, searches for them, and searching, blundering more than searching, reenters the physiological-psychological substrate from which each state dependent voice cries out from its module.

But the walls between identities have become latticed, softened in the seismic shaking of the turning process. The words and passions mingle. The energies, skills, and beliefs from one part of our personality combine with others. The bereft I-sense casts about in the babble of voices, trying to make sense of the noise.

We can look on the entry of fragmentary personality centers by a roving "I" as acts of integration. In these moments, neural reorganization brings personality components closer together. The minimal "I" left over in the final stage of the turning drama hears the voices and learns the vocabularies, and by so doing transfers them from limited to less limited personality systems, moving them from small state dependent loci to larger ones. Identity states acquire each other's vocabularies in the fluidity of the turning process when the walls of possibility fall.

For a brief time afterward, in the blessed relief, one experiences the special wholeness that mystics have attributed to an influx of the divine. In these moments, wisdom is potentiated. A person's face changes; expressive lines appear; gestures, posture and bearing alter. Character is cooked in the high stress heat of the turning.

A wisdom crisis well resolved produces a deep organic rearrangement that brings with it a new way of coping with the circumstances that drove the withdrawal. However, the new attitudes and understandings held in the newly revised and stabilized root dualisms overseen by the passing rules do not float freely. The body anchors them. The breakthrough represents the outer face of a somatic reordering.

As the turning crisis abates and the root dualisms stabilize, a new cast of part-personality modules moves into position under the aegis of a new set of passing rules. In the biggest upheavals, they produce new balances between the solitary and social tendencies, between approach and separation and withdrawal and return, between the sense of time and space, closedness and openness, hope and fear.

**82**
### *The Ox-herding Pictures of Kaku-An*

The *Ten Ox-Herding Pictures of Kaku-An*, a Chinese Buddhist document of the 11$^{th}$ century Sung dynasty, gives a particularly clear view of the wisdom journey from an Eastern perspective.

The first picture "Searching for the Ox" shows a youngish man in the open countryside. It depicts the early stages of the withdrawal.

Kaku-An comments:

*"His home is ever receding farther away from him, and byways and crossways are ever confused. Desire for gain and fear of loss burn like fire; ideas of right and wrong shoot up like a phalanx."*

*"Exhausted and in despair he knows not where to go, He only hears the evening cicadas singing in the maple woods."[28]*

In the second picture, the young man sees the ox. He chases after it. A long time passes before the ox-herder can tame and ride the ox, but by picture six, the man is riding on its back playing the flute.

The painting shows a kind of graceful competence, not arrogantly framed as power over the animal, but spontaneous, relaxed, in the moment, and it shows the young man playing the flute as he rides the ox at his ease. He makes music, relaxes and travels on. Toward what goal?

Next, we find the little man living in a hut, his hermitage in the woods. The ox is gone. He experiences deep withdrawal. Kaku-An calls picture seven "The Ox forgotten, Leaving the Man alone" He comments: *"Though the red sun is high up in the sky, he is still quietly dreaming, Under a straw thatched roof are his whip and rope idly lying."*

By picture eight a transformation has occurred, but we don't see it. The world is shrouded behind a circle of white light. Kaku-an captions it "The Ox and the Man Both Gone out of Sight." There is some tension here, some extremely energetic exchange proceeds in the field of light. But what is actually happening in this mystical moment? The standard explanation for this kind of symbol is that the white circle represents the ineffable; the state cannot be represented; it has gone beyond pictures or words. Moreover, the accompanying poem seems to confirm this:

*All is empty—the whip, the rope, the man, and the ox: Who can ever survey the vastness of Heaven?*

But the next two lines veer unexpectedly:

*Over the furnace burning ablaze, not a flake of snow can fall: When this state of things obtains, manifest is the spirit of the ancient master.*

Whence the burning furnace? Does the white circle represent a blazing energy field? What is the fuel? Does the little ox-herder generate the energy? Who is the ancient master? Kaku-An does not say.

We can conjecture that the circle of white light conceals the transformation of the ox-herder. Kaku-An paints the circumference of the light as vibratory, not firm. A collision of complex physiological wave fronts occurs. Confusing tidal currents butt up against each other. Each generates its own set of hopes and fears; each stirs emotions in the body (as in the James-Lange theory of emotions) and deals with the alarm reactions to those changes.

The incandescent storm surge goes on with an intensity reflecting the transforming power of the moment. Training in meditation in the Eastern spiritual traditions makes these moments tolerable to their practitioners. They are less inclined to frame their experiences in the language of torment than Western spiritual seekers in the same situation.

In the next picture, "Returning to the Origin, Back to the Source", a blossoming tree appears. The exteroceptive senses begin to stir. The painting shows a world coming to life. The tree is flowering vividly red. There is a wide flowing stream. Herbs grow among rocks. A clear sky shines above. A tremendous sense of relief and peace. The poet comments:

*Sitting alone he observes things undergoing changes. With his senses alert, clarified of delusion, exteroceptive, he is happily part of the world. No longer concerned with gain and loss, he's attuned to the life outside and around him.*

Strangely, unexpectedly, Kaku-an calls this attunement a mistake: "To return to the Origin, to be back at the Source—already a false step this!" Why? Because the journey must not end in the hermitage. Peace is not the goal. The withdrawal has its limits. It serves to nurture the energy for continued journeying.

Walt Whitman touches on this same sensibility in his *Song of the Open Road*.

*Allons! we must not stop here, However sweet these laid-up stores, however convenient this dwelling we cannot remain here, However shelter'd this port and however calm these waters we must not anchor here, However welcome the hospitality that surrounds us we are permitted to receive it but a little while.*

In picture ten the ox-herder, now a full-bellied, hairy middle-aged man, smiling and carefree, strides to the city with a great pack on his back.

He returns to the world. The artist calls the picture "Entering the city with bliss bestowing hands." He walks into the marketplace. Kaku-An informs us, "He is found in company with wine-bibbers and butchers, he and they are all converted into Buddhas."

The transformed man carries a big pack. He has supplied himself with everything he needs, everything that counts. All his abundance is portable. It does not weigh him down.

But look. Entering the city, he meets a young man leaving for the countryside. Perhaps they exchange a few words. Or only share a glance. But with "bliss-bestowing hands" a brief encounter counts for much. Have you met a bright spirit on the road yet?

## 83
### *Failed Wisdom*

Wisdom turnings can fail in various ways. They may not complete themselves. Events may stymie the seeker on the return leg. The world may not be ready to receive the gift gained in the turning. Black Elk, a late nineteenth-century Ogalala Sioux, painfully recollected his own failures as a shaman this way.

"To the center of the world you have taken me and showed me the goodness and the beauty and the strangeness of the greening earth, the only mother - and there the spirit shapes of things, as they should be, you have shown to me and I have seen. At the center of this sacred hoop you have said that I should make the tree to bloom. With tears running, O Great Spirit, Great Spirit, my grandfather, with running tears I must say now that the tree has never bloomed. A pitiful old man, you see me here, and I have fallen away and have done nothing."[29]

When our love turnings fail, we lose our lovers. When our wisdom turnings fail, we lose the part of ourselves that nurtures our creative élan. We do not get the gift of wisdom unless we endure self-transformation. Self-transformation leads to it. When we haven't reached into our souls deeply enough to discover our sources of strength, we cannot change. Nor can we truly accompany others through their changes. There's too little of ourselves left over to give. We have not gone deep enough to discover more.

Wisdom turnings break down in different ways. Some people never want to go on the inward journey in the first place. Others never want to come back. For some, identity confusion dominates the withdrawal and deepens as the withdrawal proceeds. What presents as a clinical depression is often a stuck turning. I've known people so firmly anchored in the opinions of others that, when interoception was strong and the sense of the presence of others stretched too thin, they felt they had no choice but to deny themselves or go mad. They scurried back to the consensual world. They lost courage. They couldn't take the first step on their own. They came back to live on the surface of life with empty posturing.

I have known others who went down and hung in bravely in withdrawal, but could not make progress. Over the long run, stress and irresolution got to them and they became ill. Others made the turning but suffered on the way back because every step of the return terrified them and they refused the help that was offered, or in their terror couldn't see it, and out of fear of self-assertion or fear of rejection clung to themselves inside themselves and wallowed in fantasy or became cynical, angry and depressed. It's hard to live with yourself when you refuse your own powers.

Some people start back resolutely but never get far. They may have bad luck. Or die on the way. Or the world resists them and they cannot stand the conflicts, especially when they come from the persons closest to them. Some almost get home, but find it too hard to cross the return threshold. The world does not want them. It is hard to keep going then. Great disappointment can follow even a well-received return.

But many have come back invigorated, better suited to their world, more purposeful, more useful, clearer in intention, happier, less compromised and less compromising and yet more adaptable. They have made great contributions, some hidden, and some known.

\* \* \*

## End Notes

[1] Llinas. The I of the Vortex. p. 10

[2] William James, Psychology, The Briefer Course. Chap 15

[3] Loewald. *Playing and Reality*. Tavistock. 1971. p.65

[4] Kierkegaard. Concluding Unscientific Postscript p. 280

[5] ibid p.284

[6] Hobson. *The Dreaming Brain* p.110

[7] Polich J., Craido J. R. "Neuropsychology and neuropharmacology of P3a and P3b". Intl J Psychophysiology. 60(2): 172-185. May 2006.

[8] Mann describes Castorp's sensory world this way: "He had not heeded the silent entry of the tenth month, but he was arrested by its appeal to the senses, this glowing heat that concealed the frost within and beneath it. It was a sensation which, to anything like this degree, he had never before experienced, and it aroused him to the culinary comparison which he made to Joachim, of an omelet en surprise, holding an ice concealed within the hot froth of the beaten egg. He often made such comments, talking headlong and volubly, as a man does in a feverish chill. But between whiles he was silent; we shall not say self-absorbed, for his attention was presumably directed outward, though upon a single point. All else, whether of the animate or inanimate world, swam about him in a mist—a mist of his own making, which Hofrat Behrens and Dr. Krokowski would doubtless have explained as the product of soluble toxins..." Thomas Mann. *The Magic Mountain*. Knopf 1960. p. 227

[9] C. S. Lewis. *A Grief Observed*. 1961. Harper 1989. p.15

[10] ibid p. 15

[11] ibid p. 47

[12] "Suppose that the earthly lives she and I shared for a few years are in reality only the basis for, or prelude to, or earthly appearance of, two unimaginable, supercosmic, eternal somethings. These somethings could be pictured as spheres or globes. Where the plane of Nature cuts through them—that is, in earthly life—they appear as two circles (circles are slices of spheres). Two circles that touched. But those two circles, above all the point at which they touched, are the very thing I am mourning for, homesick for, famished for." ibid p. 36

[13] ibid p.23

[14] In this condition Lewis cried out against God: "What chokes every prayer and every hope is the memory of all the prayers H. and I offered and all the false hopes we had. Not hopes raised merely by our own wishful thinking; hopes encouraged, even forced upon us, by false diagnoses, by X-ray photographs, by strange remissions, by one temporary recovery that might have ranked as a miracle. Step by step we were led 'up the garden path.' Time after time, when He seemed most gracious He was really preparing the next torture." ibid p. 42

[15] Joseph Campbell. *The Hero With a Thousand Faces.* 1949. Meridian Books. 1967. p. 17

[16] ibid p. 246

[17] ibid p. 246

[18] Julian Jaynes. *The Origin of Consciousness in the Breakdown of the Bicameral Mind.* Houghton Mifflin. 1976. p. 85 Jaynes maintained that "the presence of voices which had to be obeyed were the absolute prerequisite to the conscious stage of mind in which it is the self that is responsible and can debate within itself, can order and direct, and that the creation of such a self is the product of culture. In a sense we have become our own gods."[18]

[19] That the shamanic séance was a source of wisdom is argued well by Joseph Campbell who wrote: "The Way of Suffering of the shaman is the earliest example we know of a lifetime devoted to... the serious use of myth hermetically, as *marga,* as a way to psychological metamorphosis. And the remarkable fact is that the evidence points irrefutably to an achievement—at least in many cases—of a perceptible amplification of the individual's horizon of experience and depth of realization through his spiritual death and resurrection, even on the level of these first primitive explorations. The shaman is in a measure released from the local system of illusions and put in touch with the mysteries of the psyche itself, which lead to wisdom concerning both the soul and its world; and he thereby performs the necessary function for society of moving it from stability and sterility in the old toward new reaches and new depths of realization."[19]

[20] Mircea Eliade, in *Shamanism,* describes the Eskimo shaman's journey this way: "The term most commonly used in referring to a shaman is 'one who drops to the bottom of the sea.'" Eliade goes on to say, "For the bottom of the sea is the abode of the Mother of the Sea Beasts, mythical formula for the Great Goddess of the Animals, source and matrix of all life, upon

whose good will the existence of the tribe depends. This is why the shaman must periodically descend to reestablish spiritual contact with the Mother of the Sea Beasts."[20] Mircea Eliade. *Shamanism.* Princeton Univ. Press. 1972 (first published 1951) p.294

[21] I call the agencies through which our actions carry our intentions into the world our virtues. By virtues I mean strengths. Among them I count our morals, ethics and values, because they are the muscles we use for aiming, releasing and guiding our deeds into the world. Let's look at each in turn:

When I speak of morals I mean sentiments coming from the heart. Not everyone is endowed with the same moral sentiments in the same proportions. We specialize.

When those sentiments reach our heads and get thought about and articulated and shared, they become ethical maxims. Ethics belong to the ethical mind. Moral sentiments belong to the moral heart.

We only get direct access to the moral heart inwardly. It is part of our inwardness. A value is the ethical articulation of a moral sentiment created in inwardness. But it takes energy to bring it into the world in deeds – as part of our outwardness.

The better we practice and refine our values, the better we get at owning them as ours, and this allows us to develop them as skills. The stronger our virtues, the more virtuous we become. Virtue then is the strength to act upon our choices skillfully. But skills require instrumental thinking and instrumental thinking comes from the technical mind. Our best paths to the future, therefore, bring inwardness and outwardness together. We thrive only when the technical and ethical minds grow strong together.

[22] Toynbee, A Study of History, 1962

[23] St. John of the Cross, quoted in Evelyn Underhill. *Mysticism.* Meridian Books. 1955.[ first published, 1891] P. 391. In his great study of spiritual consciousness, Evelyn Underhill described the Dark Night as a paring down process: "As in [the earlier stage of] Purgation [when] the senses were cleansed and humbled, and the energies and interests of the Self were concentrated upon transcendental things: so now the purifying process is extended to the very centre of I-hood, the will."[23]

[24] T. E Lawrence. Seven Pillars of Wisdom

[25] Michael Gazzaniga. *The Social Brain.* Basic Books. 1985. p. 6. Drawing on experimental data from split brain patients, he writes that "I argue that the normal person does not possess a

unitary conscious mechanism in which the conscious system is privy to the sources of all his or her actions. I want to build the argument that the normal brain is organized into modules and that most of these modules are capable of actions, moods and responses."

[26] With the materials neuroscientists have now assembled we can model oscillatory processes in the brain that coordinate modular input through the thalamus, and keep frontal con-sciousness shifting many times a second between many foci of awareness, a spinning carousel of consciousness, perhaps polling or handshaking with the different modular centers in succession or simultaneously, each dabbing its colors and forms on the screen, and spinning so fast as to seem to us to be unified, supportive of our evolved illusion of a constant identity.

[27] Quoted In Erikson. Insight and Responsibility. W. W. Norton. p. 147

[28] In Suzuki *Manual of Zen Buddhism*. Grove 1960

[29] John G. Niehardt. *Black Elk Speaks*

# 4. CONVERGENCE

### 84
### *Real-Life Skills*

If human nature was simple we would have the rudiments for a good life before us. But we are conflicted creatures by nature; we hold intrinsic contradictions. Parts of us war with each other. We generate self-directed perturbations. We constrict our own possibilities. *But not beyond all recognition.* And the normative, the abiding foundation in what we recognize is vitally important. It shows us what we can become by living fully.

Stated simply, the great aim of personal growth, and of the methods truly allied to our needs, is to return us to the field of action where we truly meet each other through a widened understanding of rhythmic love, and truly encounter ourselves in a deeper experience of our inwardness through wisdom. We truly meet each other because in love the deeds produced in turning points weave the interpersonal world together in a rich and variegated fabric moving out in wider circles of affiliation. We truly encounter ourselves in our inwardness because in our freedom in the heart of wisdom we are truly alone.

To live well, therefore, a person must engage in four deeds energetically: withdraw when going inward, return when going outward, come close to others when seeking intimacy, and separate when seeking solitude. In wisdom, the leave-taking has to be genuine, the inward journey has to be real, and the sacrifice of outward comity has to be actual. The same standards apply to love: the approach has to be genuine, the intimacy has to be real, and the separation has to be actual. The intensification in both the closest and most distant moments must engage our most powerful hopes and desires.

## 85
### *Mapping and Stage-appropriate Action*

The better mapped we are on the legs of love and wisdom, the better we will understand what we need to do each step of the way to keep life vibrant with meaning. Mapping helps us find stage-appropriate action. The rule of thumb for mapping is to make your actions stage-appropriate so that you can fully engage your turning points when they come.

The most needful thing is to possess the discernment to distinguish the major turnings from the mini turnings that lead to them. The mini turnings, though they too involve choice and bring meaning, and make a difference, are less deeply transformational. Mistaking the mini for the major is to mistake the shadow for the form, or the backwater for the current. To traverse the great turnings well, we have to recognize the turning of turnings when it comes and comport ourselves appropriately in it with an eye to attaining our time-sensitive freedom.

I count three principal mini-turnings leading to the big turning of turnings, the one's Ken Wilber, following Hegel in the Phenomenology of the Spirit, treated as the great dualities of life. The first requires us to recognize the split between self/world, the second between ego/body, the third between self-regard and self-loathing. Wilber wrote of these,

"At the Existential Level, man imagines himself separated from and therefore potentially threatened by his own environment. At the Ego Level, man fancies that he is also alienated from his own body, and this the environment as well as his own body seem possible threats to his existence. At the Shadow Level, man even appears divorced from parts of his own psyche–from the point of view of the spectrum of consciousness, these terms all refer to the same process of creating-two-worlds-from-one which repeats itself, with a new twist, on each and every level of the spectrum."[1]

The three basic dualisms weave into all dramas of love and wisdom. In my view, we can stabilize the dualisms, but we cannot overcome them for long without deceiving ourselves.

The stable dualisms tend to fall apart in crisis times. Falling apart is what makes them adaptive. We must know when not to resist it. That takes mapping too. Our work is to put them together differently once they decohere. That's when lovers change their aims and seekers after wisdom reject their current understandings.

## 86

It takes an odd but important kind of strength to accept the paradox that comes with holding your intentions firm while accepting the impermanence of the self that does the holding. That firmness-in-fluidity gains you a point of purchase in the 'now'. From that place, you can apply enhanced moral leverage to your actions.

This amplified power in the moment I think of as the existential equivalent of leverage, on an analogy to the mechanical advantage one gains in using the simple tools, the screw, the inclined plane, the pulley, the lever.

FIRST CLASS LEVER
EFFORT
LOAD
FULCRUM

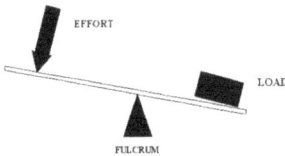

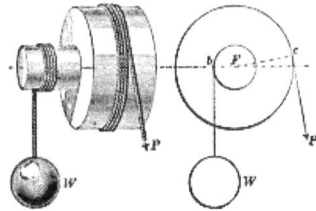

If we can muster intention appropriately and make choices timed to the turning moment, a small implementing gesture, itself a slight perturbation to the ongoing rhythms, can initiate a bifurcation that brings amplified effects into the world. This amplified power does best when expressed rhythmically. Our sensory-motor grace, mental agility, firm intention, imagination and courage come together in the leveraged moments. (I explain this more fully in # **118-121**)

Only in the moving 'here' that rides on the fleeting 'now', can you focus the energy of your intentions to a point. Stage-appropriate action brings personal power. Moreover, personal power is inherently ethical because it rests on choice.

**87**
*Wounds to Love and Wisdom*

In an imperfect world, mapping shows the actions we can and cannot take. It points at the deeds we have not the strength to perform along the legs of love and wisdom. Mapping shows us our wounds and reveals the pain in them and plots the misdirection they cause us. Face it. Nobody gets through life unscathed. We carry our injuries along with us. Our wounds create leg-specific breaks in our approach/separation and withdrawal/return patterns. The breaks interfere with our actions. They keep us from fully engaging our natural inclinations. Most importantly, they undermine our capacity to give and receive.

With good mapping, we can learn to see how our wounds stop us in action in specific places on the legs of love and wisdom. Experiencing our wounds *as wounds to action*, we will become less likely to mistake the sensory changes accompanying them as threats to our well-being in themselves. We'll know where we're at. Even in our failures, we will have achieved more meaningful transactions with life.

I stress meaning here because meaning deprived people are *always* poorly mapped. They do not know whether they are coming or going in love or wisdom. They do not clearly experience the alternation of both legs within each dynamic. Failing to experience the power of turning points, they cannot really connect the legs of love with each other. In wisdom, each step seems like an independent process, parts of an unconnected series of "moves" going nowhere with no clear purpose. Approach, separation, withdrawal, and return don't fit together as sequences in a series.

In the tensions of modern life, the modules of the mind that shape behavior defend themselves against encroachment. Each draws on its own style of aggression. As if one could foreclose on one aspect of the polarity love without distorting the other. As if one could separate without approaching. As if one could redesign wisdom by excising the need for withdrawal from the need for return, or vice versa. From this misapprehension people lose contact with the desirable movement of one phase into the other. Lacking authentic turning points, they never acquire the skills to manifest their best intentions. They do not appreciate their strengths or know their weaknesses.

Eventually our conditioned fears dispose us to favor one side of the oscillation over the other. To the extent that our experiences in any one sensory zone nurture and please us, we favor that zone and reinforce it. However, when we encounter harsh treatment, and the harsh treatment repeats, and we associate it with sensory cues at certain sensory distances, we avoid going there. It follows from this that approach, separation, withdrawal, and return develop characteristic shapes differently in each of us. Those whose energies are dominated by the interpersonal may have difficulty withdrawing into themselves. Those whose energies are dominated by the internal may have difficulty in the interpersonal world. You and I are different. We endure different blocks to action.

## 88

We accommodate ourselves differently to the visceral realities of our gains and losses in love and wisdom. We react differently to the sensory immediacy of other persons, to the different circles of sensation, to the ring of solitude and silence in our lives, to the stress of change, to our experiences with belonging and estrangement and giving and receiving. We have faced different obstructions and aggressions along the way; life is hard for us in different ways; we have different needs and different motivations. The peculiarities of the way we learn to move through the four distances, our state-dependent associations with each, the emotions they evoke in us, the balance of inner and outer senses employed, the kinesthetic memories of our early caregivers, especially the anxieties transferred to us by our parents at specific distances, condition us to live certain lives and to seek certain loving relationships.

These basic differences cut close to the bone. They give us our temperaments, and, with these, we come upon the contrasting ways we respond to opportunities and danger.

That is to say, our capacities for love and wisdom, our eagerness for them and our outgoingness on the legs of return and approach are idiosyncratic. We each go down a path of possibilities at our own rate and readiness. When our wounds are deep and festering, we lose our way and customarily hide our confusion and dislocation from ourselves. We compartmentalize and quarantine our injuries. Our psychological defenses make our memories selective and unreliable. The wounded areas grow shadowy in our consciousness. As time passes we can hardly depend on our sense of our own pasts, so clogged with compensatory fictions, false regrets and phony enthusiasms have they become. Moreover, these memories of explanations become as solid as memories of events. They much more readily spring to mind because we repeat them endlessly to ourselves in order to justify ourselves to ourselves. They form our behavioral repertory, our *modus vivendi*. They instruct us how to deal with estrangement, rejection, disappointment, loss, betrayal, confusion, anxiety and distress.

People who get hurt too often and too hard, who carry many wounds, may after many sharp losses and failed turnings give up on love and wisdom altogether.

## 89

The wounds to love are particularly onerous because their pains come from the fear of abandonment, and abandonment grows from inherent infantile dispositions. On the other hand, the wounds to wisdom have ignorance as their source, not abandonment, and not a happy insouciance, but a fearful uncertainty of "where the next blow will fall". Rabbi Nachman of Bratslav considered the wounds to wisdom worse than the wounds to love, because we experience them alone. In his view, a person living alone is more likely to break into fragments than is a person living in a family, because in solitude the warring inner factions fight each other, while in a family people do the fighting.

The most perilous wounds to both love and wisdom hit during turning points. This is the worst possible timing for trauma, for in the turbulence of a change we are more vulnerable to injury, particularly in childhood when the modular components of the personality are still forming.

## 90
### *Strong and Weak Legs*

We have weak and strong legs in love and wisdom. Weak legs lead us into diminished or pathological turnings. When the withdrawal leg is weak, for example, it does not support a full turning toward return. Under pressure, we may turn back toward home too soon because turning back feels better than going aahead. For some of us, the inward path toward wisdom evokes terror and doubt, while for some it evokes delight.

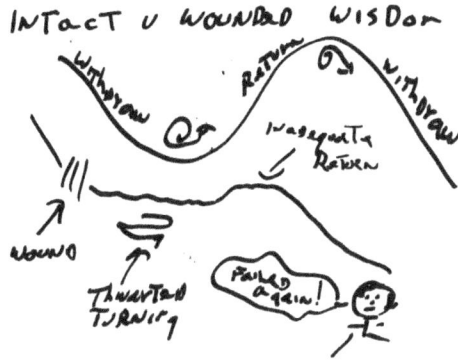

Some of us prefer the depths and are very eager to leave the daily round. We love to wallow there tracing castles in the air, and we fear return and avoid coming back to bestow the products of our inwardness on others. We may be tempted to turn again to the withdrawal stage before we complete the return because we find reentry difficult. Others eagerly rush back from a brief inwardness waving their plans and projects, ready for validation, comfortable to implement any plan as long as it makes them feel wanted.

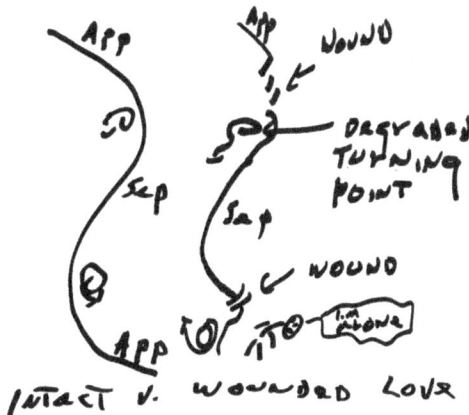

Some have an easy time getting close to each other. They're more comfortable on the approach leg, more realistic there, more in the moment. Separation, even mini-separations, may be hard for them. Therefore, they consider the approach the essential part of love, indulge in fantasies of eternal felicity in the embrace of the beloved, and consider separation cowardly or dissolute. Others separate easily and come back refreshed and seemingly open, but are wary of intimacy and fear dependency and hedge their bets. These people consider separation the key to successful relationships. They want to maintain boundaries, take responsibility for themselves and avoid codependency.

If the weak leg is separation and you've never been able to face up to solitude in love, you may cling to the other out of fear, but in your delusion call it devotion and never understand why the other keeps trying to scrape you off. Alternatively, you can believe truly that love is a chase in which your goal is never to be caught, but never give up the chase. You get strength in the subtleties of separation. Approach terrifies you. On the weak leg in love, for instance, you are shocked by the unexpected rebuff (or approach) of your partner.

The shock itself has physiological roots and generates a stress response. The stress can throw you out of your sustaining fantasy, but without giving you the means to come to a direct perception of the real situation. That's why so few people deal well with disillusionment in relationships. On the strong legs, when they are at their best and feel more capable and freer, they indulge themselves out of proportion to the rhythmic play of the whole. The larger pattern gets lost, and with it their sense of the meaningfulness of their lives.

On the weak leg, as we experience it ourselves, we are likely to have a harder time accepting that the changes we go through are parts of a meaningful design. We suffer from tunnel vision because we cannot discern the overall design. We may feel the presence of the Dark Night of the Soul, for example, but only as a timorous anxiety that throws us out of the darkness that if persisted in bravely could bring us to a turning point. For some people identity confusion overwhelms them with terror in the movement to the turning point. During the withdrawal, for instance, they so firmly anchor themselves to consensual validation that they cannot stand being thrown onto their own resources. It evokes great anxiety and confusion. Identity confusion disturbs them in intimacy where the needs and rhythms of others enters their consciousness and whips them away from their hold on themselves. Here too not the whipping away but the holding too tight is the problem.

## 91

Abrupt, premature turnings don't bring much growth because they are weak in intention. In them, we haven't accumulated the depth of experience that comes in a real withdrawal to generate a self-transforming turning point. I knew a woman who went mad consulting the *I-Ching*. She had several turning points a day. She stopped seeing friends. She didn't leave her apartment. She was going through too much. But her changes of fortune were limited to tossing coins and interpreting the oracles. Being swept from elation to despair and back by the toss of the coins, she had no time or energy to do anything else besides tossing the coins for another reading. She did not experience real turning points because her changes did not change her.

The more out of balance the weak and strong legs, the more the turnings degrade and the emptier they become. Without memorable turning points, as I said, we cannot really connect the legs that lead to or from the real dramas of life.

## 92

Since on the weak leg we are more rigidly defended, more out of touch with the natural flow of experience, more awkward and ill-at-ease and fearful, we become less competent on it, less realistic, and more inclined to create compensatory fantasies to cover up our lapses. These lapses influence sensation first. How we see what we see – and what we can see at all change. Presentness turns to absence, awareness to distraction, inwardness to outwardness, outwardness to inwardness.

Since the fears and defenses are specific to the weaker leg, they produce state dependent experiences. We don't grasp our real situation. All we can detect, as if from the corner of our eye, is how deftly we have scooted away, how smoothly we have shifted from one personality part to another with hardly an inkling we have done so.

Too much fantasy makes our love lives weak because it keeps us out of touch with our partners' needs, and our own. Because the fantasies and actualities are intermingled on the weak leg, our experience becomes harder to recollect and harder to make use of later. Our sense of time on the weak leg is foreshortened or elongated because it is built on the impoverishment of experience and not checked and shaped by realities. In this state of confusion, our hopes and fears become less realistic. We can come to believe practically anything: that love will always make us suffer, that we must accept it as a sacrificial, selfless clinging, that this may not be its own nature, but it is our fate decreed by a cruel God; that we are extraordinarily gifted in ways that we are not; that we are strong on the weak leg and weak on the strong leg, that we are loved when we are abhorred, or hated by our parents who love us deeply, that we are giving when we are taking, that our malice and thievery are justified.

These illusions lead us into all kinds of conflicts, drawing us away from present perception, hiding the sensory cues that would give us information for appropriate distance regulation. Finally, we hardly know where or who we are. Being out of touch, the unexpected rebuffs (or desires) we receive from our partners shock us. We live like the Chaplin tramp who dances attendance on an indifferent girl without even noticing how aloof she is. We are so enthralled; we are inattentive to the real state of affairs.

The less we use the weak leg, the fewer chances we have to strengthen it. Moreover, when the inevitable perturbations come, we are easily tossed off our pathways. If the impinging forces are large, even the strong legs can break and when they do, our transformational moments become even more inaccessible.

## 93
### *Insight in Turning Points; Signs of Turning*

To my observation, the turning moment does not bring insight, unity, relief, beauty, meaning or gratitude. They come, but they only come later. The insight is like the exclamation point at the end of a sentence. Only afterwards, during implementation, on the next long leg, little by little, do the insights come – and only if you keep making the changes.

Insights feel strong and important. However, the feelings inside the insights can deceive us. Felt shifts make unreliable indicators of change. People feel all kinds of things under all kinds of circumstances. They can feel they are flowing when they are stuck. They can feel awkward when they move gracefully. Hysterical people determine truth by their feelings.

Experiences of inrushing energy, of an urge to do, of exaltation, do not signify completed turnings. Transformational turnings require evidence of actual transformational events, not just sensibilities.

I have observed in myself, and among my counseling and biofeedback clients, three relatively trustworthy indicators of change. They occur in close succession, usually over a few days. One is a special dream, another is the restoration of a significant lost memory and, third, and most difficult to explain, is a synchronistic event that confirms the change.

The powerful dream has been well described by Jung. The restored memory, called abreaction in psychoanalytic theory, we understand as an event following the release of physical tension, particularly the unlocking of bracing patterns in the musculature. The restored memories let us put the past together differently. During personal paradigm shifts, when the model changes, the temporal sequences leading to it shift as well.

The third sign, the synchronistic event, is harder to grasp. It comes unexpectedly, as if from beyond the causal realm. But it strikes you as significant, right and meaningful. Someone important unexpectedly shows up. That which was lost is found. A person you are thinking about thinks about you and makes contact. The world seems to be affirming you for mysterious reasons. Jung called synchronicity "an acausal connecting principle."[2]

Where does the synchronicity come from? Is it really *acausal*? No. The straightforward explanation is that it registers in consciousness with a special emphasis: you see what you could not see before. We take the shock of recognition for a synchrony. If your voluntary control over your autonomic functions, on the borderline between what you think you can do and what you are reasonably sure you cannot do, shifts, for instance, if the boundary moves, your range of freedom expands. You encompass more of what before had been autonomic and unconscious. You experience new freedom as a "synchronicity".

You can liken the experience of synchronicity to the experience of a fellow who trudges off to work each day, his eyes averted, face to the pavement. He does not look up because his expectations are so low. Coming his way is a woman, her eyes averted too. She never looks up either. Then one day something changes. They both happen to look into each other's eyes. He smiles, she smiles back, they are attracted to each other, and in that moment they make a recognition. They see each other. A great deal travels on eye contact. *That's* the wondrous thing: they get a sense of their possibilities. Model change helps them see the possibilities better – or gives them new possibilities.

Here a small starting gesture, performed at a moment of readiness, something seemingly trivial, like the glance and focus of the eyes, produces great consequences. Opportunities open.

On their twentieth anniversary, the two tell their children about the mystery of their meeting. They call it a miracle. Something God arranged. The scales fell off their eyes. They got a new way of seeing.

## 94
### Full and Empty Turnings

I call turnings full when they are conscious, lit by intention and empowered by a precisely timed starting gesture. I call turnings empty when we have them without knowing we are in them, or when we are not aware that an opportunity has come. Moreover, this opportunity sets the groundwork for the future.

In a full turning, one accepts reversal of fortune. In an empty turning, one resists it. In a full turning, a person makes choices from the heart of his or her vulnerability. In an empty turning, the person denies or resists the vulnerability, making authentic choice unattainable. In a full turning, a lover realizes the consequences of his actions on the beloved. However, in an empty turning, love lacks consideration. The heart of caring fails. A full turning in wisdom serves others. An empty turning in wisdom seems to serve oneself, but really does not.

Empty turnings fail in five ways:

1) When the turning comes it is entered, but without powerful affirmation and without clear consciousness. The turning resolves, but it is empty. Moving mechanically through our moments of highest opportunity, we generate only a weak fanning of possibilities. We come to rest on a low adaptive peak, with little potential energy for a next move. Environmental, cultural and technological forces drive the changes. The per-

son, by long habit entrained to these forces, really follows the line of least resistance. However, he does not know it. He calls himself realistic.

2) The turning is never fully engaged. Instead, one precipitates a premature turning from a shallow place along the way and leaves the scene too soon. Love stays love, but goes from approach to separation too early, in various combinations. Wisdom stays wisdom but goes from withdrawal to return too soon, or vice versa. The reversal in a premature turning is defensive in nature, a flight from the full force of life.

3) Another kind of failure comes from delay. We stay in withdrawal too long, or in approach in love, fearing separation. We fear the turning because we know that different actions will be demanded of us afterward. We will have to confront the insubstantiality of our fantasies. We may be shamed before others for our weaknesses. We hang back from the transformational opportunity until it passes. Then we enjoy a depleted, late turning.

4) Where in a full turning point there is one great moment of reversal, in certain troubled turnings we tremble indecisively. We go back and forth. We endure reversals and then reversals of the reversals. A rhythmic vacillation ensues, a tremor of intent. The competing personality parts struggle for dominance. Hours or days pass, or longer. We cannot establish order around a new set of passing rules. We neither change nor remain the same. Instead, we oscillate around a center, unable to enter, unable to leave, unable to retreat, unable to advance. We endure the experience of the Dark Night of the Soul continuously. Repeatedly we flirt with despondency. We cope with interminable fear. Finally, the disorder gets familiar. It wears us down and numbs us out. In love, Mr. Numbnuts cannot commit and cannot break up. He "can't live with her, can't live without her." Something from outside eventually resolves the impasse. Someone, something, or some series of events chooses for us. Last of all, death makes the choice. If we are spiritually inclined, we may even delude ourselves into thinking, when we re-

view our lives, we were Taoists and "went with the flow" – but we did not. We went in the direction imposed by the concatenation of forces.

5)  We cross over. A crossover is an abrupt and inappropriate jump from love to wisdom or wisdom to love.

It happens close to a turning point when the temporal patterns are most easily disrupted. For example, there you were writing the great American novel, now you have suddenly fallen in love. Who has time to write? However, when the intimacy gets too hot you jump back, you are writing your novel again. Crossovers are defensive shifts: They let one escape from oneself just when the going gets tough. Instead of hanging in there, one jumps away. The defensive crossovers relieve the short-term physical stresses but they undermine the integrity of the process. Crossovers are usually made by jumping from the weak leg of one dynamic to the strong leg of the other. People who agonize over why they cannot experience deep love, or people who never have a satisfying creative accomplishment, may be avoiding the transformational turnings by crossing over.

## 95
### *Favoring the Convergence of Love and Wisdom*

In the strongest full turnings, (particularly in the signal events of life) love and wisdom converge. When love and wisdom converge, you make better, more transformational choices. The deeds that follow on them work their way into the world better and achieve outcomes that are more favorable.

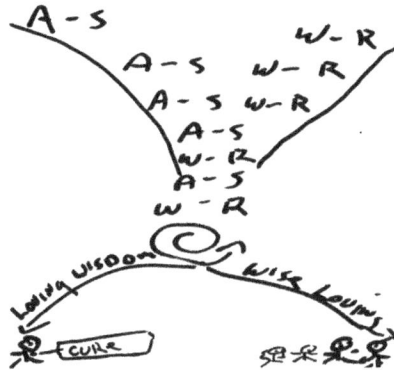

**Wise Loving/ Loving Wisdom**

How do I explain convergence? Our loves and wisdoms are not mutually exclusive concerns. They entail each other at decisive moments. Destiny altering turning points happen when love and wisdom move closer, as they can in many permutations and combinations.

An Hassidic rabbi, Moshe Leib of Sasov, explained how he came to understand the convergence of love and wisdom. Sitting in a tavern, he overheard a conversation between two Russian peasants. He gets his insight when one says to the other, "you don't really love me; if you really loved me you would know what I need."

Knowing what the beloved needs is a wisdom function; wanting the beloved to have what he or she needs, and offering it, is a function of love. To know what the other needs requires solitude in love. Without this core of solitude, we could not distinguish between our lovers and ourselves. We'd have no way to know what to give or how to give it, or how to receive with gratitude. Without the wisdom in love, we'd entangle ourselves in incompatible relationships, painful dependencies, cold estrangements, lies and betrayals. We'd eat each other alive.

And wisdom needs love in it. Without love, our inventions are cold. Without love, the withdrawal lacks consequence, lacks a reason for being. A surge of love in wisdom encourages us to return to the world out of longing for others living in it, or longing for na-

119

ture. Without the presence of love in wisdom, the desire to serve is missing.

From convergent turnings of love and wisdom we get a deeper sense of self-worth, and from that we learn to estimate the worth of the world differently, we view it through the eyes of service and develop management styles that recruit our efforts to live peacefully in new ways.

However, we have great difficulty achieving full convergence. Strong cultural forces drive love and wisdom apart.

## 96
### *Signs of Convergence*

My clinical studies with biofeedback clients show that the convergence of love and wisdom happens strongest in illness and healing settings when most is at stake and when the largest numbers of oscillators interact. When resonant sub-systems, the fractal units of love and wisdom, touch, they shape the turning process in special ways.

When love and wisdom move into a convergent relationship, we experience giving as receiving and receiving as giving. We readily understand the pleasure in giving; at times one gives with such evident joy that it makes the receivers act of acceptance a giving itself. Receiving well sparks the wisdom in love because it knows that receiving poorly hurts the giver. If we examine ourselves, we will see that giving-as-receiving occurs when approach/separation and withdrawal/return come into phase with each other. Their common dynamics let the two systems come into resonance in one or more of the common frequency bands in human nature.

You can experience the convergence in the span of a heartbeat or a breath length or over the course of a 90-120 minute rest/activation cycle. You can experience it in a night of dreams. Or in a waking day spent in powerful connection with a beloved. You can have seasons of love and wisdom linked primordially to the effects of day length on melatonin flows. There's always a time course. And the convergence experience always fades away.

When many people living under similar socioeconomic pressures endure correlated turnings with overlapping and interacting approach/ separation and withdrawal/return pulses, world changing social movements can develop. This kind of togetherness manifests first in personal intimacy, in the closest circles of life. It uses eye contact, hand gestures, touches, expressions and vocalizations to carry meaning. From there it gains momentum and spreads out in wider arcs of affiliation.

In certain revolutionary situations, brief periods of *bonhomie* can provoke historical tipping points. The young Wordsworth experienced this spreading wave of values in his travels in Revolutionary France in 1789. He wrote in *The Prelude* that

*A benignant spirit was abroad*
*Which might not be withstood, that poverty*
*Abject as this would in a little time*
*Be found no more, that we should see the earth*
*Unthwarted in her wish to recompense*
*The meek, the lowly, patient child of toil...*[3]

I deal with the historical consequences of conjoined turnings in a later chapter.

## 97

Persons who only experience non-convergent love turnings never learn from love. They cannot recognize the interplay between their social and private selves. Persons who only experience non-convergent wisdom turnings live coldly. They never understand the real worth their deeds play in the lives of others. They do not extend their ethical powers.

A convergent turning, even one incompletely realized, produces personal growth, adds something in the world, and touches others. Following the turning, it issues out on a meaningful next leg.

## 98
### *Yin/Yang: The Circle of the Opposite*

I can make the convergence even clearer by illustrating it with the Yin/Yang symbol. On one side of the central curving line, in the field of white, you have a black circle floating, and on the other side, in the field of black, a white circle floats. We can use them to symbolize how love touches into wisdom and wisdom touches into love.

You know you are close to the heart of a wisdom turning when the circle of love opens in the deeps of solitude, and you know you have grown deeply intimate at the turning of closest approach when the circle of solitude opens up in it.

We experience the circle of the opposite as a surge of love in the heart of the solitude of wisdom, and the solitude of wisdom in the heart of love.

Together the circles with their opposites (imagine the symbol folded over on itself making a helix with the circles aligned) suggest the image of the Caduceus or the Shamanic world axletree around which wind our two most characteristic human enterprises.

The circles of the opposite are the pivotal moments, the creative nexus in both dynamics.

Perhaps love can manifest in wisdom in extreme withdrawal because then we touch a resonance beyond the intimate, personal, social and public distances. There we immerse ourselves in the greater dance of nature in the cosmos. With our sensitivities to gravitational pulls, electromagnetic spectra, day/night and annual/seasonal cycles, in those moments we experience the world as a living presence. Here perhaps the sensory grounds for withdrawal-return and for approach-separation touch. And contact with nature following the turning point in deep withdrawal may evoke memories of home, of life with others. The twined snakes on the axletree have long stood as symbols of healing and wholeness.

As an experiential feature of life, the circle of the opposite shows the closeness of a single turning moment better than the Caduceus. The energies in the circle of Yin and Yang rise, fall, and bear on each other in describable and quantifiable ways. The *I Ching* uses the solid or broken lines of the trigram to represent these processes. The Caduceus, on the other hand, better represents a series of related turnings, an extended process, an historical unfolding. It appeals to us as a Western symbol of continuity and directionality in change.

Perhaps the symbols have deeper meaning too. Both show waveforms. Both show the phenomenological interactions between two dynamisms. The curved line in the Yin-Yang symbol, its waveform representation, shows a kind of embrace between love and wisdom.

If in the brief period of delicate balance we can integrate the circle of the opposite into the turning without annihilating the main thrust, we will amp up the energy of transformation, and its results will become more life affirming. The wisdom is wiser when love falls in the heart of it. The love is more loving if touched by wisdom in the loneliness of intimacy. This *touch of the other* also occurs during the smaller turbulent passages in the principal mini-turning points of love and wisdom. At these moments of vulnerability, love and wisdom trade words, images and concepts. The experiential uncovering of the circle of the opposite, therefore, marks a high point in restoring meaning to turning points, and as does the shamanic axle-tree, it opens a route to moral endeavor based on exploration of connected realms of life.

Notice that the circle of the opposite is *surrounded* in the color field; it has no direct access to its own color. It works as a window, not a door, a dynamic tension, an energizer, a seed crystal we must treat with care. If you fall too deeply into the circle of the opposite and give yourself over to it, you may suddenly drown to love and be resurrected in wisdom, or the reverse. The purpose of the turning will be thwarted then. When love wells up in wisdom, it can be experienced as something fearful and if we act upon this fear, the circle of the opposite becomes the source for regressive tendencies. It can intensify and bleed out and take over. We can flee to symbols of the mother for comfort, undermine the turning point in deepest withdrawal, and abort the return. This crossover would not only involve a premature turning, it would require a jump from withdrawal to approach instead of return. The return would not be enriched and enlivened by love, but dominated by it, forfeited to it. For the same reasons, the serpents on the Caduceus never touch. Crossover is a danger. If we take the circle of wisdom, when it appears in the close approach phase of love, to be a signal to withdraw, a similar danger rises. Instead of separating with an enlivened, enriched sense of the need for boundaries and solitude and good judgment in the dance of love, the love song stops. The dance ends. The relationship is sometimes sacrificed in the crossover to wisdom. Something like that must have happened to Kierkegaard when he broke his engagement. He fell into the circle of the opposite. Wisdom became his escape hatch.

To get the power of the circle you usually must resist the temptation to cross over from one dynamic to the other. However, there *are* those rare moments when the jump is right, rare moments when you climb through the hole, as through the *Sipapuni*, the Hopi passageway between worlds, and experience transformation full on. You can switch planes then, change paths, revolutionize your life, and with commensurate action, change the world.

## 99
### *Krapp's Last Tape*

The strongest dramatic representation of the loss of convergence I know is in Samuel Beckett's play *Krapp's Last Tape*, a play without turning points. Krapp, the lonely protagonist, practically a bum, listens to a tape he made 35 years ago when he broke up with his girlfriend in order to write his Magnum Opus.

*"... Thirty-nine today and sound as a bell, apart from my old weakness, and intellectually I have now every reason to suspect at the... crest of the wave—or thereabouts. "*[4]

Krapp eats bananas and listens to his younger self.

*"Spiritually a year of profound gloom and indigence until that memorable night in March, at the end of the jetty, in the howling wind, never to be forgotten, when suddenly I saw the whole thing.*

*The vision, at last. This I fancy is what I have chiefly to record this evening, against the day when my work will be done..."*

On the jetty, in the storm, he has an intimation of Truth. It hovers before him ready to be captured in a great book he is destined to write. As the young Krapp starts to expand on his epiphany, the old Krapp shuts off the tape recorder. He fast-forwards, cursing under his breath, until he gets to

*"... my face in her breasts and my hand on her. We lay there without moving. But under us all moved, and moved us, gently, up and down, and from side to side..."*

Then he rewinds a bit and plays:

*"She lay stretched out on the floorboards with her hands under her head and her eyes closed..."*

The epiphany on the jetty regarding his Magnum Opus and the encounter with his beloved are part of the same incident.

*"Sun blazing down, bit of a breeze, water nice and lively,"* he continues. *"I noticed a scratch on her thigh and asked her how she came by it. Picking gooseberries, she said. I said again, I thought it was hopeless and no good going on, and she agreed without opening her eyes."*

They broke up, he to write, she perhaps to enjoy her other lover. The Magnum Opus never appears. Missed turnings, lost love, failed convergence. No next long leg. Long term despair. Krapp puts a new tape on the machine and starts recording:

*"Just been listening to that stupid bastard I took myself for thirty years ago, hard to believe I was ever as bad as that. Thank God that's all done with anyway. (Pause.) The eyes she had!"*

Later he adds:

*"Nothing to say, not a squeak. What's a year now? The sourcud and the iron stool."*

## 100

The health of the rhythmically restored body gives us the apt-ness and dexterity we need to work the levers of change gracefully. We use the hand of giving-that-is-receiving to pull the lever. William Blake exalted this power in his poetry. The Blakean body, born in the fires of convergent turnings, enters the fields of endeavor armed with arrows of desire. Those arrows show the fusion of love with aggression.

**Blake Jerusalem etching**

*Bring me my Bow of burning gold!*
*Bring me my Arrows of desire!*
*Bring me my Spear! O clouds, unfold!*
*Bring me my chariot of fire!*

*I shall not cease from Mental Fight,*
*Nor shall my Sword sleep in my hand,*
*Till we have built Jerusalem*
*In England's green and pleasant land.*

## 101

When new values and motivations percolate into the world, they rise up biologically, pulsing through the tissue of life, first stirring in intra– and intercellular communications, then extending their reach into sensorimotor physiology and from there into voli-tional behaviors. Social and political commitments come from them. Can we directly discern the physiology of "percolation"? Can we tell when it is happening?

How can we know when the historical lever has moved? We experience giving-that-is-receiving on wide social scales as in the genuine ethical movements underway today. There is another sign that may seem at first sight contradictory: the restoration of creative competition, the affirmation of the agonal component, which I take to involve the fusion of love with aggression. (Harold Bloom sees the *agonal* in all great art.)

**102**

It will take powerful rhythmic resilience to cope with the changes we are going through today. But that's the only way we can meet the almost unremitting perturbation and complexity of technological civilization vigorously, without losing our hold on love and wisdom. The needful thing is to stay afloat in the turbulence. That means having the strength, at times, to welcome the chaos and still pursue authentic turnings— and to recognize convergence when it comes. You get the strength and courage for this first from understanding the temporal patterns in human endeavor. From our present vantage point, we can accept the following rules of thumb as parts of human nature:

1) Intensification and reversal are inherent in human endeavor.

2) The opportunities for growth increase tremendously at crossroads in crisis times.

3) To traverse turning points with greater knowledge and power is to gain special skills of timing, a kind of primordial canniness to see change coming and to meet it on the fly.

4) Love and wisdom converge only at singular moments and rarely for long stretches of time. It is hard to foresee these moments, but easy to recognize them when they come. To have clean and strong intentions then delivers your presence into the world with special force. Then intention spreads across the cortex, attains freedom and leaps into the world.

I have insisted on freedom for the first time here. But do we have it? Where does it come from? How does nature produce it? To these ponderations we now turn.

### *End notes*

[1] (Ken Wilber. *The Spectrum of Consciousness*. Theosophical Publishing House. 1977. p.199)

[2] The three confirming signs must happen in quick succession over hours or days or a week or two at most. If you search for them after the window of opportunity closes, you are likely to find something-or-other, anything, and make it into a confirming sign. After a real turning, however, the confirming signs find you.

[3] William Wordsworth. *The Prelude*. 1850 version

[4] Samuel Beckett. *Krapp's Last Tape*. Grove.

# 5. ON TIME-SENSITIVE FREEDOM

### 103
### *Peripetia*

Because it is full of reversals, life takes on an essentially dramatic character. Its presiding quality is change of fortune. But dramas and changes of fortune, we will see, have no meaning if we lack freedom of choice and action in them. Without freedom, moral choice is empty. Aristotle called this reversal in tragic drama its *peripetia* .

"Reversal of the Situation," he explained, "is a change by which the action veers round to its opposite..." Or as Plato wrote: "In the seasons, in plants, in the body and above all in civil society, excessive action results in violent transformation into its opposite." (*The Republic*)

This same veering dynamic, flung up from deeper physiological venues where the chemical rotors of the organism turn, (described in # **13**, **26**, **32**) finds its way into the drama of love. And it makes wisdom tentative and thrilling. Desire and satiation, belonging and estrangement, attraction and repulsion spiral round and round. Though we repress our awareness of it from fear of our own and each other's inconstancy, life inclines us to understand our changes of fortune dramatically, which puts the problems of freedom and determinacy onto center stage. Theatrical representation, ritual, dance and dramatic performances in every culture exemplify this. They have from the beginning.

According to Aristotle, the reversal in a *tragic* drama can provoke recognition. "Recognition, as the name indicates, is a change from ignorance to knowledge, producing love or hate between the persons destined by the poet for good or bad fortune." The recognition in classical drama echoes the transformational moments inhering in real-life turning points. Character is tested in Aristotelian tragedy by this crisis of recognition. At its best, it simultaneously changes not only what the central character knows, but also the action of the drama swirling around him. That's why Aristotle adds: "The best form of recognition is coincident with the Reversal of the Situation."[1] When they happen together, a sudden influx of knowledge can change the course of the action. In this moment of discovery, Oedipus plucks out his eyes.

In Greek tragic drama, a flaw in character, a blind spot, marked by deficiency or excess, by overweening pride, *hubris,* chained the protagonists to the wheel of their fates. A peripetia occurs that the hero can neither resist nor use for the sake of freedom. He spins in the clutches of nemesis, chained, as was Ixion, to his wheel. That's the tragedy. Interestingly, the gods punished Ixion for resisting the opportunity to improve his life when given the chance at a turning point. That was his *hubris*. Don't block the way to your turning points.

**Ixion bound to the wheel**

## 104
### *The Signal Events*

Life does not toss peripetieas promiscuously across our paths. The transformational opportunities come only when we meet three boundary conditions:

1)  Our biological systems are stressed out to the max,
2)  Our life situation teeters on the edge of chaos, near a tipping point.
3)  Love and wisdom are at stake.

These conditions produce the crucial turning point dramas in life. They give us the chance to change, and to get free, but only if we energetically engage them. I call these *built-in* peripetieas the signal events of life. Each generates a destiny-shaping turning point. They come at critical developmental stages, not unlike those described by Erik Erikson.

Among the signal events I include the trauma of birth itself, the first separation from the mother, weaning, leaving home, first love, marriage, childbirth, encounters with life threatening illnesses, creative accomplishment, discovery, making a home, experiencing combat, uprootedness, divorce, finding one's calling, change of profession, loss of friends, death of parents, death of a spouse, confrontation with evil, retirement, and finally the way we face our own dying moments.

Each developmental drama presents us with a crisis that only love or wisdom, as we now understand them dynamically, can resolve.

## 105

The signal events rarely run smoothly. Typically, family, cultural and environmental factors disrupt them. Events hasten, retard, reschedule, or give them unexpected twists.

Even though the transformational opportunities may have been building for years, when they actually arrive they come on us urgently and pass quickly. In seconds, we must decide what to do.

We decohere, then we come together again. But the only repatterning forces that take hold are those we have earned by living on the long legs coming into the turning. And they only come to us if we endure the legs authentically. Our own perseverance, our habits of life, gives us the moral energy to carry ourselves through our turning moments. Moral energy, and the courage that sustains it, form the core elements around which the repatterning builds.

Deep biological alterations mark the life-stage changes made in the signal events. During the turning process, clashing personality parts struggle for dominance in the body. Each fights for its own style of access, contesting the use of organ systems. These struggles later etch themselves on the face as character. They live in the body as posture, movement and gesture. They shape bearing and attitude. They present themselves in action as the character of our deeds, and in basic vitality as health or illness.

## 106

When Holmes and Rahe studied the triggers for illness in more than 5000 patients admitted to Veterans Administration hospitals, they discovered events in the previous year that correlated with subsequent illness. The top ten correlations:

Death of a spouse
Divorce
Marital separation
Jail term
Death of a close family member
Personal injury or illness
Marriage
Fired from work
Marital reconciliation
Retirement

Notice first that meaning and health are associated, and second that the events provoking illness all came from the turning dramas in the signal events of love and wisdom — marriage, death, work, abandonment, change of status (though the manifest illnesses were more likely to surface later in life when long-term chronic stresses had taken their toll.) In every item on the Holmes-Rahe scale, a person suffers a rough break to the rhythms of life. Each interruption centers on a crisis in love or wisdom or both.

However, not everyone dealing with these challenges got sick. What made the healthy people different? Did they negotiate their crises differently? Did they live a different drama, comic rather than tragic? Were their rhythms resilient enough to favor resolution over nemesis? Did they more successfully navigate their turning dramas? Did they achieve "Recognition coincident with the reversal," as Aristotle recommended? Did this inflow enable them to direct their new knowledge to beneficial ends? Did freedom manifest?

## 107

How can we have freedom? you may object.

*-If the environment entrains the body, if the effort to live is enmeshed from beginning to end in causal webs, if frequencies pull on each other, and biological systems run on cascading information flows, the turning points must spin into life as a vast summing of antecedent causes. We are driven creatures, behaving under compulsion even in what we take to be our most significant, changeful moments.*

We need freedom. Love and wisdom have no moral force without freedom of choice and freedom of action. If we are not free agents at crucial moments, our choices ride on the changes in our bodies brought on by the causal texture of the environment. The world moves the body.

Without freedom, our turning points become self-indulgent flights of fancy. Why care, why swing into action, why be brave in the midst of fear, if our hopes and fears, and the choices and decisions that come from them, are caused before we ever get to them and then pass on through on their way somewhere else predetermined?

Emerson wrestled with this problem when he wrote in his essay *On Fate*: "Man is not order of nature, sack and sack, belly and members, link in a chain, nor any ignominious baggage; but a stupendous antagonism, a dragging together of the poles of the Universe... the lightning which explodes and fashions planets, maker of planets and suns, is in him...Forever wells up the impulse of choosing and acting in the soul. Intellect annuls Fate. So far as a man thinks, he is free."[2]

But freedom of thought is not the same as free will or freedom of action. It may lead to them, but not necessarily. As we experience it, freedom falls into a hierarchical arrangement. In each, a higher freedom depends on one lower down. Free to fancy is one thing, free to dream another, free to conceptualize, free to intend to do something, free to choose it, decide it, or free to do it are further marches still. They do not all have the same force in the world and they are not always accessible or lined up. A break at any point terminates the effort. Only the full array brings intention cleanly into the world.

Emerson does not distinguish between grades of freedom or the occasions on which one rises to another. He suggests that sheer strength of will makes the difference. But strong men bang their heads against stone walls too. What we need are not strong skulls, but doors in the walls.

## 108

Scientists have found these doors. They tell us that at least on some occasions indeterminacy exists in the world. Can we use indeterminacy to bolster our argument for freedom?

To know whether we had freedom on a particular occasion, and to gauge our levels of responsibility in it, we'd have to figure out whether there was a through-line of indeterminacies leading in rapid succession from membrane related and inter-cellular functions, to neural pulsing, to muscle movements and thence to fully intended behaviors. If the passage of energy rides on the interactions of many oscillators, it may be that a narrow window for freedom opens up only when the system moves into specific temporary alignments.

Love becomes love, according to this line of reasoning, because freedom blossoms in its turning points as an emergent property of their complexity. Entrainment, interference, resonance and dissonance between the oscillators and their perturbing forces come together to create this complexity. Likewise, wisdom becomes wisdom from the indeterminacy in its turning points.

## 109

Sounds good to me. But not to Daniel Dennett. In *Freedom Evolves*, he maintains that the determinacy-indeterminacy controversy makes no practical difference to choice. He tells us that

"An indeterminate spark occurring at the moment we make our most important decisions couldn't make us more flexible, give us more opportunities, make us more self-made or autonomous, in any way that could be discerned from *inside or outside*, so why should it matter to us."[3]

"...the issue is not about determinism, either genetic or environmental or both together; the issue is about *what we can change* whether or not our world is deterministic."[4]

Dennett advises us to act in such a way as to improve our odds.

"If you have the gene for phenlyketonuria," he explains, "all you have to do to *avoid* its undesirable effects is stop eating food containing phenylalanine. As we have seen, what is inevitable doesn't depend on whether or not determinism reigns, but on whether or not there are steps we can take, based on information we can get in time to take those steps, to avoid the foreseen harm. There are two requirements for a meaningful choice: information and a path for the information to guide."[5]

Of course, you have to want to take these steps, and according to him, that is what we do "if we are normal."[6] If you don't, you're abnormal. You may be in denial. So at least part of Dennett's argument rests on culturally relative standards of normality. But how "normal" are his standards? When Dennett argues that a normal person falling down an elevator shaft should act in such a way as to increase his odds, is that normal? "Perhaps in some of the worlds in which he lands, he survives... There is some elbow room...he may at least improve the odds by taking whatever actions are necessary, and thereby, with some luck, find himself in one of the vastly many possible worlds in which he lives."[7]

## 110

The happy insouciance with which he puts the case starkly contradicts the horrifying story he tells. What a view of life he offers us! – Hurtling down an elevator shaft, encased in a machine, out of touch with nature, we are hapless victims crashing to our deaths. We have only seconds left to think. How are we to come to terms with our situation? Though Dennett would have us roll up into a fetal position to improve our survival chances from one in a million to one in a hundred thousand, mightn't a thoughtful person choose instead to relax, not bother to calculate the odds, let the odds play out while he spent his moments coming to peace with himself? Which is the saner choice?

Where do we find Dennett in his example? Is he in the elevator with us, using his last breath to advise us? Is he dealing with his own mortality, sharing his insights as Socrates did in the prison house? No way. From the breezy tenor of the language, you can tell he is standing outside watching the elevator fall. We're the poor saps inside. How can he offer us advice as the steel car hurtles by? He must be talking to the graduate students by his side. No wonder he's so light hearted. He's an outside consultant.

## 111

It's the same in the game of *Prisoner's Dilemma.* In a curious way, it mirrors real-life circumstances. The story behind the game is that two prisoners are being interrogated. The interrogator offers each a deal separately: if you rat on the other guy, you go free and the other guy gets ten years in jail, but only if he hasn't ratted on you. If you both rat on each other, you both serve three years. Tough choices. The prisoners are supposed to sit there in the interrogation room and consider the options. A whole generation of American scholars has modeled altruism on this game.

All kinds of troubled thoughts must go through the prisoners' heads. "Why should I stay quiet? If my partner squeals on me, I go away for ten years, if I squeal on him the most that I can serve is three years. And if he hasn't squealed, I might go free. Of course, if neither of us squeals, we both go free. How can I take the chance of his not squealing? I'd be risking ten years." The researchers consider these options in their scholarly papers. They use computer simulations to play the game thousands of times over trying to determine the best strategies. They never consider what prisoners would really be thinking in those circumstances. Wouldn't they think, *"Who are these interrogators? How can I trust anyone who offers such a crazy deal to follow through on it? Will they tell my partner what I really said? What will they want from me next?"*

Computer simulated play of Prisoner's Dilemma shows that the winning strategy over many iterations is to start out altruistic and stay that way until the other prisoner betrays you, then betray back once and return to altruism. That's how to win in the prison world. That's what the experts have given us as a model of our lives. But don't forget: the prisoners may not even be guilty. For all I know they may be confessing under duress like the detainees in Guantanamo.

In its basic mythos the game envisions a world of prisoners in which only certain privileged researchers are free. They either conduct the interrogations or watch them from behind the one-way mirrors. Who are the prisoners? Us. The great bulk of the population (who we take as the "us") the tiny elite "us" of the experts considers the "them". We don't even get to be an "us" in prison. We're kept in solitary confinement. In ancient Rome, in the gladiatorial combats that provided their version of Prisoner's Dilemma, Dennett and his students would be watching the entertainment from the Emperor's box.

## 112

Spinoza dealt with the problem of freedom in his *Ethics*. He maintained, "Men believe themselves to be free, simply because they are conscious of their actions, and unconscious of the causes whereby those actions are determined..."[8] If we became conscious of those causes, he argued, we would know we were not free.

However, he still found freedom in the world, because nature as a whole was free. There was nothing outside to determine it to be other than it was.

By viewing our actions in the context of the freedom of Nature as a whole, Spinoza believed that we would be happy. And he thought we were capable of it because ours was a rational universe and, once understood, we would acquiesce in its order. "Our understanding or intelligence, that is, the best part in us," he insisted, "will entirely acquiesce in this, and will strive to persist in this acquiescence. For in so far as we understand, we can desire nothing other than what is necessary, and we cannot entirely acquiesce in anything other than the truth."[9] To do so is to live life "sub specie eternitas", under the aspect of eternity.

By participating in God's plenitude (for Spinoza God and Nature were coterminous), our acquiescence would be meaningful. It would make us happy. Moreover, our happiness, as he saw it, would give us energy to become active rather than passive participants in the freedom of the whole. That would be our freedom.

Stuart Hampshire, a philosopher who has written extensively on Spinoza, recognized that the power of what you could call *positive acquiescence* "by itself converted an external cause into an inner ground of affirmation or action."[10] Nevertheless, its relationship to freedom still puzzled him. "Perhaps this picture of the free man as self directing, as an integrated mind with a continuous controlling reason, is so far a clear one. But the notion of freedom itself is still unclarified..."[11]

<br>

## 113
### *The Third Attribute of the One Substance*

Spinoza saw the universe as one, single and whole. It was composed of one unified substance. That single substance manifested under an infinite number of attributes. But we knew of only two of them, thought and extension. God knew all of them. As Spinoza put it: "... mind and body are one and the same thing, conceived first under the attribute of thought, secondly under the attribute of extension."[12]

Interpreters disagree whether Spinoza believed we were capable of knowing more than the two attributes. I would say we have already taken on a third in the *space/time continuum*. Time itself has a place alongside thought and extension. In our chronobiological inquiry, we have seen it at work in the unfolding of our turning points.

With time as the third attribute, the wonder of life and its beauty and truth come to us in the deep *synchronicities* we experience. These occur across all scales of being in ways Spinoza overlooked, for though he aimed for "vital choice of action" he did not locate action in a temporal flow. He disregarded the special qualities of time because he viewed time like space, as uniform, infinite and eternal, with a consistent texture everywhere and no special "spots of time" in it. By viewing time as a manifestation of change in extension he relieved himself of concern over finding the special moments of accessibility to change, (events would be *scheduled* if his determinism was consistent). But our relativistic physics helps us see the world differently. We have come to conceive of time itself as malleable and grainy. Space we understand as curved and bumpy. Even the pure vacuum bustles with energetic particles twinkling in and out of existence.

## 114

Physicists today have a much less confining view of causality. They know that the causal web is nothing like the billiard ball cosmos upon which Spinoza based his "geometrical reasoning", and that mathematical proofs are tautological. Reality in its billion-fold branchings, whether in the neuronal traffic in the brain or in the molecules of a heated gas, or in the intergalactic water maser clouds, or in the behavior of crowds, produces emergent qualities that make the whole different from (and not predictable by), the sum of its parts.

Indeed, biologists no longer dispute the presence of indeterminacy on the molecular scale in membrane events. They have to use quantum theory to account for the reactions between enzymes and their substrates. They treat many exchanges in cells statistically. Prigogine and Stengers posit indeterminacy at bifurcation points in biological systems.

"We expect that near a bifurcation, fluctuations or random elements would play an important role, while between bifurcations the deterministic elements would become dominant." They go on to explain how these bifurcations occur in systems far from equilibrium. They take place when "the amplification of a microscopic fluctuation occurring at the 'right moment' resulted in favoring one reaction path over a number of other equally possible paths. Under certain circumstances, therefore, the role played by individual behavior can be decisive."[13]

What do the authors mean by individual behavior? Individual behavior of *what?* Quantum events, atoms, molecules, human projects? Where do the microscopic fluctuations come from? Prigogine and Stengers refer to James Clerk Maxwell, the great 19th century English physicist who established the basic laws for electromagnetic fields, who called them 'singular points' and treated them as accessways to freedom *in the human realm.* They quote him saying

"...the rock loosed by frost and balanced on a singular point on the mountain-side, the little spark which kindles the great forest, the little word which sets the world to fighting...Every existence above a certain rank has its singular points: the higher the rank, the more of them. At these points, influences whose physical magnitude is too small to be taken account of by a finite being, may produce results of the greatest importance. All great results produced by human endeavor depend on taking advantage of these singular stages when they occur."[14]

According to them, Maxwell believed we could take advantage of a readiness in the world, that we could meet it with a human readiness timed to the moment.

## 115

Do these indeterminacies point to freedom? That is the philosophical puzzle. They may not. As we have seen, Dennett and others argue that indeterminacies form parts of a vastly more complex *deterministic chaos.* We live in a world far too complex for us to get our minds around. We can never comprehend its causal texture – except probabilistically. The causal web is too vast and various and too easily tweaked by small changes in initial conditions to allow us any assurance we have freedom of choice or action. Our control over our fates is minimal, stochastic, and uncertain. We cannot know whether we are free or not.

But there is another possibility. Perhaps we can initiate biological membrane events. As unlikely as it sounds on first hearing, perhaps at certain moments we can find a way "in". If the accessible moments had relevance in the human realm, we might be able to raise cellular indeterminacies to freedom.

You would be right to point out that this is a hope, not an argument, and that I haven't explained how our intentions can penetrate cellular processes to give volition an "in". I would have to demonstrate how the phenomenological and physiological realms meet, touch and interpenetrate. Mind would have to touch matter – not an easy case to make!

However, our approach to the biophysics of the turning process will point us in a promising direction. It will show that indeterminacy on the smaller scales _can_ move to larger scale events. At certain moments, under special conditions, we can pass volition from one indeterminacy to another, and along the way bring human freedom into the world of events. Reality comes together in such a way as to allow actual human beings, set in their habitats, to find and wield freedom at significant moments in their lives. The key is "at significant moments." Something shifts in the texture of reality then.

In the field that surrounds and penetrates us, turbulence produces indeterminacy and indeterminacy turns to freedom. I will soon explain how, for biological and biophysical reasons, we acquire mechanical advantage at those moments. With this leverage, we can cultivate the opportunities in our turning points and volitionally restructure our fundamental ways of being in the world. Love and wisdom provide those moments. Made refulgent by freedom then, they turn toward each other; they develop convergent tendencies. They grow together. Wisdom seeks love and love seeks wisdom.

To make these claims clear we'll have to go once again into the thicket of neouroscience, biology, biophysics and philosophy. Bear with me.

Research shows that biological systems produce indeterminacies at their bifurcation points. This we accept. I will now argue that intention can raise certain of these indeterminate bifurcation points to freedom.

Recent findings in neuroscience support this way of thinking, particularly research in neuroplasticity, complexity theory, quantum neurobiology and neural Darwinism. With their help we will try to bring phenomenology and physiology closer together.

## 116
### *A New Take on Freedom*

I used to think the main constraints to my freedom were in me, not in circumstances. I didn't see that the textures of the fields of endeavor themselves changed. I didn't understand that the readiness of the situation was the main thing and that it affected each of us differently, you in your "now" and me in mine. I didn't know how to deploy myself in the now, or understand that down to a fine-grained texture, there were freer and more determined times, bowls of attraction with steep sides and strong suction that you couldn't escape no matter how hard you tried, and shallower ones in which even small deeds could have great consequences. To me determinism was largely psychological. I believed I could not break my chains because my will was weak. I was ignorant of timing. I couldn't seize the moment because I did not know how to figure out *which moments could be seized.*

**Me, lost**

## 117

An indeterminacy that turns to freedom, freedom that manifests then slips away. Strange notions. To make them useful we will have to temporarily put aside the 'all or nothing' view of freedom and make freedom time-sensitive. We need to treat it as something that comes and goes. This implies that in some settings we are free and in others we are not because something in the structure of reality itself shifts. When that happens, the field that surrounds us (and that is in us) changes and freedom opens up or closes down.

James Crutchfield and his collaborators made this argument when they wrote, "innate creativity may have an underlying chaotic process that selectively amplifies small fluctuations and molds them in to macroscopic coherent mental states that are experienced as thoughts. In some cases the thoughts may be decisions, or what are perceived to be the exercise of will. In this light, chaos provides a mechanism that allows for free will within a world governed by deterministic laws."[15]

What I want to know is whether *I am the one* who "selectively amplifies" the small fluctuations in my brain and raises them to freedom, or whether the indeterminacy just happens to configure itself in a way that I call a decision that I flatter myself I have made freely?

Some very smart people argue that while indeterminacy is a characteristic of the physical universe and lurks in the quantum realm, freedom is an attribute of living beings only and needs consciousness, intention and action to manifest meaningfully. Even if indeterminacies occur at critical junctures and open paths to freedom, we have to take those paths, and the taking requires consciousness and intention. And where does that come from?

In this fractured world, who has clear consciousness? Where does it sit among the modules? How close can we actually come to experiencing the moments when indeterminacy opens to freedom if we are in chaos then? Where is the "me" that can ride them into meaningful choices and actions?

What we are looking for – unlikely as the successful search promises to be – is an access to freedom that connects the smallest neural fluctuations with the broadest, most meaningful intentions, choices and deeds. To demonstrate that we have real freedom, freedom to make meaningful choices with creative historical consequences, then, we have to demonstrate not only that the wholeness is free on some ultimate (and perhaps remote) level, as Spinoza saw it, but that it contains within it numberless eddies of freedom – indeterminacies within an undetermined whole – closer up and accessible to us. The most important of these opportunities we are supposing must occur in our brains. But how do they liberate our intentions and actions from the causal web? Do we dare whisper the magic words, *Quantum effects*?

## 118

Werner Loewenstein, a biologist researching intercellular communication, sounds the proper cautionary note: "One instinctively shies away from that weird, counterintuitive particle world, in the hope that all the answers may lie in the macroscopic realm. But they don't, and if the perspective of a quantum-coherent cellular ground substance stands the test, then it is at the quantum level of protein-water organization in the brain where we must seek the physical basis of mind."[16]

His interest in protein-water organization ties into our treatment of the thermal effects of sun on water in early evolution. The vortex movements that established some of the rhythms of primordial life pulled protein chains through water, their hydrophilic and hydrophobic regions reacting to the combined influences of electrogenic and kinetic forces. In primordial times, they whirled round in the oceanic vortices, great and small. The "handedness", the chirality of proteins may still preserve traces of this gyre motion.

Loewenstein maintains that protein/water complexes carry quantum coherent information into cell signaling processes. From there they shape our mental life. From there, the hierarchy of freedom ascends to intentions and actions.

## 119

Can we make the case that our *attending* in and of itself, our conscious awareness, has the power to provoke bifurcations that issue out into choices that produce affirming actions? Can we treat our intentions as neurological events? Would indeterminacy rise to freedom if they were?

Loewenstein makes this case with a quantum computer analogy.

"Now a computer of this sort—a device subject to the laws of quantum physics—is what Penrose postulates to be operating in the brain. This view does not disregard the classical digital brain activity. The two computational operations would exist side-by-side in our brain, though at different levels of energy and matter. We will deal with the two-level aspect further on, but first this question: What may be the substrate for such computations? This may seem an adventurous question to ask at this early stage, but it strikes at the heart of the matter. So, let us at least write down the *a priori* requisites for that substrate. This will be useful in sieving through the proposals that are already on the table.

"The requisites are easily specified: (1) the substrate must be insulated from the cellular-sap, or else, as we have seen, the coherent quantum state gets drowned in the gross; and (2) it must provide intercellular continuity in order to allow a multicellular quantum-coherent state. As to the identity of the substrate, all we can usefully do at this time is to weigh some general possibilities in this a priori light.

"We start with a possibility Penrose has raised, that the microtubules—the 140-angstrom-wide tubes found in virtually all sorts of cells criss-crossing their interior—might be the substrate. This seems unlikely because they do not fulfill condition 2. While they have attractive dialectric features that might conceivably enable them to serve as *intra* cellular conduits of quantum coherent waves, it seems unlikely they could do so as *inter*cellular conduits; all the available evidence indicates that they do not traverse the cell membrane.

"However, this does not exhaust the possibilities. There are other dielectric candidates in the cell that show more promise. They form part of the cytoplasmic ground substance, a net stretching through the interior of cells that is made of finer filaments than the microtubules. That net, or at least a major component of it, fulfills requisite 2 and bids fair for requisite 1. Indeed, this is no ordinary net—its mesh holds organized water! Or perhaps I should say the mesh organizes the water, because the protein constituents make hydrogen bondings with the surrounding water, ordering it. But it's not the information in the molecular order I am harping on, but that in the quantum order—the information inherent in the quantum states of the hydrogen atoms. This information changes in an oscillating electric field. Hydrogen atoms flip between quantum states in such a field. They can act as logic gates—bits of information can be registered by the quantum spins of the electron and proton in a hydrogen atom."[16]

Here we have frequencies and synchronization on the quantum level. They move to the macro level in the dance of hydrogen bondings in protein/water structures caught in the harp of the microfilaments, just as we described it earlier (see # **15, 25**.) Note that the information conveyed is oscillatory. As such, it can establish resonant relationships with other oscillators. In addition, these may link to systems in the macro realm operating in the ways that Winfree modeled in the heart. If quantum indeterminacy can enter living tissues in molecular events, and by doing so generate bifurcations in biological processes, and if the precise moment these occur *is* the route by which organic life communicates quantum effects to the macro world, and it happens in the human brain, these *can* produce events on the macro scale.

## 120

One can reasonably argue that, over time, evolution favored certain indeterminacies in biological processes over others – ones that counteracted dysfunctional automaticity. Natural selection used naturally occurring indeterminacies. These indeterminacies undergird the cellular chemistry underlying our primal approach-separation and withdrawal-return behaviors.

Randomness becomes highly relevant to survival on the edge between order and chaos where the essential energy exchanges of life occur.

Stuart Kauffman, the complexity scientist, makes these situations vitally important.

"...A working hypothesis, bold but fragile, is that on many fronts, life evolves toward a regime that is poised between order and chaos. Borrowing a metaphor from physics, life may exist near a kind of phase transition." [17] He explains "the space of relevant possibilities of the biosphere – its phase space – cannot be prestated. Thus the biosphere is creative in a way we cannot prestate...you do not yet know how to pick out the relevant collective variables...as the variables that will matter to the unfolding of the biosphere." [18]

On these adaptive pathways, living organisms evolved to use indeterminism as an *opportunity,* an occasion to increase variability and biodiversity in order to attain greater fitness in development and behavior. In human evolution, natural selection turned the subcellular random events on the edge of chaos into the pathways to freedom that we now enjoy in the power of mind that lets us love and seek wisdom.

We shall see later that the advances in the evolution of the human brain in its period of most rapid growth did actually follow this route. For now, we can surmise that selective processes raised indeterminacy into freedom in order to overcome dysfunctional automaticity – that volition proved advantageous, and when play and intention opened new adaptive niches, freedom explored them.

**121**

In the near chaos conditions of our turning points, when we are most vulnerable to perturbation, special time-limited opportunities for self-directed change open up. Multiple oscillators move into phase with each other (review findings in # **8-10, 17, 54**.) When they reach their turning moments together we get access to freedom.

Always one particular bio-oscillatory system will be up front, serving as the fundamental to which the other bio-oscillators relate as unisons, overtones or undertones. The front face might be breath or heart rate or the flight of ideas or a tremor or a slower activation rhythm in the ultradian band. It might be a committed desire, a thought or an intention. It could happen in a dream or during an insight in waking life. The opening could come during a chance meeting or in deep intimacy with a beloved, or after great concentrated effort. Doors open to freedom then, but they quickly shut in the aftermath of the reversal. You can miss them.

*The puzzle we must now solve is why, after all this trouble, and with so much meaning at stake, we almost invariably miss our moments of freedom and let them pass by?*

* * *

## *End Notes*

[1] Aristotle. *Poetics*..(xi 1-2

[2] Emerson. Fate. In Selected Essays. Penguin Classics. 1982. p. 373-4. First published 1860.

[3] Daniel C. Dennett. *Freedom Evolves*. Viking 2003. p. 136

[4] ibid p. 160

[5] ibid p. 156

[6] Ibid p. 169.

[7] Ibid p.88

[8] Spinoza. Ethics. III.ii. note

[9] *Ethics*. IV Appendix. xxxii

[10] Start Hampshire in *Spinoza – a Collection of Critical Essays*. Ed. Marjorie Grene. U. Notre Dame Press. 1973. p. 310

[11] ibid p. 302-3

[12] Spinoza. *Ethics*.III.ii.note

[13] Prigogine and Stengers. *Order Out of Chaos*. Bantam. 1984. p. 176

[14] James C. Maxwell. *Science and Free Will*. Quoted in Prigogine and Stengers. p. 73

[15] James Crutchfield, et al "Chaos" Sci Am. Dec 1986 p. 49.

[16] Loewenstein. *The Touchstone of Life*. Oxford U Press.1999. p.332

[17] Stuart Kauffman, *At Home in the Universe*. Oxford, 1995. p. 26.

[18] Stuart Kaufmann. "What is Life." In Brockman ed *The Next Fifty Years*. P.138.

# 6. THE READER'S OBJEC-TIONS, AND MY RESPONSES

## 122

*Your arguments don't hold up. Their premises are flawed. It's obvious that we love more than one person at a time and that we approach some while separating from others, and some who we approach are separating from us. All kinds of contradictory trends are mixed together. There's no overriding rhythm. Our depth processes suffer all kinds of halts, breakdowns and reversals. While we head inward in some dimensions, we are turning outward in others. The phantasmagoria of our inner lives is as complex as our sensory experiences. And all these movements in love and wisdom ramify out in a vast scattering of happenstance on happenstance that have no pattern whatsoever. Young, then old, then dead. That's all we can depend on. And not even that, if we die young. The evidence you presented from the sleep/waking cycle, the rest/activity rhythm, brainwave synchronies, monthly and seasonal behaviors, and more, don't apply. In real life, every one of them is interfered with. Chance occurrences are always sweeping in and driving us away from our intentions with no way back. Who we meet, what we hear, what happens to us, who our parents are, and the events that formed them, they're the most important factors, and they're all out of our control.*

*The biological oscillations are real. I'm not denying that. They have reversals in them by definition. Organisms run on oscillating systems at many different frequencies. But only on low levels. And love and wisdom do rest on biological foundations rich in oscillation. But only on low levels, in cells and tissues. You make an unjustifiable jump from molecules to conscious experiences. Quantum events have nothing to do with meanings, values or the rich content of our turning points.*

149

*Which brings up an even bigger problem: I don't experience the turning points you describe. I don't feel mine coming. I don't remember them later. If we have turning points, they're just descriptions after the fact. Real life events don't converge on turning points. What you call turning points are not dynamisms that drive events, they're words, concepts, retrospects. They don't turn anything.*

*They're not dynamisms in nature. They have no pulling power. Nothing draws us toward them. They're not fields of attraction. They're not whirlpools. You can call any moment a turning point depending on what happens later.*

*If the turning points were real, wouldn't we remember them forever? Why would the meaning bestowing moments be hidden from us?*

*And if turning points did have pulling power, if they were real, which goes against all the empirical evidence, given all the confusion and complexity of life, there'd be many in play whirling around us, and they'd annihilate each other. How can you even speak of a turning point when so much is going on? You want us to make dangerous commitments to an untested approach to life on flimsy evidence. You say take the existential plunge, celebrate reversal, accept the intensification and chaos. You're prescribing madness!*

**You don't experience your turning points because you have been dulled to them by culturally learned psychological defenses. Our culture can't deal with rhythmic reversals as a primary characteristic of life. We have been trained to the expansive, accelerative, unidirectional sensibility. The ladder of success always goes up. Everything has to grow in a linear way. Explosion is our mythotype. The Big Bang and the inflation afterward shoot the arrow of time ever forward. We're always moving away from our point of origin. We're always further from the center. We have great difficulty seeing the turning points in life as anything other than unfortunate crises threatening failure.**

**But you are right about your main point: No rhythm in life goes unperturbed for long. We are seriously interfered with on the way to love and wisdom.**

*- You hardly mentioned this. Were you sparing us?*

It was a literary choice. I focused on love and wisdom because I wanted to introduce you to the natural goodness in nature first. But I knew we would have to plunge into cruelty, hatred, greed, envy, competition, oppression and economic exploitation soon enough.

Terrible things can happen on the way to love and wisdom. And to call them "perturbations", a value neutral term, does nothing to instruct, strengthen or even console us. The word accounts well enough for the biophysics of the forces impinging on our homeostatic rhythms. But it's too abstract to use to describe the *enemies of love and wisdom*. Because that's what really concerns us here.

And the enemies are the aggressions we launch against each other's deepest needs and values. And against ourselves. You do it. I do it. We are the indispensable carriers of aggression. It travels from person to person.

### 123

Aggression has been discussed in great detail in our Western civilization. But not by us nowadays. We don't look for it in ourselves inwardly through experience. We let experts tell us about it. And the experts base their findings on dog urine markings, dispute resolution among bonobos, chimpanzee politics, rape among mallard ducks. Or criminal behavior, mental illness, or on the psychodynamics of deviance as seen from outside. Or we attribute the viciousness in human nature to original sin.

*-You've suddenly gone from light to dark. The good world looks evil now.*

No. Not really. I mean to show that all the powers in human nature come together to form a whole. In fact, they're forged together in the furnace of the sun. Our evolutionary history creates in us the same "fearful symmetry" that Blake wrote about in the tiger.

## 124

*Tiger! Tiger! burning bright*
*In the forest of the night,*
*What immortal hand or eye*
*Could frame thy fearful symmetry?*

*In what distant deeps or skies*
*Burned the fire of thine eyes?*
*On what wings dare he aspire?*
*What the hand dare seize the fire?*

*And what shoulder, and what art,*
*Could twist the sinews of thy heart?*
*And when they heart began to beat,*
*What dread hand? And what dread feet?*

*What the hammer? What the chain?*
*In what furnace was thy brain?*
*What the anvil? What dread grasp*
*Dare its deadly terrors clasp?*

*When the stars threw down their spears,*
*And watered heaven with their tears,*
*Did he smile his work to see?*
*Did he who made the Lamb make thee?*

*Tiger! Tiger! burning bright*
*In the forest of the night,*
*What immortal hand or eye*
*Dare frame thy fearful symmetry?*

When Blake asks, "Did he who made the Lamb make thee?" he wants us to understand that the tiger is not the enemy of life, but its compressed core. As Alfred Kazin commented: "Never is he more heretical than in this most famous of his poems, where he glories in the hammer and the fire out of which are struck the 'deadly terrors' of the Tiger. Blake does not believe in a war between good and evil; he sees only the creative tension presented by the struggle of man to resolve the contraries."[1] The creative tensions form the whole. We are deeply dyed with aggression, but also incandescent with love.

And limned in wisdom. The strands are twisted together. It takes skills of self-examination to unweave them so that we can understand ourselves. We must be the hand that dares to seize the sinews through the fire. Until we do, we won't grasp the raw creative energies on which human life thrives. Without twisting and untwisting the sinews we'll have no way of finding, furnishing or living in the real world of experience.

Blake's poem is a cry of hope, not despair. As Mark Schorer commented: "The innocent impulses of the lamb have been curbed by restrainings, and the babe is turned into something else, indeed into the tiger. Innocence is converted to experience...The tiger is necessary to the renewal of the lamb."[2] Now we are ready to deal with aggression.

### *End Notes*

[1] The Portable Blake. Viking Press. 1946. p.43
[2] Mark Schorer, William Blake, The Politics of Vision. Vintage. 1946. p.214)

# 7.AGGRESSION 101

## 125

Konrad Lorenz considered aggression an appetitive instinct. It had its own rhythms. It "speaks up when it has been silent too long, and forces the animal or human to get up and search actively for the special set of stimuli which elicit it and no other hereditary co-ordination."[1]

He observed this rhythmicity in the threat display behaviors in geese, and identified a contrapuntal instinctual dance, one pulse a dance of appeasement, the other a dance of aggression.

**Graylag Goose Threat Display/Appeasement Dance**[2]

## 126

Lorenz insisted that aggression was the basic support for social life. It preceded all ties of affection—in fact caused them. Enmity was the root and trunk upon which later expressions of altruism grew. "Personal bonds belong to the aggression- inhibiting, appeasing behavior mechanisms..."[3] The aggression/appeasement system – one side advances, the other submits – left little room for a primary erotic instinct. Appeasement was the source of bonding behavior. "Intra-specific aggression," Lorenz argued, "is millions of years older than personal friendship and love."[4] Evolution, he concluded, had "chosen, of all unlikely things, the rough and spiny shoot of intra-specific aggression to bear the blossoms of personal friendship and love."[5]

**127**

The aggression-appeasement system worked within species between members of a social group. Without group membership, there were no grounds for appeasement. The great benefit of the appeasement system was that it affirmed belonging. And belonging is necessary for the well-being of social animals, including human beings.

"That indeed is the Janus head of man," Lorenz concluded. "The only being capable of dedicating himself to the very highest moral and ethical values requires for this purpose a phylogenetically adapted mechanism of behavior whose animal properties bring with them the danger that he will kill his brother, convinced that he is doing so in the interests of these very same high values. Ecce Homo!"[6]

For Lorenz, wider inter-communal aggressions presuppose an "us" that defends against a "them". The "themness" of the "them" follows from the "usness" of the "us". And the "usness" came from the aggression/appeasement dance.

The "us and them" mentality produced a condition Lorenz described as "militant enthusiasm" – a "specialized form of communal aggression", that could

"be elicited with the predictability of a reflex when the following environmental situations arise. First of all, a social unit with which the subject identifies himself must appear to be threatened by some danger from outside...second...is the presence of a hated enemy from whom the threat to the above 'values' emanates... third...is an inspiring leader figure... fourth... is the presence of many other individuals, all agitated by the same emotion."[7]

## 128

Tough positions. Cold and hard. A bitter controversy flared over Lorenz' analysis. Leon Eisenberg, a psychiatrist and professor at Harvard Medical School, wrote that

"To believe that man's aggressiveness of territoriality is in the nature of the beast is to mistake some men for all men, contemporary societies for all possible societies, and, by a remarkable transformation, to justify what is as what needs be," adding that "Pessimism about man serves to maintain the status quo. It is a luxury for the affluent, a sop to the guilt of the politically inactive, a comfort to those who continue to enjoy the amenities of privilege."[8]

In rejecting Lorenz, most researchers argued that there were no grounds for treating aggression as an appetitive instinct. In their view aggression functioned as a kindled drive that moved along pre-structured, definable biochemical pathways. It flares, shoots, fires, and then abates. People do not go spoiling for a fight when they haven't been in one for a long time. Nor do they build up an appetite for war from too much peace, or for misery from too much happiness.

John Hurrell Crook arguing *contra* Lorenz, in an interesting collection of essays in *Man and Aggression* edited by Ashley Montagu expressed the consensus view that "aversive behavior is a response to undesirable or harmful stimulation and persists until the individual flees or until the stimulation is removed... According to this account, aggressive behavior is non-rhythmic and lacks an appetitive phase..."[9]

Ashley Montagu himself went so far as to deny the reality of all instincts.

"Everything he is and has become he has learned, acquired, from his culture, from the man-made part of the environment, from other human beings... his brain, far from containing any 'phylogenetically programmed' determinants for behavior, is characterized by a supremely highly developed generalized capacity for learning; that this principally constitutes his innate hominid nature..."[10]

## 129

The heat of the reaction against Lorenz had a great deal to do with his politics. His critics reminded their readers that Lorenz willingly participated in the intellectual life of Nazi Europe. His Ecce Homo!" remark carried echoes from Nietzsche that did not sit well with people still pondering German war crimes. Lorenz denied the accusations.

What actually happened?

Freud and Lorenz were both in Vienna in 1938. Freud, 80 years old, ill with oral cancer, lived in danger of being swept into the Holocaust as a Jew and enemy of the Third Reich. Lorenz was a young man of a prominent family whose father was an avid Nazi supporter and propagandist for the Anschluss ( March 12, 1938.) Lorenz with family help pursued career advancement.

To advertise his availability, he published an article in a German journal deploring racial degeneration, making the point that mankind, like other domesticated animals, was losing its instinctual strength through 'self-domestication'. He saw us undergoing a kind of genetic drift to decadence that included an attraction to mixed race breeding and hence to ugliness and demeaning sexual selection.

Even his sympathetic biographer, Alec Nesbitt, maintains that the genetic decay paper was 'angled' at Germany's Nazi masters and was "clothed in Nazi terminology" to promote his career as a natural scientist."[11]

It worked. On September 2, 1940, Lorenz was given Kant's chair in psychology at the Albertus University of Konigsberg. But the war on the Eastern front pushed him into military service in the German army as a medical officer posted to a psychiatric unit in Poznan in the Polish Corridor. Poznan was an S. S. center. Himmler visited there in October 1943 and spoke to the assembled SS officers, "whether nations live in prosperity or starve to death like cattle interests me only in so far as we need them as slaves to our *Kultur*; otherwise it is of no interest to me."[12]

In 1974 Lorenz told Nesbitt he was unaware of the German atrocities until 1943 when by chance he saw a transport of Gypsy prisoners passing through Poznan, an odd failure of perception for a future expert on aggression considering the prominent position Poznan played in the SS role in conquered territories. According to Nesbitt, Lorenz was transferred to the German army in Vitebsk in White Russia where he was captured by the Russians and taken prisoner on June 24, 1943.

During his imprisonment, Lorenz wrote his first book on aggression.

Freud's fate ran a different course. With the help of American and British diplomats, he received exit visas for himself, his family and staff, paid the various bribes and taxes, had his bank accounts confiscated, and went into exile in London. He provided money under Austrian guarantees to care for his four elderly sisters who were too frail to travel. In a final humiliating gesture before his departure, the Gestapo required Freud to sign a document stating he had been treated with extreme fairness by the authorities. To this official communication he appended a remark wise in Jewish irony: "I would recommend the Gestapo to anyone."

The money he provided for his sisters was stolen by the authorities. The sisters were exterminated in the gas chambers. Freud died in London in 1940.

## 130

The passions surrounding Lorenz' politics and findings have abated. Academic opinion has now established a comfortable middle ground, acknowledging that aggression has both instinctual and learned components.

E. O. Wilson posited seven original triggers for aggression: "...the defense and conquest of territory, the assertion of dominance within well-organized groups, sexual aggression, acts of hostility by which weaning is terminated, aggression against prey, defensive counterattacks against predators, and moralistic and disciplinary aggression used to enforce the rules of society... Aggressive behavior is in fact one of the genetically most labile of all traits. In short, there is no evidence that a widespread unitary aggressive instinct exists."[13]

He tells us "most kinds of aggressive behavior are perceived by biologists as particular responses to crowding in the environment."[14]

In our conceptual scheme, Wilson's seven triggers function as responses to disruptions of social rhythms. Overcrowding was a particularly strong example. His analysis confirms our view that the ties between aggression and homeostatic physiological functioning are deeply set in biology, that these triggers are efforts "to gain control over necessities—usually food or shelter—that are in short supply."

Next, Wilson tries to persuade us that "although the evidence suggests that the biological nature of humankind launched the evolution of organized aggression and roughly directed its early history across many societies, the eventual outcome of that evolution will be determined by cultural processes brought increasingly under the control of rational thought"[15]

To support it, Wilson argues that when instinct and learning come together "genetic biases can be trespassed, passions averted or redirected, and ethics altered..." The redirection, in his view, would incline us toward a social harmony that he associates with "...the human genius for making contracts." He expresses great confidence that it can "continue to be applied to achieve healthier and freer societies."[16]

The learned components will come to dominate. Big claims for the taming power of reason!

## 131

It is easily observed that the contracts Wilson praises are hardly free of "genetic biases". Many contracts are nothing more than papered-over instances of coercion by the dominant over the subordinate parties. They're full of aggressive territorial and dominance interests often hidden inside a surface geniality and insidious modesty of which Blake writes:

*Now the sneaking serpent walks*
*In mild humility.*
*And the just man rages in the wilds*
*Where lions roam.*

The interests and passions of competing factions do require alliances and coalitions, compromise, delay and renunciation. However, it is wrong to suggest that reason has triumphed in these contracts. Or that its successes have come at the expense of aggression.

It makes more sense to see the social contract as a late formulation of a long – practiced sociability in small communities where the subscribers to the contract knew each other well – they knew what to trust or suspect. Passionate avowals make promises real, not the contracts that record them. The passions are necessary to those promises because words are slippery and the future is unknown; we do not like to hear our partners say, "That was then, this is now." These reversals hit us emotionally.

In our mass culture, the contractual arrangements that supposedly secure belonging lose their passionate avowal. We are strangers to each other. We sign on to all kinds of agreements. Some even contradict each other. One contract can cancel another. Obligations can change at any time. Read the fine print.

## 132

Why do we suppose that in large-scale settings teeming with stress-producing perturbations, we are more disposed to promise keeping than promise breaking? Both the creative and destructive kinds of aggression maintain equally close ties with reason. That our cognitive powers tame us for social life tells half the story. Here's the other half: Once our ancestors had the brains to figure out the long term advantages of increased cooperation, they also understood the advantages of subterfuge and deceit. Even primates put out false signals for food because they know it will send the others off in the wrong direction. And certainly, our forbears made promises they never meant to keep. They erected systems of belief they didn't believe in themselves. They made treaties they never meant to honor. And they wrote social contracts that made some people more equal than others. As Tolstoy put it,

"Some do not believe in anything and are proud of it. Others pretend to believe in what for their own advantage they have persuaded the masses to believe in beneath the guise of faith. The rest, the great majority of the population, accept as faith the hypnotism exercised over them and slavishly submit to everything demanded of them by their non-believing rulers and persuaders."[17]

## 133

Wilson's compromise won't work in human life for many reasons. Any kind of stress can be registered as a threat to homeostatic functioning. Why limit ourselves to seven triggers for aggression? Why not 700? Why not an endless series of them? Civilized aggression seems to be able to respond to anything, go everywhere, penetrate all behaviors, and serve all masters. A million things can bring it on. We can give almost any insult meaning. Fathers throw their babies against the wall. Aggression is almost infinitely recruitable – recruitable precisely because it is triggered, not pre-structured and lying in wait.

## 134
### *Freud v Lorenz on Aggression*

Freud accepted the interplay of instincts, but saw them differently. Where Lorenz placed the "blossoms of personal friendship and love" on the "the rough and spiny shoot of intra-specific aggression," Freud gave them each their own rootedness. Neither had priority. They rose through the tree of life together. In a late formulation in *New Introductory Lectures on Psychoanalysis* (1933), he wrote:

"Our hypothesis is that there are two essentially different classes of instincts: the sexual instincts, understood in the widest sense-Eros, if you prefer that name—and the aggressive instincts, whose aim is destruction... But it is a remarkable thing that this hypothesis is nevertheless felt by many people as an innovation, and, indeed, as a most undesirable one which should be got rid of as quickly as possible."[18]

Later he asserts that "Luckily the aggressive instincts are never alone but always alloyed with the erotic ones. These latter have much to mitigate and much to avert under the conditions of the civilization which mankind has created."[19]

They were terms in a dualism "... both kinds of instinct would be active in every particle of living substance, though in unequal proportions..." Which makes us inherently problematical, conflicted beings, bound to suffer Given our real situation he considered "a belief in the 'goodness' of human nature... one of those evil illusions by which mankind expect their lives to be beautified and made easier while in reality they only cause damage."

The alliance was really a fusion of instincts, as Freud understood it. The fusion gave aggression a role in sustaining erotic ties against harsh intrusions from the world. To illustrate the fusion of aggression and Eros, Freud most often focused on the rough and tumble of sexual foreplay, and on masochism and sadism. But he also noted its broader implications by observing, "Every instinctual impulse that we can examine consists of similar fusions or alloys of the two classes of instincts."[20] Animal studies of breeding season aggression, in recent decades, show the rhythmic alternation of aggressive and erotic behavior in mating rituals. Scientists have analyzed the underlying biochemistry for many social mammal species: "When the breeding season approaches and the gonads recrude, aggression increases with a corresponding elevation in testosterone levels."[21] We find an even more excruciating transitional expression in the agonal orgasm that occurs under acute stress in humans, as documented in cases of asphyxiation with erection and ejaculation.

We can conjecture that in nature aggression, when it fused with Eros, defended the primordial social rhythms of approach and separation from disruption. This made distance regulation a response to attraction and desire rather than fear and aversion – a relationship much different from what Lorenz proposed. It protected life rhythms.

Therefore, aggression is not prior to or a cause of love. Love has its own power as a motivator of action and as a counterforce to aggression. But what lets them come together in an instinctual fusion? And how does defusion tear them apart? In the defusion of Eros and aggression, Freud saw the working of the "death instinct". This he associated with the thermodynamic tendency of all things to tend to increasing disorder, arriving at a final heat death.[22]

Insofar as it appears disorderly, we can perhaps liken the depatterning to the dispersal phase of the dispersal-aggregation rhythm as it expresses in body tissues, particularly in neural cells that lose their beat, or to the breakdown of the passing rules in its phenomenological face. Rhythmic derangement is in fact the diagnostic criterion for an epileptic seizure when evaluated through an EEG. And we live in a rhythmically deranged time. So in our lives aggression defuses from Eros frequently, drawn out, tempted, or swept away by all kinds of challenges and threats.

That the aggressive instincts played a primordial conservative role, Freud acknowledged in *The Ego and the Id*: "The rudimentary creature would from its very beginning not have wanted to change would, if circumstances had remained the same, have always merely repeated the same course of existence. But in the last resort it must have been the evolution of our earth, and its relation to the sun, that has left its imprint on the development of organisms."

Why the sun? Freud understood that the solar rhythms of the days and seasons were embedded in our organisms by evolutionary selection and we seek to conserve them. Freud goes on to say that

"The conservative organic instincts have absorbed every one of these enforced alterations in the course of life and have stored them for repetition; they thus present the delusive appearance of forces striving after change and progress, while they are merely endeavoring to reach an old goal by ways both old and new."[23]

## 135

We can draw the following inferences from our brief analysis of the biophysics behind aggression and love:

o   Between oscillatory instincts, on whatever time scale they move, resonances and dissonances naturally spring up as they move in and out of phase with each other.

o   If aggression were a rhythmic function, it would have its own hunger-satiation cycle. It would act on us according to the frequency and amplitude of its oscillations. In doing so, it would fall into or out of phase with love and wisdom. But as a triggered drive, aggression has no rhythm. It acts as a perturbation on love and wisdom. Its effects vary with its intensity and the timing of its moment of impact.

o   Only aggression specifically recruited by love and wisdom to protect love and wisdom can safeguard homeostasis. It defends rhythm from perturbation by establishing a defense perimeter against intrusions.

o   Approach/separation and withdrawal/return are resilient in the face of perturbation. They link to internal clocks; the clock cycles (assuming they have not been destroyed by deeper disturbances) help restore tweaked rhythms. Lorenz insisted that human evolution didn't have time to evolve these safeguards. But he was wrong.

### 136
### *Distance Regulation*

Now we can add something important to Lorenz' argument. I stated earlier that researchers consider overcrowding a principle trigger for aggression. Yet many species overcome crowding with distance regulation behaviors.

In *On Aggression*, Lorenz argues that distance regulation behaviors get their energy from aggression or aggressive display. The aggressions flare up when violations occur to ongoing approach/separation rhythms. Lorenz called distance regulation "a very simple mechanism of behavior-psychology that gives an ideal solution to the problem of the distribution of animals of any one species over the available area in a way that is favorable to the species as a whole."[24, 25]

That is to say, aggressive behaviors evolved as responses to the disruptions of rhythm whenever overcrowding became a critical infringement on homeostasis. When overcrowding threatened the food supply, interfered with breeding patterns or impinged on the regular ordering of social distances, aggressive displays were stimulated.

To the extent that aggression fought to restore the biological rhythms needed for homeostasis, it played a conservative role in species chronobiology. However, at the same time, aggression exerted force as a perturbation, not as a rhythm itself, rather, like a lion tamer's whip lash, it snapped to keep the lions dancing. It worked as a perturbation for the sake of rhythm.

The evidence from animal behavior I interpret to show that under certain conditions (like overcrowding) aggression reinforces distance regulation to conserve the rhythms of dispersal and aggregation-the basic food gathering, sleep/waking and safety/defense rhythms. Aggressive displays keep daily dispersal/aggregation rhythms going by organizing the individual rhythms of approach and separation between individuals into systems of distance regulation.

These larger systems maintain Edward Hall's fourfold division into intimate, personal, social and public distances. Hall explains, "in addition to territory that is identified with a particular plot of ground, each animal is surrounded by a series of bubbles or irregularly shaped balloons that serve to maintain proper spacing between individuals."[26] These bubbles conform to the four social distances.

## 137

We can liken the aggressive distance regulation system to a patchwork quilt of relationships. In the quilt, some occupy big patches, others smaller ones. The most dominant males and females and their allies command the largest, most central spaces and have the freest access to and through all other spaces. By ranking order, certain individuals can enter others' spaces. Some can come in but others must stay out. The closest contacts are allowed between mother and child, then between siblings, then mates, and then out to the broader social interests that regulate the whole group. The primordial territory, then, is the quilt of social distances regulated by both aggression and affection. And the quilt is dynamic. It changes. Births and deaths change the quilt, as do aging and wounds and illnesses.[27]

The patches may change, but they remain contiguous. They touch each other. Their borders are active and alive. Each patch is part of the whole. The sum total of the social space that everyone constructs in the daily dance of dispersal and aggregation, however it changes, always comes out to the same unitary value—one. "We are one group, we are whole. The quilt as a whole is us". Here aggression, affection, and the rhythms of nature come together to create the experience of *belonging*.

## 138
### *Belonging and Community*

Human beings need community. Whether we shun or embrace it, community poses every kind of test of the worth and durability of deeds. It is the place where a person first knows belonging or estrangement. Here he or she wins or loses the basic contests for excellence. Of all venues, it follows, community stands as the one best suited to teach us to catch the quick of aggression on the wing, to feel the unfolding of one's own aggressive energies, and to tame or use them for creative projects.

From aggression in natural communities, we taste our first actual alienations, usually starting with childhood rivalries and rejections. Even in small social groupings, not everyone will have good standing. Given the variability in human nature and the depths of our wounds, some will get subordinated roles or will be pushed aside.

Natural communities are not idyllic places. They harbor injustice and provoke malice. Nevertheless, they are central for allowing us to live full lives. When viewed as the necessary link between person, nature and culture, community, tough as it may sometimes be, serves as the most reliable agent for transforming the militant enthusiasm of the Us versus Them into the agonal Us, the "us" who struggle together to *create* not destroy.

The full play of our humanity depends on life in community. Community gives us a social territory of the size we were equipped by evolution to move in, to feel at home in, to behave viscerally, autonomically, vegetatively, and spontaneously in. Our sensory circles and social distances evolved to work in natural sized communities. It is where our personal and social epigenetic releasers get their clearest stimulation. Here the four primordial social distances receive their signals, providing the structural web from which, by lines of affinity, community life can grow.

For millions of years, the distance regulation system was folded into the broader, older aggregation-and-dispersal rhythms responsive to the day/night cycle. The primordial social order, the patchwork quilt, formed a network of relationships based on approach/separation rhythms fitted to the four social distances that kept the structure of aggregation/dispersal going through the daily round.

Depending on need and on the personalities of the members, all kinds of coalitions could form and dissolve. Defense, food gathering, healing, shelter building, hunting, tracking, traveling, keeping the juveniles in line, alliance building, and inter-band relations. Each would bring different leadership arrangements to the fore, not so much one top male or female replacing another, as adjustments in the whole scheme. The patchwork quilt was dynamic, always reforming, never finished.

## 139

Nor can you simply decide to reject belonging. The need for it presses in on you. Abandonment we experience as a root fear. We carry it from infancy. Even in favorable circumstances, the natural rhythms of love and wisdom will themselves awaken the fear of abandonment. It stirs at the extremes before their reversal points. Abandonment always hovers close. The possibility never leaves. When someone dies, the fear stirs. Change provokes it.

Belonging is our hedge against loneliness and despair. You cannot choose to reject belonging. Even voluntary exile will not snuff out the visceral pull. The presence of others will draw us into new relationships.

Our intimacies strengthen belonging through language affinity. The very words we use when we ponder our choices are not of our own devising. Human speech emerges and develops only in community. 'Community' and 'communication' have the same root. Language is the tie that binds our hopes and fears together. We require language to become real to others and maybe to ourselves. Here with language and higher cognitive abilities, a distinctly human, one could say a normative human social life, develops.

## 140

More insightfully than any scholarly writer I know, Hannah Arendt probed the need for belonging. Reflecting on her own experiences as a refugee from Nazi Germany, she wrote in a 1943 article:

"Man is a social animal and life is not easy for him when social ties are cut off. Moral standards are much easier kept in the texture of a society. Very few individuals have the strength to conserve their own integrity if their social, political and legal status is completely confused... we lose confidence in ourselves if society does not approve us; we are – and always were – ready to pay any price in order to be accepted by society."[28]

Without community life, love and wisdom, though they maintain their existential and creative dimensions, lose ethical and political force. To gain strength of character in the absence of deeds is an empty pursuit. To have only a private life creates enormous frustrations.

## 141

Community is the natural venue for values. It is where love and wisdom thrive. When we withdraw and return, it is always from and to community.

Ethical transmissions require community life because moral behavior comes into the world more through example than precept. Exemplary deeds need to be witnessed. The witnessing makes community the conducting medium for ethical examples. Our power to solve shared problems gains force when exemplary behaviors travel through real communities.

## 142

We have a primordial desire for cozy belonging. The larger the group, the weaker our gut-level identification with it. Something in our makeup resists the anonymity and leveling of mass culture, rejects distant rule, and resents absentee ownership. We are biologically predisposed to break larger groups into smaller units. And if we cannot find real communities, we opt for false ones.

1) We attach ourselves to subgroups, to smaller scale natural social entities, within which natural dominance associations still form and function – a family, a home site, neighborhoods, churches, work associations, circles of friends – and then we feel wanted. But these secondary institutions compete for loyalties and put pressure on each other. So a complex web of conflicting loyalties forms, eliciting different behaviors and requiring complex splits in consciousness.

2) Our sense of belonging gets more abstract, more diffuse, diluted. We allow ourselves phony belonging; we accept abasement by buying worthless products, "theme park" substitutions for life, fake adventures, false memberships and trivial connections.

3) Those who cast their lot with humanity against the debasement of belonging, and try to protect "themselves

against the loss of the object by directing their love, not to single objects but to all men alike" commit a folly. Freud wrote that "a love that does not discriminate seems to me to forfeit a part of its own value, by doing an injustice to its object; and secondly, not all men are worthy of love."[29]

4) Many people in the prosperous world try to find belonging by acquiring the portable symbols of belonging. They make possessions, wealth and money their goals. But these expedients drain belonging of meaning. The infantile fear of abandonment pushes through to the surface and drives us to measures that are even more desperate. The main one being the Us/Them mentality.

5) The last most desperate and catastrophic way to secure belonging is in the Totalitarian state, a peculiar 20th century invention whose ideology promises to explain everything to people for whom nothing makes sense. Arendt argued that the crucial support for totalitarianism comes from loneliness. In the 2nd edition of *The Origins of Totalitarianism* she wrote that "totalitarian domination as a form of government is new in that it... bases itself on loneliness, on the experience of not belonging to the world at all, which is among the most radical and desperate experiences of man." The totalitarian state offers a very complete kind of belonging, and people long for this, but they can only get it by accepting, as an unshakable principle of organization, that the world is divided into a superior Us and inferior Them. In this world, sacrifice for the sake of the Us is the highest civic duty. The "us" has to prove its "usness" repeatedly. The slightest deviation arouses suspicion. People spy on each other. Only the ideology is pure. A secret disaffection drills its way into the self until finally we grind the "us" down so far that one suspects oneself of being one of "them". In Stalinist days that frame of mind, heated by terror, made people confess to thought-crimes or testify against each other or testify against themselves, or even commit suicide on request.[30]

* * *

## *End Notes*

[1] *On Aggression* p.83

[2] Ibid p.177

[3] ibid p. 132

[4] As Lorenz saw it, The personal bonding that comes from redirected aggression shows the "... ingenious feat of transforming, by the comparatively simple means of redirection and ritualization, a behavior pattern which is partly motivated by aggression, into a means of appeasement and further into a love ceremony which forms a strong tie between those that participate in it." ibid p. 167

[5] ibid p.45

[6] ibid p.233

[7] ibid p. 263-4

[8] Leon Eisenberg

[9] Man and Aggression p.151.

[10] Montague. *Man Against Aggression* p. 15

[11] Nesbitt. *Konrad Lorenz*. Harcourt Brace. p. 81

[12] Quoted in Shirer. *The Rise and Fall of the Third Reich.* 1959. Fawcett Crest. p.1224

[13] E. O. Wilson. On Human Nature. Bantam 1979. p. 104

[14] Wilson *In search of Nature*. Island press. 1996. p.85

[15] Wilson. *On Human Nature*. Bantam. 1978. p. 119

[16] Idea of Nature p.90-94

[17] Tolstoy "What is Religion" p.102 in *A confession and other Religious Writing*. Penguin classics.

[18] Freud. *NIL*. Norton. p. 103

[19] Freud. (New Introductory Lectures. p.111

[20] ibid p. 105

[21] Stewart. "Brain ACTH-endorphin neurones as regulators of central nervous system activity." In *Peptides.* 1980, Scriptor.

[22] We often see aggression restless in the net of love. Defused aggression discovers new projects. This in turn further impacts the rhythmicity of love and wisdom on hormonal and neurological levels by altering the time course of the stimulating signals in the cultural environment. This reorganization of our priorities changes our awareness of who we are, what we want and what we care about. Arguably, the fusion of aggression and Eros acting as a conserver of rhythm predates the epigenetic signals for defusion; defusion points to

more recent functions. Someday DNA analysis will be able to tease out the sequence of events. My guess is that defused aggression serves in our line of descent as an enforcer of distance regulation and of broader territory and dominance arrangements and attaches to property and wealth. With the help of defused aggression these become prominent features in our species social life. Once human cultures develop, the liberated aggression can be sent even farther.

[23] ibid p.31. Under the influence of his friend Fleiss, Freud, for a long time did believe biological rhythms were the main shaping forces in human nature, but he gave up on the notion for lack of evidence, and after a long personal struggle ended his relationship with Fleiss. Freud's biographer Ernest Jones reports a humorous conversation he had with Freud: "I asked him how Fleiss managed when one attack of appendicitis occurred an irregular number of days after a previous one. Freud looked at me half quizzically and said: "That wouldn't have bothered Fleiss. He was an expert mathematician, and by multiplying 23 and 28 by the difference between them and adding or subtracting the results, or even by more complicated arithmetic, he would always arrive at the number he wanted."

Freud may have turned his attention away from the structural role biological rhythms play in human nature too soon, not recognizing that the cultural-historical role of aggression begins when it develops an enthusiasm for projects that push it beyond rhythm driven resource exploitation, bonding and distance regulation. Once it is freed from its rhythmic context, aggression can become a general purpose impulsion to action, and then a new kind of assertion emerges: one without an inherent goal, one that can serve many goals, an instrumental rather than an emotional or ethical expression. We would expect to find it in the service of a project working as a means not an end.

[24] Lorenz ibid p.35

[25] Many other zoologists have covered this same ground and confirmed the connection between distance regulation and dominance and territorial behaviors. "Individuals of many species, but by no means all, maintain a space around their bodies within which they repel approaching individuals...Repeated encounters give rise to hierarchies, the so-called 'peck order' first described from flocks of chicken... Territoriality is thus a special case of spatial defense not easily separable from the maintenance of personal space." John

Hurrell Crook. "The nature of territorial aggression". In Montague ed. *Man and Aggression.* p.152 E. O. Wilson, the father of sociobiology wrote recently, "Humanity is decidedly a territorial species. Since the control of limiting resources has been a matter of life and death through millennia of evolutionary time, territorial aggression is widespread and reaction to it often murderous."[25] Consilience. Sarah Blaffer Hrdy who studied distance regulation in female primates, drawing on the work of Hausteder, wrote: "for each infant at birth, the world can be divided in two: between those females who may approach the mother with impunity and attempt to pull the infant off, and those females who would not dare to! So fixed would be the resulting system that even if the mother dies prior to maturity, the daughter still fits into the hierarchy at the same spot the mother did..." Hrdy. *The Woman That Never Evolved.* Harvard U. 1981. p. 113. In primate bands the daily rhythms take on the qualities of a dance across geographical features. And beyond the daily ballet there are slower dances tuned to the seasons that keep the troops moving in step with the availability of food sources and population pressures. As the range changes or is relocated, the dominance arrangements change. Franz de Waal gives many examples of the variability of the dominance scheme among chimps in captivity. In his book *Chimpanzee Politics* he writes: "Real dominance is the sum total of aggressive encounters. The outcome of such encounters is not 100 per cent predictable, particularly because chimpanzees have such a strong tendency to form coalitions. Incidental reversals in the real social hierarchy are far less rare than with other animal species. That is why the chimpanzee hierarchy is so often termed 'flexible' and 'plastic'." De Waal. Chimpanzee Politics. p. 89

[26] Edward Hall. *The Hidden Dimension.* Anchor, 1969. p.10

[27] Each band has its own memories of alliances and enmities, its own local history, its own learning and schooling. And these arrangements are maintained with varying strictness or leniency, sometimes tested, defied, challenged, lightly taken, ignored or vigorously defended, depending on the current leadership, the individuals involved, the space requirements, the environmental resources, the traditions of the band and the daily and seasonal rhythms of life. (Perhaps the most aggressive individuals keep the social web together not so much as leaders, leadership may be another function, but by

keeping it from becoming lax and disrespectful and so non-functional)

[28] "We refugees." *The menorah journal.* Jan, 1943. Quoted in Hannah. *Arendt The Recovery of the Public World.* p. 10) Arendt maintained that "a life without speech and without action... – and this is the only way of life that has renounced all appearance and all vanity in the biblical sense of the word – is literally dead to the world; it has ceased to be a human life because it is no longer lived among men." Quoted in Hannah Arendt the Recovery of the Public World. p. 157."This revelatory quality of speech and action comes to the fore where people are with others and neither for nor against them – that is, in sheer human togetherness." Human Condition p. 160. She believed that in the modern world we had lost the forum for self-revelation. "Men have become entirely private, that is, they have been deprived of seeing and hearing others, of being seen and being heard by them... The end of the common world has come when it is seen only under one aspect and is permitted to present itself in only one perspective." ibid. p. 58

In the modern West, in Arendt's opinion, "the public realm has almost entirely receded, so that greatness has given way to charm everywhere; for while the public realm may be great, it cannot be charming precisely because it is unable to harbor the irrelevant." ibid p. 47

Instead of personal engagement we have conformity. "It is decisive," she explains, "that society, on all its levels, excludes the possibility of action... Instead, society expects from each of its members a certain kind of behavior, imposing innumerable and various rules, all of which tend to 'normalize' its members, to make them behave, to exclude spontaneous action or outstanding achievement." ibid p. 37

[29] Freud. *Civilization and its Discontents.* p. 49

[30] She wrote "Loneliness, the common ground for terror, the essence of totalitarian government, and for ideology or logicality, the preparation of its executioners and victims, is closely connected with uprootedness and superfluousness which have been the curse of modern masses since the beginning of the industrial revolution and have become acute with the rise of imperialism at the end of the last century and the breakdown of political institutions and social traditions in our own time... Even the experience of the materially and sensually given world depends upon my being in contact with other men, upon our *common* sense which regulates and

controls all other senses... What makes loneliness so unbearable is the loss of one's own self which can be realized in
solitude, but confirmed in its identity only by the trusting and
trustworthy company of my equals. In this situation, man
loses trust in himself as the partner of his thoughts and that
elementary confidence in the world which is necessary to
make experience at all." Arendt. *Totalitarianism.* P.475-477

# 8. STRESS AND AGGRESSION

## 143

**D**uring Paleolithic times the relationship between aggression and stress changed. I will next explain how we evolved anatomical overlays to the stress response along fronto-limbic tracts that widened the scope of social life by forestalling the bifurcation of stress into aggression. With hardwired modifications to the stress response in place, our understanding of threat itself changed, and our capacities for empathy and sympathy grew. I conjecture that these suites of behavior entailed enhanced facial recognition skills. As we let more signals in, more meanings could be shared. They increased our social tolerance, extended our capacities for natural belonging, and widened our curiosity.

## 144
### Fight or Flight

The stress response releases thousands of neural, hormonal and cellular signals throughout the body. They drive up the vital signs, stimulate the cardiovascular system and shift the blood flow from the periphery to the core. The stress response blocks digestion in favor of quick sugar metabolism, diminishes immune competence and tips the neural balance toward sympathetic and away from parasympathetic dominance. It postpones ongoing vegetative cellular functions in order to prepare the body for emergency action. In both stress and aggression, the arousal interrupts ongoing activity. They both make us turn away from a natural flow to acknowledge and usually tighten up against an intrusion. In both responses, the same enzyme, dopamine-beta-hydroxylase (DBH), catalyzes the conversion of dopamine to norepinepherine. Up to this point both processes run on the same chemical cascade. Jerome Kagan speculated that reduced sympathetic reactivity on this pathway promoted anti-social behavior. "Because DBH level is under genetic control," he wrote, "it is reasonable to speculate that some extremely aggressive boys with low anxiety over asocial behavior have inherited a chemistry that makes it easier for them, given an environment that permits aggression, to become aggressive and antisocial."[1]

The way we differ in the unfolding of this branching process shapes our aggressive natures. It makes us different from each other, particularly with regard to how we love, how we ponder our situation and how we place ourselves in the world – with what energy, resolve and resistance to intrusion.

How and when and for what purpose do stress and aggression diverge?

## 145

Usually stressors end before they rouse us to aggression. Consult yourself. When the stress hits, it comes on fast and usually quickly passes. We reconnoiter, understand and try to account for the stimulus. What you thought was a gunshot turned out to be a car exhaust backfiring. Knowing this, you go back to what you were doing.

However, in the instant of arousal, as it quickly passes, the germ, the first possibility of an aggressive response stirs. The body takes on a posture of vigilance during this period. And vigilance is the outer face of the inner realization that "this interruption matters".

The longer the person stays vigilant, the surer he or she is that something important is at stake, and the more seriously the body prepares itself for action. If we take the stimulus to be a threat, rightly or wrongly, we will make a defensive or aggressive response. We'll fight, flee, or freeze. The interpretation we make, based on a true or false assessment of the facts, may draw out an appropriate or inappropriate response.

What do we take as threats? What arouses vigilance? Distance regulation responses still operate inside us. Boundary crossings, trespasses of personal space, infringements of prerogatives, perceived insults and injustices carry important meanings. They do so, underneath it all, not because someone is in your face, but because they challenge (or are perceived to challenge) one's position in the reigning territorial and dominance system—still, in some shadowy way, reflected in the primal distance regulation ordinances of the culture. That is why you cannot shake the president's hand without an express invitation to enter his or her space.

Therefore, the stimuli that distinguish aggression from other expressions of the stress response turn on position, status, prestige, possession, ownership wealth and dominance.

## 146

Since both stress and aggression start with the interruption to a flow, the switch to aggression initially entails a *bifurcation from the stress re*sponse.

Researchers have described this transition across an enormous range of physical, chemical, biological, social and even economic systems for many species of animals. (I treat the evolutionary and historical significance of the stress/aggression pathways below in # **202-210.** I chose to divide the argument in two parts. If you are curious to follow the physiological modeling now, jump ahead.)

Type-A personalities, to illustrate the shift in human beings in the context of pop psychology, bifurcate early from stress to aggression. They have a low set point. According to Friedman and Rosenman, who did the original clinical research on Type A personality,

"… most Type A subjects possess so much aggressive drive that it frequently evolves into a free-floating hostility." Type A's, according to them, pit themselves in a fight against time. Friedman and Rosenman wrote: "Type A Behavior Pattern is an action-emotion complex that can be observed in any person who is *aggressively* involved in a *chronic, incessant* struggle to achieve more and more in less and less time, and is required to do so, against the opposing efforts of other things or other persons."[2]

177

As Friedman and Rosenman argue: "It is the Type A man's ceaseless striving, his everlasting *struggle* with time, that we believe so very frequently leads to his early demise from coronary heart disease."

The struggle against time I take to be a struggle against natural rhythmicity. Its distinctive cause, in our technological civilization, is the pervasive neglect and consequent inaccessibility of natural rhythms.

In their denial of temporal flow, Type-A personalities are culture-heroes of a sort: they reject and rise up against the basic rhythms of life in pursuit of new projects and "higher" goals.

When the struggle with time is colored by hostility, the resulting disorder thrusts a dangerous restriction into meaning. The type-A's life becomes obsessive and narrow, lacking in empathy and sympathy. As Kenneth Pelletier noted, "He is hopelessly myopic, concentrating always on today's achievement and spending little or no energy considering the far more important question, 'What's it all for?'"[3]

The narrow focus predisposes the Type-A person to resist reversal. He struggles against his own turning points. Linear progress, as for techno-enthusiasts like Ray Kurzweil, becomes the mark of a victorious struggle against time. It takes a stand as a fight against unforeseen contingencies. Despite the best analysis and planning, the unforeseen keeps coming up.

## 147

A body of research shows that testosterone in the presence of high noradrenaline (but not in the presence of adrenaline) is pulled away from sex and put in service of aggression. Instead of functioning as a signal to the DNA in sexual arousal, testosterone in the presence of noradrenaline works in the synapses as a neuromodulator.

Noradrenaline makes testosterone available for the "fight" response by blocking the receptor sites for 5-HT, a serotonin precursor. Noradrenaline, by occupying its receptor sites, impedes the transmission of serotonin across the synapses. It prepares the receptor sites to respond to testosterone instead, evoking quickly kindled aggression. Katherine Thompson, a Canadian scholar, in a very thorough and well-thought-out online review article states "many studies have been undertaken that implicate 5-HT, a metabolite of serotonin in modulating aggression. In general, increased activity of serotonergic synapses inhibits aggression: studies in male rhesus monkeys showed that those with low levels of 5-hydroxyindoleacetic acid (5-HIAA) were more aggressive. Furthermore, fluoxetine, a serotonin reuptake inhibitor, tends to decrease aggression in both animals and humans. Strong evidence suggests that these effects are mainly mediated via 5-HT1A and 5-HT1B receptors."[4]

Very likely, the aggressive response to noradrenalin evolved early in mammalian life for use in territorial defense, protection of the young and in other situations where quickly kindled aggression would have served survival interests. However, in humans, it has a much wider range of uses.

## 148

After it branches off from the stress response, aggression follows one of three behavioral routes, each with its own biochemical correlates.

1)  The aggression can bind to sexuality. In this kind of aggressive response, the testosterone does not retreat from sexuality. The breeding season rituals, some involving aggressive displays or combat, show this clearly in animal behavior.

2)  The aggression liberated by noradrenaline triggers the "fight or flight" response with its primordial origins in individual and troop defense and in the maintenance of dominance and territorial relationships in band-sized groupings.

    This second kind of aggressive response is important for our discussion because it breaks its bonds with rhythmicity. Fight or flight temporarily interrupts ongoing circadian and ultradian rhythms. When the occasion ends, the body returns to rhythmicity. Crimes of passion, fits of temper, acts of vengeance, bouts of jealousy and malice fit into this category. When the triggered aggression uses up its charge, shame and guilt replace it.

3) The aggression of persistent derhythmization is triggered by permanent cultural territory and dominance rules, by ambition, conflicts, by money problems, by threats to status. In acquisitive cultures where people pursue wealth and property as a prime aim in life, the thresholds for aggression are regularly crossed.

Many people condition themselves to bear up under the strain. Some even thrive on it. For many, the bifurcation from stress to aggression has a happy face. They take it as an opportunity to advance, to compete, to feel triumphant. They take on the stress/aggression bifurcation in the Hegelian sense. It becomes their struggle for recognition.

Moreover, for a long time the human body can accommodate itself to chronic stress without serious consequences. Decades pass until something breaks down. Our organ competencies, our resilience and resistance eventually fail. The chronic aggravation of body sites exhausts specific organ systems. Stress-related illnesses appear, multiply and interact. We enter what Hans Selye called the "exhaustion phase" of the General Adaptation Syndrome. The stress symptoms we experience develop into chronic diseases. The diseases overlap. Aging becomes a balancing act between competing ailments. Many researchers consider stress related illnesses the principle causes of death for humans with long life spans in high-tech cultural settings.

### 149

We have constitutional differences in our personal stress triggers and responses. Our sensitivities and set points for aggression vary with temperament. According to Jerome Kagan, these temperamental differences run along the shy/bold, inhibited/uninhibited, endangered/safe spectra.

Certain temperaments do well in specific historical settings. Others do poorly. What we shy away from, what we face, what we rationalize and defend against, all have stress components that cross from physiology to meaning insofar as they color our hopes and fears and twist our capabilities. The way we deal with stress predisposes us to suffer certain wounds in life (See more detailed explanation in # **87-88.**). Our avoidances, defenses and over-compensations along the legs of love and wisdom make us fail in specific ways.

However, our defenses keep us from suffering the full recognition of how the aggressions that caused our wounds have ruined our chances for fulfillment. I suppose there is some advantage to palliation here.

I have mentioned our wounds. The deepest most dreadful wounds come early. They fall along the legs of love and wisdom at different stages along the way. Our wounds, rightly understood, tell the story of how and where we were stopped or beaten back on the way to love or wisdom. Traumatic events, patterns of neglect and abuse, many of them culturally relative, like the timing of toilet training or weaning, open wounds. A young child probably experiences these as sudden, incomprehensible insults. Being torn from the mother, getting lost, being slapped away, confronting an angry father, or the experience of gnawing hunger – all of these can wound us. Chronic deprivations wound us as deeply as sudden traumas. When later events occur on the same places along the legs of love and wisdom, the wounds are aggravated. Our sensations signal them. We feel them coming.

To ease the pains and to keep ourselves from losing hope, we develop defenses. Anna Freud did a clear exposition of them. We repress memories and feelings, we project our conflicts onto others, we deny, rationalize, split ourselves into walled compartments and more. We keep despair at bay, but our defenses cause gaps, absences, lacunae that depress what little momentum we have to move to our turning points.

Finally the wounds scar over. We find we've been bent by life. We cannot stand tall or move gracefully.

As I mentioned earlier, our defenses, along with our habitual aggressions, fears, passions and anxieties dispose us to favor one side of the oscillations of love or wisdom over the other. The one-sidedness disrupts the patterns of movement in approach/separation and withdrawal/return. We stop seeing them as phases in a polar dynamic. Existential anxiety warps us in action until, by lifelong habit, the side that is blocked grows shadowy in consciousness. When we favor one side over the other of a slow oscillation whose period we cannot easily sense, the shape of the whole pattern is hidden. I referred to this condition in our quick dialogue on hiddenness (in # **122**). Our lopsided psychological perspective obscures the turning points, particularly in the signal events of life. We miss the big turnings. We turn back from them. Or enter a turning point too soon. Or bog down. Or follow the line of least resistance. We're all scared. The more cowed we are, the further we fall from our possibilities.

## 150
### *Venues for Aggression*

As it bifurcates from the stress response, human aggression comes up against love and wisdom in four venues. They roughly correspond to Hall's four social distances.

The closest in, the most intimate venue for aggression, is internal. We experience it in the aggression we turn against ourselves. Guilt is a kind of aggression against the self. It can tear us apart. However, internal aggressions can also give us moral fiber, courage for action and staying power. To break addictions takes aggression of this sort.

The second venue of aggression manifests in our dealings with the persons nearest to us. With them, we fight in the intimate social distance, through touch and feeding, in harsh words, in sexual rejection, in coldness, and in cruel or negligent parenting. Here we hurt and are hurt by those we love.

The third venue of aggression stirs in public and political spaces. Here people *gang up* against each other in larger groupings. You find third venue aggression in neighborhoods, schools and in voluntary groupings of all sorts. It expresses in workplace conflicts over money, job responsibilities, promotions and firings. Here as affections turn to enmities, pecking orders form.

The enmities and amities in the third venue greatly influence the forms of civility in a society. They welcome us or drive us away. We belong or are excluded.

To some extent, groups need to defend themselves against the deviant tendencies of their own members. The problem is deciding what constitutes deviant behavior. Since the standards generally lie beneath the threshold of awareness, the monitoring of deviancy becomes a "fine art," an art often abused by those with vengeful motives, who make themselves into the gatekeepers of belonging in groups. Usually only a few group members take on this gatekeeper function. Those who most fear having the group turn against them, pursue the job most keenly. They run the welcoming committees. They reject individuals whose gifts or personalities threaten to change the group's standards of "usness". They fear any changes that may make the group less congenial to them. In their insecurity about belonging, they stymie real Philia for the sake of false Philia.

Sometimes the gatekeepers ostracize troublemakers or sociopaths. More often, however, they get rid of the hapless ones, the victims of bad luck or mental illness. The gatekeepers also find troublesome the community's most creative individuals. They threaten the gatekeepers merely by being different, because, by their quirks, they widen the standards of belief and affiliation. By seeking the freedom they need to make their creative contributions, they alarm others.

Third venue aggression appears wherever the "us" forms in social units larger than the nuclear family. In order to belong, we must deal with group aggressions. We can all identify the third venue incidents from our lives. We have all been harmed by others. We react defensively. We close down. We harm ourselves. We trouble our waters, overturn our relationships, and with our senseless disputes spoil the small rhythms of life.

There is plenty of shame to go around. We are envious and withhold praise. We break promises. We neglect and abandon people who have been important to us. We forget those we should remember and remember those we should forget.

## 151

I was a wild young man. I had trouble connecting with groups. I arrogantly supposed I did not need to belong to anything. I was fine on my own. When I first read Allen Ginsberg's poem *Howl* I pitied the poor suckers

*...weeping and undressing while the sirens of Los Alamos wailed them down, and wailed down Wall, and the Staten island ferry also wailed, who broke down crying in white gymnasiums naked and trembling before the machinery of other skeletons...*

I did not realize I was becoming one of them. *Howl* ends with a terrible image. I have since seen it played out for real, sometimes

183

with me the protagonist. Ginsberg describes the creative person standing

*before you speechless and intelligent and shaking with shame, rejected yet confessing out the soul to conform with the rhythm of thought in his naked and endless head, the madman bum and angel beat in Time, unknown, yet putting down here what might be left to say in time come after death...*

Now I can catch my own destructive and self-destructive tendencies on the wing. They are shame based. I have been on the receiving end of other people's aggressions and I have overreacted. Even a small gesture that I interpreted as a rejection when I was hoping for intimacy roused my ire, and I have turned my back and walked away never to return from persons and situations I loved. I am too thin-skinned. That's been my ruination.

The aggressions I have given and received have not braced me up or made me manly. They made me fail. For a long time I hid this from myself by rejecting my rejecters the moment before they, in my judgment, were about to reject me. When I could no longer fool myself with this maneuver, I got frustrated and despondent, agitated and depressed. I turned my aggressive energy against myself. I dismantled my own creative life. My modular parts tore each other down.

## 152
### *The Fourth Venue: War*

By any measure, we must count war as the greatest enemy of love. As a fourth venue for aggression, it exceeds all the rest in numbers of wounded and dead, of populations in flight and cities destroyed, in the conjoined tragedies and mutual annihilation of the possibilities of life. War is the place where the greatest enmity converges to destroy the most love.[5]

War is not driven by hate. Money, ambition and statecraft organize the destruction. As Stephen Spender wrote of WWII:

*The guns spell money's ultimate reason*
*in letters of lead on the spring hillside.*
*But the boy lying dead under the olive trees*

*Was too young and too silly*
*To have been notable to their important eye.*
*He was a better target for a kiss.*

History shows that the greatest threats to cultural life, and the most heinous acts against love in war, come from aggression kindled in cold deliberation. Hatred is just a tool exploited by policy makers to rally the masses. Hatred plays a minor role in strategy because it lacks patience. You need an invigorated enmity based on reason to steer aggressive campaigns. Reason can take cognizance of shifting circumstances and learn from experience. Hate negates reason. Enmity sharpens it.

As Iago, the great destroyer of Othello's love for Desdemona, explains to Roderigo: *"If the balance of our lives had not one scale of reason to poise another of sensuality, the blood and baseness of our natures would conduct us to most prepost'rous conclusions. But we have reason to cool our raging motions..."*[6]

Iago's hostility to Othello stood on firmer ground than hate, because reason had freed it from its rhythmic rounds of intensification and abatement (the mirror image of love) that made it strategically unreliable as a driver of policy. To produce a great destruction, enmity needs to link reason to will and shift away from hate. It has to be able to wait. That is why Iago insists that real destructive intent works *"by wit, and not by witchcraft; And wit depends on dilatory time."*

The cool calmness of the policy-driven mind can be clearly seen in the great murderers of history. Often they are family men like Rudolph Hoess, commandant of Auschwitz, who during the Nuremberg trials insisted that his wife and children led a fine and normal life at the camp. He expressed pride that he had enormously improved gas chamber technology by introducing Zyklon B through the showerheads.[7]

When self-awareness and self-examination fall into disrepute the search for meaning dies.

To theorize that we get to self-restraint by way of reason, as Wilson does, is to ignore the evidence of history. It's an argument that goes back to Darwin. Paul Bloom, in an interesting short essay on the need for a psychology of morality, distinguishes the influence of reason on self-restraint from the argument for altruism and other social instincts.

"In *The Descent of Man*, Darwin sought to explain human morality in terms of a general increase in human intelligence – one that enabled us to transcend the emotional reactions of our primate ancestors to appreciate the very notion of ethical behavior, of a code of morality that can be applied in a fair and objective manner. [William] James had a different view, which he defended in *Principles*: that the unique aspects of human nature are merely the result of the addition of social instincts, such as shyness and secretiveness that other animals lack."[8]

I side with James in this and present my arguments in # **137** and **138** and later in more detail in # **196** and **197**.

## 153

In each venue, unleashed aggression upsets a particular balance between approach/separation or withdrawal/return, sometimes both simultaneously. In personal venues, as we have seen, aggression resets the rhythms in Eros by rerouting testosterone. In similar ways, though with different biochemical reactions, it shifts the balances within Philia or Agape. So too with the inward-outward movements of wisdom. The perturbations shift the balances between our interoceptive, exteroceptive and proprioceptive sensations by which we draw the maps of our lives and store our findings in state-dependent memory.

Aggressive intrusions upset the rhythms of love and wisdom by disabling one leg more than the other, by pushing us abruptly from leg to leg or by stalling us in one place too long. The long, comfortable approach stages we enjoy in normal family life can be suddenly curtailed. We cannot think straight, we cannot decide things, we get stuck inside ourselves. Our highs and lows, our strengths and weaknesses, our timidity and assertiveness, no longer meet our expectations. We cannot prepare for contingencies. They no longer fall along the usual legs of our social and solitary rhythms. Life as a whole becomes a contingency.

## 154

If we respond vigorously to the aggressive perturbations that fall along the legs of love and wisdom, if we react appropriately where the culture sets up its roadblocks, if we get mad when we're expected to get mad and submit when we're supposed to submit, then our sensitivities, interests and competencies will coincide with cultural expectations. We will conform to the culturally approved balances between inward and outward knowing. We will readily manage the repertory of recruitable aggressive behaviors we need to succeed. We will be well adjusted. Our ambitions will be apt for the perturbations we are most likely to meet in the course of life. We will find it relatively easy to take on the culturally approved goals as our own. We will be good citizens, able to accommodate cultural values to our consciences, so we can pursue success without equivocation.

Surely, we are not meant to accommodate ourselves to *all* culturally approved values. There are times when we have to say No. And times to say Yes too.

Rhythm breaking aggression can serve good purposes. It can focus us on danger, or motivate us to seize on opportunity. It has adaptive value not only as a rhythm-protector but also as a force to reorient our priorities.

Part of its virtue comes from its triggerable nature and its recruitability. In order to be recruitable, a drive has to be separable from its objects; its separability is what makes it recruitable. It would be harder to recruit hunger for anything beside the desire for food than to apply aggression to the desire for gain, to justice, artistic achievement, sports, politics or business.

The great heroes were drawn to their deeds by all manner of recruitable aggressive energy. When the written record of history begins, the myths of heroic endeavor no longer exemplify the primordial use of aggression as a restorer of rhythm. They become Promethean. The heroes become thieves of fire, exploiters of new ideas, creators, opportunists, conquerors. Heroic aggression can no longer achieve its ambitions in simple conflicts that can be resolved by distance regulation.

Konrad Lorenz called this kind of wider aggression the 'aggredi'.

"What is certain," he wrote, "is that, with the elimination of aggression, the *'aggredi'* in the original and widest sense, the tackling of a task or problem, the self-respect without which everything that a man does from morning till evening... would lose all impetus; everything associated with ambition, ranking order, and count-less other equally indispensable behavior patterns would probably disappear from human life."[9]

Are we talking about one aggression here, differently used, as Freud believed, or are we supposing that there are seven kinds of aggression, each with a separate origin attached to different aims, to use Wilson's arithmetic? Or some other specific number? Is the aggression of a chess master fundamentally different from that of a battlefield general?

Is there something more than raw recruitability that lets aggression serve as a protector *and* a disrupter of love, a boundary maker or violator, a form giver or form breaker? Is there an agonal element in the personality that drives creative individuals to surpass their peers and predecessors? It will not be easy to determine this. The good and bad aggressions interact in many ways. They sometimes turn into each other.

## 155

Once aggression relinquishes its conservative role as a preserver of the rhythms of life, we enter a new cultural milieu. Here aggressive interests link with reason. Sometimes they even *require* an alliance with reason. Moreover, they can get it because both reason and aggression are general-purpose non-appetitive, non-oscillatory functions with indefinitely wide areas of operation. Both have an instrumental character. They are both motivated to respond to problems. Stress arouses both. Innovation, assertion and creative expression flow from both.

Without planning, foresight, will, initiative and determined efforts to surmount obstacles (including those posed by other people), large scale ventures cannot move forward.

The likelihood is that evolutionary selection designed reason and aggression to help each other. Nature gave them routes by which to cooperate and even mesh. Far back in the human past, when the chemical cascade to aggression first came under partial voluntary control, the testosterone accumulated at the erotic/aggressive bifurcation point could (to some limited extent) be directed one way or the other by ideation, fantasy, imagination and intention. Reason and aggression made deals, negotiated their relationship, during the delay in the branching of the stress response.

\* \* \*

## *End Notes*

[1] Jerome Kagan. *Galen's Prophecy.* Basic Books. 1994. p.53

[2] Friedman and Rosenman. *Type A Behavior and your Heart.* 1974. Knopf

[3] Kenneth Pelletier. *Mind as Healer, Mind as Slayer.* Delta. 1977. p. 125-127

[4] Chronic cortisol secretion makes most people avoidant. Researchers have found positive relationships between cortisol and shyness and withdrawal (Kagan, Reznick, & Snidman, 1977), and also with anxiety, depression, and introversion (Dabbs & Hopper, 1990). Dabbs, Jukovich, and Frady (1991) suspect that cortisol will thereby inhibit violence through social withdrawal and depression. Robert Sapolsky carries these researches forward in his studies on stress and depression; he sees them occurring in epidemic proportions in modern life. But for a subset of the population chronic aggression lowers cortisol secretion. Perhaps the long lasting aggressions of leaders in government and business and those found in repeat violent offenders come from the same biochemical sources: a diminished cortisol response.

[5] Of this condition Wilfred Owens, a poet who died in the trenches in WWI, wrote:

*Red lips are not so red*
*As the stained stones kissed by the English dead.*
*Kindness of wooed and wooer*
*Seems shame to their love pure.*
*O, Love, your eyes lose lure*
*When I behold eyes blinded in my stead!*

*Your slender attitude*
*Not exquisite like limbs knife-skewed,*
*Rolling and rolling there*
*Where God seems not to care;*
*Till the fierce love they bear*
*Cramps them in death's extreme decrepitude.*

*Your voice sings not so soft, –*
*Though even as wind murmuring through raftered loft, –*
*Your dear voice is not dear,*
*Gentle, and evening clear,*
*As theirs who none now hear,*
*Now earth has stopped their piteous mouths that coughed.*

*Heart you were never hot,*
*Nor large, nor full like hearts made great with shot;*
*And though your hand be pale,*
*Paler are all which trail*
*Your cross through flame and hail:*

Weep, you may weep, for you may touch them not.

[6] Shakespeare. *Othello* I,3

[7] From testimony at the Nuremberg trials: "So when I set up the extermination building at Auschwitz I used Zyklon B, which was a crystallized prussic acid which we dropped into the death chamber from a small opening... We knew when the people were dead because their screaming stopped... Another improvement we made over Treblinka was that we built our gas chambers to accommodate 2000 people at one time, whereas at Treblinka their ten gas chambers only accommodated 200 people each."

[8] *The Next Fifty Years.* Ed. John Brockman. Vintage. 2002. p. 74-85.

[9] Lorenz. *On Aggression* p.269 The unique human problem, as Lorenz saw it, was to preserve the 'aggredi' of good aggression without fomenting violence over territorial competition in the pursuit of money and property. Even with the best of intentions greed and power-lust work their way free of the 'aggredi' – and perhaps they *must break free*, given Lorenz' view of human nature, and given the eagerness for gain, the imaginative power of our malicious intentions, the ignorance and arrogance apparent in the pursuit of pleasure and, indeed, in the ordinary conduct of business. In every big project there are hostile individuals ready to take over when the opportunity arises. When things go wrong they're eager to take advantage of others. To restrain these tendencies that are resident in human nature Lorenz proposed eight sublimations. First, know thyself: "...insight into the causality of our actions may endow our moral responsibility with the power to control them." Second, sports: "...the main function of sport today lies in the cathartic discharge of aggressive urge."

Third encourage "dangerous undertakings, like polar explo-ration and, above all, the exploration of space, all give scope for militant enthusiasm..." Fourth, develop wider familiarity with other cultures. "Being friends with a few 'samples' of another people is enough to awaken a healthy mistrust of all those generalizations which brand 'the' Russians, English, Germans, etc., with typical and usually hateful national char-acteristics." Fifth, espouse causes. "What is needed is the arousal of enthusiasm for causes which are generally recog-nized as values of the highest order by all human beings." Sixth, appreciate our common intellectual property "More than any other product of human culture, scientific knowl-edge is the collective property of all mankind." Seventh, de-velop intergenerational trust "a relation of trust and respect between two generations must exist in order to make a tradi-tion of values possible." Eighth, humor. "In its highest forms it seems to be specially evolved to give us the power of sift-ing the true from the false...Laughter resembles militant en-thusiasm as well as the triumph ceremony of geese in three essential points: all three are instinctive behavior patterns, all three are derived from aggressive behavior and still retain some of its primal motivation, and all three have a similar social function..." ibid p.274-284.

# THE BODY IN
# HISTORY

# 9. FIVE HISTORICAL EPOCHS

## 156
### *The Emergence Struggle*

**W**ell into Paleolithic times the natural signal sources for dispersal/aggregation were coupled to the day/night cycle. The signals to disperse came in the morning. The aggregation came at night. The choreography moved the group as a whole.

The taming of fire 400,000 years ago had made the hearth a powerful aggregation signal.[1] The technological expression of warmth with all its drawing power probably influenced the evolution of our thermal regulation physiology (which may intrinsically correlate fire and clothing with social life and belonging on some deep organic level). Desire and longing, jealousy, remorse and compassion may well have spiraled out from the hearth: the warmth of the woman and the safety of the hearth were early civilizers.

Close proximity brings women's menstrual cycles into phase with each other. Pheromones waft between sexual moieties. Group sexual readiness influences mating customs and, through them, influences population biology and child rearing practices. As fertile periods moved into phase among the females, the sexual desires of the group as a whole must have played out in *occasions*, ritualized encounters, and festivities of all sorts. William Irwin Thompson in *The Time Falling Bodies Take to Light* describes this early human culture as "a menstrual world," whose dominant sensory and emotional pulls and physical signals were female. He writes,

"For hundreds of thousands of years the culture of women and women's mysteries had been the dominant ideology of humanity. The hominization of the primates in the shift from estrus was a female transformation. The rise of a lunar notation and the beginnings of an observed periodicity upon which all human knowledge is based was a feminine creation." He calls women's mysteries "the dominant ideology of humanity."[2]

Sarah Blaffer Hrdy reasons that because human ovulation is hidden, women set the beat for the emergence of intimacy. She reports that,

"What strengthens the pair-bond between early human men and women was not frequency of copulations, but rather the fact that ovulations became concealed. Without estrus, the only guarantee a male would have of consorting with a female near the auspicious moment of ovulation would be to consort with her throughout the month, month after month."[3]

Feminist evolutionists have described this arrangement as a sex contract: protection and meat in return for sex and vegetables. Males made themselves useful while hanging around waiting for sex.

Within the hormonal wash and the circadian and seasonal impulsions, we fit our personal social and solitary needs – and our dawning consciousness of them. Individual mini-dramas of approach-separation and withdrawal-return play out. In aggregation, we explored pair bonding and friendship in new ways. In the dispersal, during furthest separations, we honed our experiences of self-conscious solitude.

One can easily imagine how, at the extremes of dispersal, when the troop members were out of sensory range of each other, withdrawal pulses drew individuals inward. Some actually experienced a new state of mind they named *solitude*. The inward pull became its signpost.

To continue a story we started before: Picture yourself in the primordial setting. You're alone, out stalking game or foraging for greens, far from your fellows. You begin to experience the flow of internal sensations. Their heightened intensity, vividness and immediacy draw you in. You engage the flow of your own consciousness. Images, memories, voices and thoughts mill around you. You remember and review the day. Events from the previous weeks or months occur to you. You make odd and pleasing associations. You explore self-consciousness. The *you* that explores watches the phantasmagoria.

*The Watcher inside you is born.*

Seeing with the inward facing eye, you distinguish your memories from the events themselves. The memories belong to you; the events are common property. With this private perspective, you become aware of differences, and even recognize the conflicts between your own desires and the social expectations placed on you. But this awareness is private. These are your secrets. No one needs to know about them. With that act of concealment, you realize how alone you are. Solitude is born. A momentary terror of abandonment comes over you. You have your secret place, from which you can plot and plan and think, but at a cost, because your welfare is utterly dependent on the group, and yet you are keeping secrets from them

To counter the fear of abandonment, you recollect the faces and voices of others. From within the solitude you review your shared involvements. Through memory you prove your closeness to those who know and care for you. You recall the words of your parents, age-mates, and friends, the assured manner of the troop leader, the consoling intimacy of your mate – they comfort you. Many words and images come up that reaffirm your place in the group. You belong. But is it enough?

Unsettling images also rush in. Voices assault you. You need to counter them. You cannot shout them down. So you use the modularity of your personality to compartmentalize the scenes of the passing show. Over time, your mind gets good at modular organization. It learns to distribute memories, attitudes and motivations among part-personalities. By putting up walls, the mind keeps you in your incumbent personality. It tightens up your I-sense to prevent the mental turmoil in solitude from flinging you from one part-personality to another chaotically under the pressure of recollected voices, commands, incidents, hopes and fears. To keep steady in solitude, in the conceptual scheme we have been developing, you need to maintain clear passing rules **(See # 77-81.)** The movement from one memory storage area to another has to make sense. It can't be too jumpy. Otherwise, your waking life would be as unstable and chimerical as your dreams.

## 157

As Homo sapiens emerge from embeddedness in nature they develop independent, individual approach/separation and with-drawal/return patterns, with a characteristic ability to distribute memory, attitudes, cognitive styles, knowledge and even tempera-ment among different modules of personality. Each module carries with it a repertory of state-dependent associations, memories, and behaviors. These modules have doors and the doors have locks and keys. They open, first, to sensory distance information, for these reflect the basic operations of social and solitary life very directly.[4] Then they open to cultural learning.

Shaken loose from our containment in the over-rhythm of dis-persal/aggregation by changed habits of life, we develop a reper-tory of skills that lets us play different roles at different times and distances, as needed. During these role-plays, the modules pass the I-sense to each other along the legs and stages of love and wisdom according to the passing rules so that personal identity doesn't un-dergo too much slippage from one state-dependent situation to an-other.

## 158

As human culture develops, new sensitivities in intimacy and solitude arise. They break free from their embedded condition in daily aggregation-and-dispersal to build primary alliances with the approach-and-separation and withdrawal-return rhythms as distinct and independent pulses in human nature. Eventually these emer-gent patterns of behavior enter consciousness as subjects of inter-est, as values. We celebrate them. We seek them out. We come to understand them. We name them love and wisdom.

For the first time we find ourselves capable of individuation. We overcome the "participation mystique."[5]

Musical instruments, bone flutes, sculptures, abstract designs, personal adornments, appear during this era. You see individuation in the handprints outlined on the cave walls. In the remarkable cave art so skillfully executed, so full of close observation of nature, we detect the deep pondering of the mysteries of birth, life, change and death.

Mourners scatter flowers in graves. They inter implements, tools, jewelry, and food with the mortal remains – equipment for a journey to a realm of ancestors. Corpses lay in fetal positions as if waiting for rebirth. Ancestor worship, affiliation with the past, perhaps regret over the loss of loved ones, perhaps sorrow and expiation, emerge. These just touch the surface of the new skills and powers, formed and fashioned in turning points of love and wisdom, that our ancestors assimilated into art and myth, and wove into cultural evolution in the changed organization of the passing rules. These developments obviously take us far beyond the old rhythms of life shaped by aggregation-dispersion.

I call this great process the emergence struggle. It sets the tone for Homo sapiens prehistory. Over the millennia, humans come to see themselves as *conscious* agents of their own destinies and, endowed with freedom, reach out in wider circles of affiliation and deeper journeys of discovery to move from species ethology to human history.[6]

## 159
### *The Second Epoch: Shamanic Wisdom*

By 25,000 BCE, human life has moved into a wider cultural ambit. We see the rudiments of a mythology and religion emerging. It seems to be built on a spiritual encounter with a "beyond", expressing a primordial wisdom quest that surfaces in the shamanic journey form.

Cave painting, Lascaux, France, 15,000 to 10,000 B.C.

We find the story of withdrawal-return painted on the walls of caves dating back 30,000 years.

Retreating to the deepest recesses of the caves, in total darkness, lit only by tallow lamps, the shaman and his or her apprentices (who perhaps sign with their red handprints) conduct a journey to the underworld, to "the happy hunting ground", followed by a return to the group. The trek into the dark and depths of the cave symbolizes the journey. The cave paintings depict it.

The artists represent herds of animals galloping on the cave walls. Groups of crudely sketched hunters, along with departed souls and ghosts perhaps, circulate among them. The animal archetypes move in directions that follow the contours of the cave walls, overlapping each other. A continuous artistic effort lasting hundreds or even thousands of years is recorded on the same cave walls.

In many scenes, you see the shaman in their midst, a stick figure with a bird mask.

**Lascaux shaman painting**

He seeks the "animal masters," the prototypes of the hunted species. He appeals to them to come back to the world, to be born again, to reenter through the gate of life to assure the productivity of the hunt. (Attention to the birthing process of the prey animals may have been transitional to husbandry).

On other journeys, the shaman seeks special knowledge needed by the tribe. He goes down, and then he comes back up with it following the dynamics of the wisdom journey discussed earlier.

A crucial observation: *These enormous cultural developments predate major changes in tool technology.* We shall soon see why.

## 160
### *The Third Epoch: Expressive Human Nature*

During the Epipaleolithic period, between the end of the last ice age and the dawn of the agricultural revolution, we reach a new threshold of cultural life. Under the influence of abrupt climate change, we transform tool technology and revolutionize resource exploitation and living styles. We rise further from the embedded expressions of human nature, over-leaping the emergent forms practiced in the feminine love/shamanic wisdom era. Human nature enters its third *expressive stage.*

Archeological evidence shows that a great variety of new lifestyles developed. Some people lived in permanent settlements set among oak and pistachio woodlands extending along a thousand miles long narrow arc from the Levant into Anatolia. Hunter-gatherer troops of the old Paleolithic culture lived among them, interacted, and traded with these revolutionary gardeners and husbandmen.[7]

From artifacts, tools, seeds, pollen grains and kitchen middens, researchers have pieced together a story of people living in many small permanent or semi-permanent hamlets as sedentary foragers and early planters. Ofer Bar Yosef has studied the Natufian culture in the Near East (13, 000-10,000 BCE) in detail. He concludes "Natufian culture played a major role in the emergence of the early Neolithic farming communities, or what is known as the Agricultural Revolution."[8]

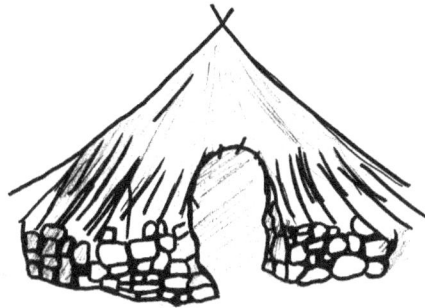

**Natufian Dwelling**

The people lived in small communities in circular stone-walled pit houses 3-9 meters in diameter, some smoothed over with plaster and roofed with wattle. They built the houses close to each other. Stone tools, microliths, arrow shaft straighteners, mortars and pestles and sickle blades have been unearthed in them, indicating an economy that harvested cereal grains (though they may not yet have planted them.) They gathered pistachios and wild grasses and apparently hunted gazelles, fallow deer and wild boar. They kept farmyard animals. The first evidence of man-dog bonding occurs at this time. In some graves, we find men buried with their arms around their dogs.[9]

In response to climate changes, "many communities in the forest belt had turned to cultivating wild grasses. Within a few generations, they had become full-time farmers... farming spread rapidly through the Levant and into the far corners of Anatolia."[10]

**161**

We're talking about major climate and geological shifts: a 350 foot rise in sea level, huge areas of dry desert now open to irrigation, warmer, wetter weather with longer growing seasons that provoked species migrations, some north to the retreating cold, some south, resulting in new ecological networks, new rhythms in them, new interspecies predation and new intra-species territorial and dominance conflicts. The world went from cold and dry to wet and warm.

Then the climate shifted again. The cold and dry weather returned for 1,500 years, a stretch called the Younger Dryas (11,000-10,300 BCE), triggered by a stalling of the Gulf Stream caused by ice jams in the arctic.

In the Younger Dryas, Brian Fagan writes in *The Long Summer*, "the landscape became drier and drier and the forests retreated far beyond walking distance...[it was] a time of intense competition for food. There are no signs of warfare, such as war casualties in local cemeteries, apparently just a quiet acceptance that food was scarce and a greater reliance on kin to stave off hunger... A permanent settlement like Abu Hureya was no longer viable in the absence of the nut harvests and in the face of a severe drought... the village was abandoned....the ancient strategy of mobility was the only option, whatever the cost."[11]

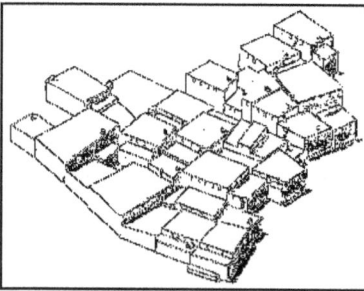

**Abu Hureya Before it was Abandoned**

Under these climatic pressures, we again had to adapt or perish, though Ofer Bar Yosef reports from his field studies that, despite the changes, "several communities maintained social relations with their original hamlets and returned there to bury their dead." He surmises, "The first experiments in systematic cultivation most likely occurred during the Younger Dryas". In his view renewed migrations led to new settlements closer to more reliable water sources.

The succession of boom times and hard times provided the conditions for rapid evolution. As William Calvin put it, we had to be men for all seasons.

## 162

The archeological evidence suggests that during these geologically abrupt transitions social groups explored many new ways of life. Multiple, tentative starts were made. Hamlets were founded and abandoned. There is evidence of trade. People occupy new environmental niches.

One finds lakeside sites with fish industry, boating, the use of nets and weirs. There are sites with seed culture. Elsewhere the beginnings of metallurgy. Elsewhere fanciful jewelry and costuming.

In contrast to Paleolithic customs, there seems to have been no standardized burial practices. There were communal graveyards, but also family burials under the floors of houses. Ben Yosef thinks these differences "indicate the existence of distinct group entities." Though nothing as grand or striking as the pyramids to come or as the Lascaux cave already painted was created, the artifacts do show evidence of great variety, signs that something new was stirring. The smallness itself, and the diversity within the smallness, tell of delicate, tentative explorations of human possibilities, with independent starts in many small settlements spread across wide swaths of territory. Groups could live in very original ways, developing their customs in relative isolation.

Where the Paleolithic culture had the hearth, the Epipaleolithic culture had the home, a settled life in a place of private belonging dug into the landscape, the ancestors buried under the floor, a building with a foundation, inhabited year round.

## 163

During transitional periods, change occurs first and more markedly in the psychosocial depths than on the material surface. The monuments, social institutions, tools and techniques come later. They culminate rather than begin transitions. The *sine qua non* is authentic turning points in love and wisdom. Under these conditions, human nature reaches new adaptive peaks in the freedom of its creative moments.

During the Epipaleolithic period, the swings in human nature seemed to get wider and come faster under resource and climate pressures. Approach-separation and withdrawal-return were repeatedly stressed and tested. Many different kinds of love and wisdom were likely to have flourished as isolated groups explored their possibilities. Solutions of unprecedented novelty appear.

I speculate that some settlements became more reliant on words, some more on tools, some more on weapons, some on cooperation, some on internal competition, some on dream wisdom, some on debate, some on dancing, some on narcotics, and some on sexual license. Groups taught and learned differently, staged the periods of their lives differently, esteemed each other differently. They had diverse belief systems, distinct brands of inwardness and outwardness, and the ethical and technical sensibilities that followed from them. They harnessed aggression differently and aimed it at different projects. Some troops used the increased cerebral processing power impulsively without periods of inward reflection. Some deliberated carefully. Some were slow and others were fast. People lived in different social worlds, perhaps even as different species, if the Neanderthal people survived into these times.

## 164

With a planetary population of under a million, groups rarely met. But when they did, what momentous occasions their gatherings must have been! Given human passions and curiosity, each word and glance and gesture observed, each social distance and border crossing experienced, must have reverberated with meaning. They experienced true intellectual curiosity. They touched the deepest unknowns: the souls of strangers. Aggression stayed close to the surface, but did not inevitably lead to open hostilities or violence. Rather, we can guess that trade and exchange, friendship, learning and intermarriage were the more frequent outcomes.

It's doubtful the people could speak to each other. There must have been many local languages built on an inherent universal grammar. However, they could understand each other in other ways, through dance, and drama; they could share drawings, and religious rituals.

Put yourself there, and bring your mind with you, (because their brains were as good as ours were, though lacking some of our conceptual equipment and cultural history.) You were perhaps more fearful and more curious than you would be now, and your sensory acuity was probably better and your passions clearer. What would you have wanted to find out from the others? Where the food was? Yes, those things, but the more crucial questions on a planet of few human souls were probably *not technical*. "Who are you, how did you get here, where did you begin, what is your story? What is the world like, as you see it? Where have you been, who are your ancestors?" That's what we would want to know.

But we could not speak to each other very well, so we had to act it all out.

Imagine playacting the story of your people's origins. Your performance, to be comprehensible, would rest on the empathy generated in your audience by your actions and expressions. You would trust in their ability to read it from your dramatizations.

Imagine discussing these performances afterward among your fellows and families in your spoken language. You would compare the stranger's stories with your own, finding similarities and differences.

Here mythology emerges from ritual enactments, pantomimes, theatre, dances, etc. It starts with human empathy, built on a foundation of the shared rhythms of nature and human life. What did you learn through empathy? That myths and rituals deal with common human predicaments, and that the creative power of the human experience comes from turning points in the signal events of life.

This catalog of stories, values and patterns of life thenceforth forms the seed stock for world mythology (and all literature indirectly). That gift the Mesolithic period gave us.[12]

## 165

Most paleontologists attribute our powers of advancement to the "problem solving" model of the human brain. Though the consensus scholarship recognizes the crucial role played by increased cooperation, food sharing, and child rearing and abstract planning in human prehistory, these essentially non-material achievements they assign to *cognitive advances* (often associated with technological progress.)

They measure the evolution of the brain by the enormous expansion of the brain case, which they attribute to the increased volume taken up by changes on the surface of the cortex. They neglect the crucial importance of the lines of communication between the expanding cortical surface and the deeper mid-brain thalamic, amygdala, hippocampus and other limbic system structures that must have accompanied it. Whole-heartedness, warmth and sensitivity, with all their emotional and intuitive elements, hardly matter to them. They rarely mention love and wisdom. The result: a misplaced emphasis on cognitive processing that obscures the critical distinction between reason and judgment and pigeonholes us as toolmakers and techno-enthusiasts.

William Calvin follows this line in *A Brain for All Seasons*. According to him, the evolutionary premium during the Ice Ages was not on specific body adaptations, like fur or fat, but on the flexible adaptability of the CNS. That's what lets us learn quickly in challenging environments.

In Calvin's view climatic pressure pushed us toward what he calls a general-purpose brain. Its greatly increased processing area helps us solve problems, acquire language and expand social interactions. Good enough.

However, the toolkit hardly changed at all during the period when the brain doubled in size. Where did all the processing power come from? Where did it go?

We are so accustomed to associating intelligence with technological advancement that we can hardly envision a cortical expansion not directly keyed to increased cognitive powers, with these powers leading directly to technological innovation. That's the reigning model. We've been stuck with it for a long time. Calvin gives it new life by accounting for rapid changes in brain volume, *in the absence of new technology*, by pointing out that improved aiming and throwing skills require an enormous number of new neurons. Once we have those neurons they become available for other purposes. Calvin overlooks the direct evolution of deeper capacities for empathy, sympathy and cooperation independent of augmented intellectual powers.

But we can look at the growth of the neo-cortex from another angle: Advancements in cognitive processing are only adaptive to the extent they rest on deepened strata of cooperation. These in turn depend on perseverance and resilience in approach-and-separation and withdrawal-and-return in the face of aggressive perturbations. What we needed more than brilliant minds capable of subterfuge, then, was *counter-aggressive restraints*. Moreover, the evidence suggests that we did evolve these restraints through better management of stress and aggression in a wider range of conditions (See # **143-149.**). Emergent love and wisdom, and the new skills that came with them, favored by evolved counter-aggressive restraints, changed the stakes in life; they gave us new goals by turning us into creatures who *cared* differently. Out of *caring,* we *created.* And our creative energy drew on aggression in new ways.

In my story of origins, the latter stages of brain evolution produced expanded frontal-limbic circuitry that freed us to use aggression in a *Promethean way*, to snatch the boon of creative moral advancement in the near chaos conditions of our turning moments.

## 166
### *Reason vs Judgment*

To understand where I'm coming from, let's step back and take a broader view of our current intellectual life. By and large, our cultural elites hold false opinions about human intelligence. They seem to assume that all the prime coping values are cognitive. With the big braincase for evidence, they assume that our moral lives can be advanced by progress in understanding the external world.

Robert Wright takes that line in *NonZero.* He writes,

"My point is just that these brains are a continuous outgrowth of something at life's very essence: a primordial imperative to process information. Given the apparent connection among information processing, sentience, and meaning, it seems fair to say that evolution by natural selection was from the beginning a veritable machine for making meaning."[13]

In his earlier book, *The Moral Animal*, he made a case for reciprocal altruism that took into account its evident abuses. Here he props up the weak reed of reciprocal altruism with the steel of intellect – more steel, less reed.

"So there you have it: the basic equipment needed for a species to hop on the co-evolutionary escalator: learning, learning by imitation, teaching, some use of tools, along with elementary grasping abilities, a mildly robust means of symbolic communication, and a

rich social existence featuring, in particular, hierarchy and reciprocal altruism…"
(p. 291)

For evolved altruism to assure civility, it would have to assert
itself when we were maddest at each other, when we were least
amenable to reason. That doesn't happen. Loveless altruism, half-
faked, half-sincere, won't pull us back from the brink. As soon as
intense group chauvinisms surface, as in our current War on Terror, goodbye altruism. Francis Fukayama, also holds on to this
false hope of altruism tempered by intellect. He assures us,

"Human nature encompasses a great deal more than male-bonded
violence. It also involves the desire for what Adam Smith called
gain, the accumulation of property and goods useful to life, as well
as reason, the capacity for foresight and the rational ordering of
priorities over the long term. When two human groups butt up
against each other, they face a choice between engaging in a violent, zero-sum struggle for dominance, or else in a peaceful pos-
tive-sum relationship of trade and exchange. Over time, the logic
of the latter choice (what Robert Wright labels non-zero sumness)
has driven the boundaries of human in-groups to ever larger communities of trust…" [14]

As if the heart and the temperament are secondary considerations, and biological tendencies toward aggression don't express
differently in different people and don't enter their intentions and
shape their actions and hide from them in the corners of their
minds in culturally relative ways. How can outward intelligence
safely shape human affairs when the roiling currents in our own
depths shape intelligence itself?

We cannot refer all matters of judgment to degrees of intelligence, or distribute them across different kinds of intelligences. If
so, by implication, people with the highest IQs would make the
most trustworthy counselors. Don't bet your life on it.

Intelligence can certainly try to solve any problem, (or anything that can be posed as a problem,) but without good judgment,
we cannot determine which problems are worth solving. Or why.
Moreover, good judgment does not come from the big brain's cognitive capacities alone, any more than it comes from caring alone.
It depends on a living balance between thinking and caring, be-

tween personal and instrumental concerns, between self-preservation and reverence for life, and between cortical and limbic interests.

## 167

Michel de Montaigne, the first modern Western practitioner of self-examination, distinguished judgment from intelligence in a way that will still serve us well:

"Judgment is a tool to use on all subjects, and comes in everywhere. Therefore in the tests I make of it here, I use every sort of occasion. If it is a subject I do not understand at all, even on that I essay my judgment, sounding the ford from a good distance; and then, finding it too deep for my height, I stick to the bank. And this acknowledgment that I cannot cross over is a token of its action, indeed one of those it is most proud of."[15]

Donald Frame, Montaigne's translator and biographer, says of his use of judgment:

"It is almost identical with understanding and with reason in the sense of right reason, but, unlike reason as wrong reasoning, it is close to the facts, responsible, teachable, scrupulous in reaching conclusions. Far more important than knowledge, it has no necessary connection with it. Its training is the heart of education. Its jurisdiction is infinite, including morals and intellect, determining not only what is true but what is good. It is the faculty that teaches man how to live."[16]

## 168

It is a mistake to conflate all knowing with outer knowing. Moreover, inner knowledge *can* be shared. This is the realization that brought Wittgenstein back to philosophy after years of alienation. He wrote in his last great work, the *Philosophical Investigations*: "Is there such a thing as 'expert judgment' about the genuineness of expressions of feeling? – Even here there are those whose judgment is 'better' and those whose judgment is 'worse'. Corrector prognoses will generally issue from those with better knowledge of mankind. Can one learn this knowledge? Yes; some can. Not, however, by taking a course in it, but through 'experience'. – Can someone else be a man's teacher in this? Certainly. From time to time he gives him the right tip. – This is what 'learning' and 'teaching' are like here. – What one acquires here is not a technique; one learns correct judgments. There are also rules, but they do not form a system, and only experienced people can apply them right."[17]

With "corrector prognoses," "better knowledge of mankind" and "right tips" we can, I will show below (see # **235, 236**), reconcile competing claims of the ethical and technical minds.

## 169

How do you know when you are getting or giving good judgment? The standards change. Today the faculty of good judgment requires:

1) A strong mind capable of sustaining a struggle in the depths to unify being kind with being true. When caring about justice is a primary motivating energy of life directly flowing from love and wisdom, judgment encompasses intelligence.

2) Judgment affirms the genuine nobility of the human aspiration for truth. But it links the search for truth to personal transformation. It accepts suffering, if necessary, but doesn't seek it. This transformation comes from a synergy of human attributes: from frontal-limbic linkages that modify the stress response, from imagination that develops into envisionment, from child's play that rises into serious and responsible exploration.

3) Without an inclination for deep pondering in long periods of inward reflection, increased processing power tends to produce a different kind of human social life, one more changeable, impulsive, brilliant, improvisational and more liable to sociopathy. Wisdom, reason and imagination need each other. Wisdom uses reason in its search for truth.

4) We can direct reason to consider problems raised by wisdom. Moreover, when reason and imagination mingle with love, qualities of temperance and judgment come into play. Without this protection, intelligence goes wild and we get into the state of mind reflected in the remarks of ex-secretary of Defense Robert MacNamara on the Cuban Missile Crisis: "We lucked out! Rational individuals came that close to a total destruction of their societies."[18]

## 170

Rational consideration *is* a direct route to certain kinds of knowledge. As a step in the process, it is indispensable. But alone it will not make the world safer or happier. Too many students of human nature push love and wisdom into the background. They obscure our access to the most important things about us. With love and wisdom deleted, the many modes of sensation and imagination, interoceptive sensory acuity, contemplation, meditation and intuitive knowing that we actually rely on to gauge our place in the world are bleached away.

This negligence I believe is willful. It rests on a hidden refusal to endure self-examination. To assure objectivity, it resolutely resists inward ways of knowing as primary sources of evidence about the world – even about the passions of human nature itself. However evolutionary, archeological, anthropological, sociobiological, demographic and biological studies of human nature have not become less subjective. Instead, they've become more inadvertently *confessional, unconsciously self-revealing. Richard* Lewontin's 1984 criticism of sociobiology pointed out that,

"Metaphors are often taken for real identity and the source of the metaphors is forgotten. There is a process of backward etymology in sociobiological theory in which human social institutions are laid on animals, metaphorically, and then the human behavior is re-derived from the animals as if it were a special case of a general phenomenon that had been independently discovered in other species."

Until we can understand our own biases through self-examination, we will be blindsided by our own self-aggrandizing tendencies. Hobbes had it right when he wrote more than four hundred years ago,

"Though by men's actions we do discover their design sometimes; yet to do it without comparing them with our own, and distinguishing all circumstances by which the case may come to be altered, is to decipher without a key, and be for the most part deceived, by too much trust or by too much diffidence, as he that reads is himself a good or evil man."

Researchers now leading the study of human nature get no training in self-examination. No wonder their writings don't have half the relevance to life you find in Montaigne's *Essays* or Machiavelli's *Discourses* or Rousseau's treatise on the *Origin of Inequality* or Hobbes' *Leviathan* or Hume's *Treatise on Human Nature*. The great students of human nature never excluded the evidence of their own lives from their inquiries.

Nowhere is the absence of perspective on personal motivations more troubling than in the study of aggression. Of all the blindnesses, those keeping us from attending to our own destructive tendencies have to be the most dangerous.

## 171

One big pitfall: evolutionary researchers must make the immediate present the terminus of the evolutionary line. The past has to add up to the present. Thousands of generations of the remote past tumble precipitously into the handful of generations of the historical record. This is a form of backward prediction. It draws conclusions before the investigation begins.

Yet, the past *does have to* add up to the present. But which part of "now" is the present? Sociobiologists living in the Middle Ages, using the same approach, would argue that the natural law had feudal tendencies. Viewed from ancient Athens, social life would be inherently polis-like. The causal sequence would be skewed by the choice of any given moment as the historical terminus. But our evolutionary thinkers (and who besides Creationists is not an evolutionary thinker) deal with vast spans of time. Ten thousand years pass in a blink of the eye. Except when you get close to the present. Then the Middle Ages, the Renaissance, the age of Imperialism, mark definitive periods. In a consistent evolutionary approach, however, all of these "termini" might as well be "now". Against the 100,000 starting generations, the hundred historical generations all count as the present because they're too close in time to reflect heritable differences from isolated breeding populations.

That's why millions of years of evolutionary selection can seem to validate our current belief that the American ethos, the final outflow of the Protestant Ethic, is the *summum bonum* of history; that success is the best indicator of a life well lived, and reciprocal altruism is the way to it.

## 172
### *The Fourth Epoch: Human Nature Conflicted*

Material accomplishments follow fast and hard on the heels of the Agricultural Revolution. From settlements of a few dozen, townships of hundreds, then thousands grow. The crossroads encampments become villages and eventually city-states. The explosion in techniques and tools that did not occur earlier, though we had the brains for it, happens now. We grow crops, keep seeds, domesticate dairy and raise meat animals. New methods of food preparation and storage appear, people brew beer and wine; they build irrigation canals, they dig water wells. Wheeled carts appear, fishing boats sail in coastal waters. Caravansaries carry produce. Thrown pots, urns and barrels are manufactured. Long distance trade begins. Together these increase productivity and encourage the division of labor, making community life more secure.

The new rhythms de-emphasize the daily outings of the foragers and favor seasonal flows of time with slower growths and longer transitions. The wise men begin to measure time mathematically through the calculation of astronomical cycles, including the phases of the moon and the passage of the sun through the zodiac. The mythology encompasses seasons, years and generations. Eons are proposed. Creation myths, legends of the origins of the people, sagas of migration, replace the Paleolithic hunting magic. The dispersal-aggregation pulse still runs day to day, but less deeply than before. Other rhythms surge forward.

Prehistorians provide two conflicting accounts of the fast changeover from small-scale gardening cultures to massive, empire building, and war-making civilizations.

The minority report, to which we shall soon return, is value centered, gender oriented and reflects the sensibility of the myths that emphasize fertility and nurturance in the emerging agricultural life. The majority account, which we will consider now, is hard edged. and concentrates on material accomplishments. It makes economic relations primary. It looks to the big historical winners, the civilizations that survived, to define success. It treats the little losers as historical debris.

Jared Diamond tells the story of the winners in *Guns, Germs and Steel.* He promotes a simple materialistic thesis: successful civilizations raise food production to support larger populations. These occur in geographical settings that favor cultural diffusion, particularly east-west movement within and between continents. The trade, invention and competition that follow encourage progress. "A larger area or population means more potential inventors, more competing societies, more innovations available to adopt..."[19]

His just-so story goes like this: the big winners of history are the early civilizations who settle the valleys of the Euphrates, Nile and Indus rivers. Alongside them some tribes and villages live, but they are conquered and absorbed. Since then small indigenous people have been pushed into increasingly remote locales.

Though Diamond does not say much about them, the archeological evidence shows that many graceful, humane and interesting societies thrived for a while in the culture areas the "winners" dominated. The point he misses is this: not every innovative group living in a pleasant spot supplied with the right flora, fauna, surpluses and trade routes, chose the path of material accumulation and trade. Many small indigenous societies pursued other goals. There are many kinds of innovation. True, the more "competitive" societies wiped them out, so for Diamond they were losers. But that does not make them either inferior or unfit. It only shows the superior *might* of the victors.

Moreover, many of the historical "failures" have changed the world for the better. Their accomplishments in language, mythology, math, alphabetics, music, drama and philosophy still shape our sensibilities. They have transformative power not because they were the conquering civilizations but because they were conquered and their contributions were assimilated.[20]

Granted, societies need surpluses of some kind to produce higher cultures. But Diamond never acknowledges that different kinds of surpluses encourage different kinds of growth. Some surpluses, based on enhanced emotional communication, spiritual striving or ecological balance with the land, create accomplishments you cannot measure by technological prowess. Moreover, material advancement can bring on unanticipated problems that civilizations can solve only by reversion, by shutting down progress and backing off, as happened when China, after exploring the coasts of Africa, dismantled their great sailing fleets.

Diamond's story of origins leaves out the skills and talents emerging from art, language and gender relations. In fact, the whole body of non-material attainment that gives us our defining human characteristics (love, wisdom and aggression in our account,) serve him mainly as a backdrop to technological and economic innovation. He goes so far as to argue that when you look for winners you have to *disregard* the aesthetic, moral and psychological qualities that make us human because over time and space they average out. He argues this on the apparently reasonable assumption that local differences would be randomly distributed over a continental mass. Even "innovativeness" he treats as "a random variable... at any particular time, some proportion of societies is likely to be innovative."[21]

In the last chapter of *Guns,* Diamond retreats somewhat from his own conclusions. In a final reconsideration, he makes room for what he ignored throughout. He calls these forces historical "wild cards". They may under some circumstances energize the transition to civilization in non-material ways. Among them he includes local cultural variations, the unpredictable influence of great men, and chance events. But "it remains an open question how wide and lasting the effects of idiosyncratic individuals on history really are". In among the "local cultural variations," presented as an afterthought, he catalogs everything that most people take to be essential to their humanity and then passes over it without comment.[22]

In his later book, *Collapse,* Diamond, in a nod to environmentalism, adds an important caveat to his material determinism "sustainably". He seems to be trying to right the balance with some kind of creative moral presence. However, he cannot squeeze into his concept of sustainability all the caring and judgment he left out of *Guns, Germs and Steel* because he holds a core belief that the evolution of human nature culminated in its technological capacities, and it leaves scant room for non-material primacy.

By ignoring the great richness of the non-material qualities in human nature, by reducing success to the capacity to innovate sustainably, Diamond makes creativity too narrow, as if aesthetic, ethical and emotional factors do not contribute to success or collapse. His work falls in with the acceleration/expansion view of civilization that underpins American entrepreneurial capitalism.

His chief error: He draws a misleading contrast between "innovative" and "conservative" societies. He assumes that a society that is conservative with respect to technological change will be conservative in all respects, that a dualistic division into conservative and innovative societies can be justified by a single litmus test.

But we have in the past revalued our values and adopted new attitudes against the grain of technological progress. We are likely to do so again.

## 173
### *Warriors, Chariots and Invasions*

The feminist pre-historians tell a different story. Diamond never mentions it, or them, or the other world historians, like Arnold Toynbee, who dealt directly with the origins and life-histories of civilizations.

In the feminist myth, the peaceful cultures of the archaic peoples suffered a catastrophic end. I'll tell the end first, then the beginnings. Into the relatively peaceful Neolithic world, armed nomadic bands intruded, some mounted on horseback or driving chariots. They came from the steppes of central Asia, pushed west and south by migrating people behind them. The migrating bands raided the agricultural settlements, some now sizeable towns. As early as 7-8000 BCE the archeological records show defensive city walls being built in Jericho and elsewhere.

In the agricultural world, the old male hunting groups became the field workers, artisans, potters, smithies, and artists. When needed, they defended the walls. They patrolled the perimeter and maintained security. A standing military presence was established.

Inside the city, food was stored in defensible warehouses. Later, bureaucrats kept written tallies. Deposits and withdrawals were recorded. The people dug wells, cisterns, and city walls to protect themselves against long sieges. Military outposts far away from the capitol guarded the frontiers. Bronze Age weaponry appeared.

Mythic tales of warfare abound. Artists depict battle scenes with heroic values glorifying conquest, endurance and harsh treatment of captives.

Cross-cultural myths depict a long war between the invaders and the settlers. One story has the displaced male hunters, now deprived of power, holding on to their prestige by forming secret societies, and from these groups misogynistic conspiracies were hatched. Another theory has it that the deposed male hunter societies successfully plotted a takeover of the matriarchal society by opening the city gates to the invaders. In time, the invaders won, settled down and ruled in a new way. As the armed defenders of the matriarchies, the warriors became the de facto negotiators with their patriarchal conquerors. Male to male agreements were made.

A struggle between the temple and palace ensued, of centralized over decentralized institutions, of hierarchy over webs of relationship, of technology over nature, of oppression and conquest over cooperation. With the take-over of the grain storehouses, the male dominated kingships secured their hold on power.[23] The sacred function of the smithy declines and iron weapons are widely supplied to standing armies under the rule of kings or warlords. The mythology changes again. Now a male sky god wars with the mother god of earth, and defeats her. The early literate civilizations in Sumer, Egypt and in the Indus valley already reflect the victory of the invaders who assimilated the indigenous culture and changed it.

In the river valley civilizations, the trend is for a centralized authority to rule. An organized male leadership class, perhaps first a council of elders, then a chief, then priests and priestesses, nobles and kings, emerges to define the customs of the clan, the laws of the state, and the relationships to the gods.

With the priestly classes comes legally enforced religious calendars, primogeniture, laws of inheritance, codes and records. The early hieratic city-states institutionalize a concentrated male political and spiritual authority. And with this system, given good climate and surpluses, and inventiveness and ambition, commercial empires trading over thousands of miles build up. Poets celebrate war as an heroic human virtue.

## 174

New institutions and technologies come to depend increasingly on advance planning, permanent administration and transmitted orders. Clear lines of communication, built on technical language and laws, on the rational capacities of administrators, carry the orders. Arithmetic for commercial bookkeeping and geometry for surveying and building develop. Natural dominance gives way to a royal or priestly class whose holdings and hierarchies eventually lead to private property. Lawgivers become culture heroes.

In the male centered successor regimes, approach-separation and withdrawal-return, long risen from their embedded condition in dispersal-aggregation, and already matured into personal love and wisdom, fall increasingly under the shaping power of property and hierarchical dominance. They find their expression in marriage moieties, inheritance laws, castes, clans, economic classes, and ownership relations. Here, bound aggression definitively breaks from its ties with love and wisdom to support social order, material culture and competitive advantage in territoriality and social dominance. These interests are advanced ahead of and sometimes in defiance of personal, familial and household interests.

Love and wisdom henceforth function *inside of* property dominance. The leading classes put wisdom in the service of invention, discovery and statecraft. They use its accomplishments for the aggrandizement of state territory, property and status. Priests and kings rule.

## 175

Starting with the hieratic states, in my reinterpretation of feminist prehistory, the exploration of the possibilities in human nature, so prominent among the small-scale indigenous Epipaleolithic societies, narrows. With their populous cities, great monuments, earthworks, armies and bureaucracies, political control takes on new roles. Aggression, still tied to the ethological rhythms keyed to habitat resources, is further liberated.

In the early empires, aggression begins to be recruited by reason (and itself recruits reason) for new purposes, from fielding permanent armies to building pyramids, to enslaving whole populations, to collecting taxes. The sheer scale of the productions, and the organized labor needed to run them, have lasting impacts on the rhythms of life, subordinating them to a cultural calendar that less and less reflects the cycles of nature.

Since then, our ways of belonging have destabilized. The deployment of attention between inwardness and outwardness has been continuously revised. Love and wisdom have taken on new inflections in both the public and private spheres.

These shifts, whatever their content, typically effect one pole of the dynamic more than another. Approach is favored over separation, return over withdrawal or vice versa. The pole that is perturbed in the greatest number of people over the widest area and for the longest time shapes the course of historical change. After each of these changes, the stress/aggression threshold has to be reset. It has to function differently within each person, and between different segments of the society.

## 176

Written history begins with patriarchal rule. The pre-invasion indigenous people are nearly blotted out of the historical record, only to reappear in a shadowy way in the sacred texts of the successor regimes. Sarah the priestess with her own grove and acolytes becomes the wife of Abraham. Significant portions of humanity begin to live in a new world ruled over by distant patriarchal gods who are abstract, absent, not made of the same stuff as humans, unlike the Great Mother, whose children were consubstantial with her. We enter the world of separation. Power relationships predominate. Men rule over women. For the first time master/slave societies develop.

In our terms, in the early patriarchal civilizations, the dominance quilt takes the revolutionary step of redefining social distance regulation by pulling it away from actual encounters between individuals and framing it instead in terms of social, economic and gender-based classes. The kindled aggression that once sustained dominance in actual personal distance disputes, by rejecting the "quilt" of social distances in favor of the hierarchical "pyramid", takes on a more abstract but no less compelling order – a patriarchal system buttressed by rules, rewards and punishments. Property, birth order, social class and hierarchical status determine our possibilities and once it is codified in laws, it becomes the system by which love and wisdom are regulated.

By opening or foreclosing possibilities for love and wisdom, property dominance comes to redefine action, and the meaning of life and the worth of deeds. Henceforth wealth, class and access determine the times and spaces in which the different kinds of love can flourish. It says who can meet whom, which liaisons are permitted, which prohibited. It establishes kinship rules and the laws of inheritance, and wisdom loses its revolutionary potential too. It becomes esoteric. Freelance shamanic voyaging is marginalized in favor of an organized priesthood.[24]

In ancient Chinese culture, the Buddhist and Taoist sages become wanderers and recluses.

### 177

All large-scale societies require a dual strategy that both invigorates and suppresses natural dominance relationships. The state needs secondary institutions built on natural dominance to support its policies, its marriage and inheritance laws, and its belief systems. However, in large-scale venues, tensions build up between the abstract, law-driven dominance system and the flaring sparks of natural dominance. It is usually resolved in favor of large-scale property arrangements, protected by the laws, priesthood and police arm of the state.

But in every civilization, the balance is set differently. So emergent natural dominance still affects property relations, influences how business partnerships form and dissolve and how boards of directors manage their enterprises, etc. In large polities, natural community harnessed to existing property and economic relations becomes a basic tool for state building. But sometimes the balances shift. In turbulent times, natural dominance can flare up and struggle against the established power structure and even win, as in the velvet revolutions that overthrew the communist regimes in Eastern Europe in the 1990s.

### 178

Patriarchal civilizations may have won, as Jared Diamond and others define victory, but feminist Prehistorians make the case that in the victory human nature lost its sensitive growing tip. In *The Chalice and the Blade,* Riane Eisler definitely treats this historical transformation as a fall from grace. She makes gender conflict central to it.

"Now everywhere the men with the greatest power to destroy—the physically strongest, most insensitive, most brutal—rise to the top, as everywhere the social structure becomes more hierarchic and authoritarian. Women—who as a group are physically smaller and weaker than men, and who are most closely identified with the old view of power symbolized by the life-giving and sustaining chalice—are now gradually reduced to the status they are to hold hereafter: male-controlled technologies of production and reproduction."[25]

In her story, male domination engulfs history in all large polities, ending "a long period of peace and prosperity when our social, technological and cultural evolution moved upward: many thousands of years when all the basic technologies on which civilization is built were developed in societies that were not dominant, violent and hierarchical." [26]

## 179
### *Women: The Chalice and the Blade*

In exploring alternatives to the myth of material technological dominance, Eisler points out that "almost universally, those places where the great breakthroughs in material and social technology were made had one feature in common: the worship of the Goddess."[27]

With other scholars, she argues that the fundamental social forms in the Neolithic gardening communities were woman centered.[28] Goddess oriented civilizations ruled the transitional period from Neolithic to the ancient river valley civilizations, and women were given a cosmic dimension associated with the periodicity of the moon. The lunar, menstrual calendar reigned It ran on the sexigesmal system, based on the divisions of the circle into degrees, also a female symbol. The calendar, the garden, the grain and the hearth were presided over by priestesses. The woman became the shaper of the larger species rhythms. "The pantheon reflects a society dominated by the mother. The role of woman was not subject to that of a man, and much that was created between the inception of the Neolithic and the blossoming of the Minoan civilization was a result of that structure in which all resources of human nature, feminine and masculine, were utilized to the full as a creative force."[29]

Women are intermediaries between the worlds, and this feminine sensibility continues down into the Mesopotamian myth of the marriage of Innanna and her consort Dummuzi, that Campbell treats as a late version of an earlier planting culture myth.[30]

For Eisler, the high point of a gender equal "cooperator" civilization – and of happiness and creativity – was the Minoan civilization of ancient Crete, originating in 6000 BCE and attaining its final form between 2000-1500 BCE.

**Minoan "Parisians" from Knossos palace**

Crete achieved "…a social order in which, to quote Nicolas Platon, 'the fear of death was almost obliterated by the ubiquitous joy of living.'"[31]

Platon, Superintendant of Antiquities in Heraklion, Crete, described

"…vast multi-storied palaces, villas, farmsteads, districts of populous and well-organized cities, harbor installations, networks of roads crossing the island from end to end, organized places of worship and planned burial grounds… the whole of life was pervaded by an ardent faith in the goddess Nature, the source of all creation and harmony. This led to a love of peace, a horror of tyranny, and a respect for the law. Even among the ruling classes personal ambition seems to have been unknown; nowhere do we find the name of an author attached to a work of art nor the record of the deeds of a ruler." [32]

In the feminist telling of prehistory, the Minoan civilization stood for something particularly hopeful in its ways of being human, of loving, raising families, displaying kinship, expressing gender relationships, developing rites of passage, growing wise, and tempering aggression. By dealing with gender, the feminist pre-historians left in their accounts what others excluded: love and wisdom.

Eisler may have been overly enthusiastic in her belief in gender-equal past civilizations, but surely she was right in thinking that gender has always been a central factor in social and cultural organization and has to be carefully considered.

## 180
### *Summary: The Five Epochs of Human Development*

o  In the first stage of its evolution, human nature is <u>embedded</u> in the natural environment. The interface between person and world is immediately responsive to signals from nature; aggression is rhythm conserving; belonging is focused on kin and troop; life is nomadic, and the body is most strongly attuned to the daily dispersal and aggregation rhythm.

o  In the second stage, beginning in the middle Paleolithic era (50-25,000 BCE), human nature becomes <u>emergent</u>; primordial approach-and-separation and withdrawal-and-return start to be experienced as love and wisdom; we become conscious of our inwardness and can distinguish it from outwardness. Endowed with a sense of self and a capacity to love, the standards for belonging change. When it does, the old distance regulation system breaks down. Aggression, no longer so tightly linked to the management of actual changes in moment-to-moment physical position, begins to explore other kinds of personal space and domination. In the course of time notions of property, ownership, status and rule develop. As they do, the ties between aggression and the conservation of biological rhythms weaken.

o  In the third stage, (20-10,000 BCE) human nature is <u>expressive</u>. Enlivened by the developing passions of the heart and mind, stimulated by the discoveries made

223

in turning points, it produces vastly different ways of being in the world. A torrent of new inventions, both material and non-material, comes pouring into the shared cultural spaces. These institutions, technologies and physical artifacts transform the environment. Their sensory worlds altered, people begin to see and sense differently. Time runs differently for them. Body attitudes change, set-points shift, sexual activity and fertility rises or falls. The duration of life stages alters. Youth is extended or contracted. This expressive phase begins at the end of the last Ice Age. Rapidly changing environmental conditions spur it on. The expressive period continues into the Neolithic era. New political forms, social contracts, creative endeavors and ecological understandings develop. Earth-altering changes are made: settled life, continuously inhabited villages, division of labor, crop culture, trade and commerce among them. During a 10,000-year span, the transformation brought in by the agricultural revolution spread through Africa and across the Middle East, north into Anatolia, west across North Africa and east into Central Asia.

o    Then under pressures introduced by these changes, our orientation shifts again. In this fourth stage (5000 BCE – 1800 AC) human nature becomes <u>conflicted</u>, constrained by culture. Human nature has to struggle for breathing room. Life is troubled, split and turned against itself. Freud caught the tenor of this tension in his late work. In *New Introductory Lectures,* he wrote, "In spite of all our pride in our cultural attainments, it is not easy for us to fulfill the requirements of this civilization or to feel comfortable in it, because the instinctual restrictions imposed on us constitute a heavy psychical burden. Well, what we have come to see about the sexual instincts, applies equally and perhaps still more to the other ones, the aggressive instincts. It is they above all that make human communal life difficult and threaten its survival."[33] In *Civilization and its Discontents,* he wrote, "Civilization has to use its utmost efforts in order to set limits to man's aggressive instincts and to hold the manifestations of them in check by psychical reaction-formations... In spite of every effort, these endeavors of civilization have not so far achieved very much "[34]

In the fourth epoch the aggressive components of human nature take on a new role: they shoe-horn us into civilization itself, not only by using the external powers of the state but by teaching us to manage ourselves from within with repression, guilt and self punishment.

o The fifth epoch is just beginning. We inhabit it. The fifth epoch is our field of endeavor. It started with the Industrial Revolution in the 18th and 19th century. In it we follow changed conditions of labor and economic regulation. We live with greatly modified signal sources for sensation, coming more frequently from technological rhythms than from nature. These reset the rhythms of life, not always homeostatically.

The fifth epoch gathered steam during the period of European colonialism and imperialism. It pushed us into two World Wars and into the atomic age, under the influence of what President Eisenhower called the Military Industrial Complex. Fifth epoch enterprise now drives our information revolution, computerized economy, biotechnology, global trade and communications networks. It creates needs, threats and promises uniquely its own. What shall we call the fifth epoch? We shall decide this in the next chapter.

Here we'll concluded by drawing from our brief survey three features common to all revolutions in human nature:

o Real revolutions result in altered distance regulation behaviors.
o These change social arrangements (and the physiological responses to them) in one or more of the four social distances.
o The changes enter economic life in altered territorial and dominance relationships.

\* \* \*

## End Notes

[1] Most of what we needed we already had. Tool use went back 2 million years, fire was widely used 400, 000 years ago. We were omnivorous, had a bipedal gait, an upright stance and stereoscopic vision. We walked long distances with hands free to carry children and manipulate tools. Our infants were born immature and required years of care and supervision.

By the beginning of the last Ice Age, the Wurmm glaciation (100,000 BCE.), considerable evidence of domesticity is found, fire rings, huts, retaining walls, clothing, fur rugs on the floors.

[2] Thompson. *The Time Falling Bodies Take to Light*. St. Martins. 1981. p.156

[3] Hrdy. *The Woman That Never Evolved*. Harvard U. 1981. p.142

[4] I first came upon the term "state dependent memory" in the seventies in Charles Tart's book *States of Consciousness*. "Memory is specific," he wrote. "The way in which information is stored, or the kind of Memory it is stored in, is specific to the d-SOC [discrete state of consciousness] the material was learned in. The material may be stored but may not be transferable to another state."[4]

Tart supported this view on the research of Goodwin and Powell. Along with Robert Ornstein, David Galin, Arthur Deikman and others, he set out to develop a program of "state specific sciences," saying "...I think it is probable that state-specific sciences can be developed for such d-ASCs [discrete altered states of consciousness] as autohypnosis, meditative states, lucid dreaming, marijuana intoxication..."[4]

[5] Where Jung saw individuation as a process involving the progressive emergence into consciousness of archetypal characters, I see it in the development of archetypal patterns of action that enter consciousness and lead in their highest form to convergent love and wisdom. In my rendering of the emergence, convergent love and wisdom produce the wider self that Jung celebrated. But the deeds produce the person, not vice versa.

[6] William Calvin, explains that "...by somewhere between 175,000 and 125,000 years ago, there were anatomically modern *Homo sapiens* around Africa. They were in the Le-

vant by 100,000 years ago (not surprising as the flora and fauna were often African). Somewhere before 50,000 years ago that spread out into the rest of Asia (and soon thereafter into Europe and Australia) in the most recent Out of Africa expansion.

[7] Similar cultural complexes have been unearthed as far away as Mexico and in the Yangtze River valley in China. refs

[8] Ofer Ben Yosef. "The Natufian Culture in the Levant, threshold to the origins of agriculture." in *International Monographs in Prehistory*, 1991

[9] Fagan, basing his narration on the excavations of British Archeologist Andrew Moore in the 1970's (on an 11.5 hectare dig in present day Syria along the Euphrates river) and on research by Gordon Hillman on the flora there, recounts the bounty with which the Mesolithic Abu Hureyra people lived during this time: "All these food sources—gazelle migrations, spring grass harvests and the bounty of nuts in the fall—gave the Abu Hureyra people a relatively predictable diet, an interlocking set of easily storable foods that allowed them to inhabit the same location for generations... A constellation of unique circumstances brought a relatively small number of hunter-gather bands like that at Abu Hureyra to an entirely new relationship with their environment and with one another. For the first time, people lived cheek-by-jowl in crowded settlements, not for a few weeks but generation after generation... Relationships between families, between kin, between young and old, became infinitely more complex. So did the spiritual relationships of people with their land, and with the oak groves, pistachio trees, band grass stands..." Fagan. The Long Summer. p. 87 -88

[10] ibid p. 108

[11] ibid p. 92

[12] Joseph Campbell called the mythology of the Paleolithic hunters "Kama" and the Neolithic planter mythology "Artha", terms he took from the Indus valley civilization. Kama represents the shaman's journey to the Animal Master for renewal of the Hunt. It is a myth from the withdrawal-return universe, depicting primal wisdom processes. Aretha emphasizes the annual renewal of the vegetation. It describes fertility as an essentially female energy. It tells the approach-separation story, in terms of seasons of renewal, in a world run on conjugal relations and mother love overseen by a Goddess.[12]
The Neolithic myths disclose different parts of human nature. "These suggest," Campbell explains, "that the obvious anal-

ogy of woman's life-giving and nourishing powers with those of the earth must have already led man to associate fertile womanhood with an idea of the motherhood of nature."[12] The shamanic form of the journey to the underworld is transformed, feminized. A goddess, Innana, Isis, Astarte, Demeter goes on a journey to the underworld to rescue a kidnapped sister or daughter where a reunion between mother and daughter occurs.

The myths follow the seasons. The Eleusinian mysteries carry traces of this transitional time into classical Greek civilization. Persephone is brought up to live six months in the light of day, and then returns to the underworld to live six months with the lord of the dead in darkness. Renewal through the gate of the female replaces the encounter with the Animal Masters. The turning point is celebrated as the mystery, the moment when the seed sprouts.

[13] Wright. NonZero. Vintage Books. 2001. P.322

[14] Francis Fukuyama. *Our Posthuman Future.* Farrar Strauss. 2002. p. 126)

[15] Montaigne. Of Democritus and Heraclitus. I.50. Donald Frame translation.

[16] Donald Frame. *Montaigne, A Biography* p. 182-4. Freud counseled this too. Phillip Reiff, in *Freud, The Mind of the Moralist*, p. 78, describes his view as follows: "The kind of intelligence the ego would ideally manifest is modest and practical, not prideful and aspiring. Equipped with a Freudian sagacity about ourselves, we do not hope to abolish the constraints from which we suffer; such hopes are moralizing and utopian, not rational and realistic." Reiff p. 78

[17] Ludwig Wittgenstein. Philosophical Investigations. P. 227

[18] From *The Fog of War*, a documentary film.

[19] Jared Diamond. *Guns, Germs and Steel*. W. W. Norton. 1997. p. 405-407

[20] When we look at the anthropological accounts of small societies a whole range of variations opens up. One can argue that tribal people live fuller lives with wider access to themselves than "civilized" people do, The lives of small societies span a larger arc of cultural diversity than do the lives of empires. The contrasts are fuller between the fierce Yanomano and the gentle Tasaday, than between Russians and Americans, between the Samoans and the Kwakiutl than between Frenchmen and Germans. Consequently we can learn more about our future potentialities from them.

[21] ibid p. 254

[22] Nor will the argument for a "nonzero sum society" work. It can be argued that reason in alliance with the technical mindset produces less, not more trust, because it blocks the routes of expression for love and wisdom as *Brave New World* shows. Without a foundation in moral sentiment, enlightened self-interest isn't elevating and isn't even self-interest. But these are the knd of justifications the winners provide for themselves.

[23] Oppenheim. Ancient Mesopotamia. p.117

[24] Campbell reports on the Mesoamerican version of the suppression this way: "The binding of the shamans by the Hactin, by the gods and their priests, which commenced with the victory of the Neolithic over the Paleolithic way of life, may perhaps be already terminating—today." Campbell. p. 281

[25] Eisler. p. 53

[26] ibid. p. xvi-xvii

[27] Riane Eisler. *The Chalice and the Blade.* Harper and Row. 1987. p. 8

[28] Marija Gimbutas, focusing her researches in Old Europe, describes it this way: "The Neolithic 'virgin' is almost as corpulent as the Paleolithic 'Venus', particularly in central Anatolia and around the Aegean. Typical seventh-millennium sculptures from Catal Huyuk in central Anatolia take the form of a massively fat woman, either standing or seated, supported by leopards. She usually either holds her hands up to her large breasts or rests them on the heads of accompanying animals...Throughout the Neolithic period her head is phallus-shaped suggesting her androgynous nature."

[28] Marija Gimbutas. *The Goddesses and Gods of Old Europe.* U. California Press. 1974. P. 152. The archeological remains of Catal Huyuk (7000-5000bce), excavated in Turkey by James Mellaart, uncovered thousands of Goddess figures. According to Mellaart, female sexuality and fertility are celebrated as high symbols of unity here and throughout the Neolithic world. Gimbutas drawing from her studies concludes that "the task of sustaining life was he dominant motif in the mythical imagery of Old Europe, hence regeneration was one of the foremost manifestations. Naturally, the goddess who was responsible for the transformation from death to life became the central figure in the pantheon of gods..."

[29] Eisler. ibid P. 236-238)

[30] Campbell. *Primitive Mythology.* p.143

[31] Eisler. p. 32.

[32] Nicolas Platon. *Crete.* Nagel Publishers. 1966. p. 15.

[33] Freud. *New Introductory Lectures* p.110
[34] Freud. *Civilization and its Discontents* p. 59g

# 10. OUR TIMES, GOOD TIMES AND TIMES TO COME

## 181

The Romantic poets were the first to recognize the fifth epoch's moral crisis. They saw it rooted in the transforming power of the Industrial Revolution. William Blake was the toughest minded of them. He etched many of his poems in acid on copper plate. In the *Four Zoas* Blake wrote,

*"And all the Arts of Life they chang'd into the Arts of Death in Albion.*
*The hour-glass contemn'd because its simple workmanship*
*Was like the workmanship of the plowman, & the water-wheel*
*That raises water into cisterns, broken & burn'd with fire*
*Because its workmanship was like the workmanship of the shepherd;*
*And in their stead, intricate wheels invented, wheel without wheel,*
*To perplex youth in their outgoings & to bind to labours in Albion*
*Of day & night the myriads of eternity: that they may grind*
*And polish brass & iron hour after hour, laborious task,*
*Kept ignorant of its use; that they might spend the days of wisdom*
*In sorrowful drudgery to obtain a scanty pittance of bread,*
*In ignorance to view a small portion & think that All,*
*And call it Demonstration, blind to all the simple rules of life.*[1]

The Industrial Revolution from its start, and in its extension in the Information Revolution, has favored outwardness over inwardness. It trains people in exteroceptive sensation, emphasizing the distance senses of sight and hearing. It teaches us to rely on outward measures of progress. The balance between inward and outward sensation tells us how we fit into the world. It gives us crucial information about whether we are alone or in company, intensely involved with another person or only casually in touch, and all shades in-between. It tells us whether time goes slow or fast, whether space is open or closed, opportunities nearby or far away, hope internally generated or dependent on outside sources, truth universal or consensual. It supports our notions of whether dream-time is real or illusory, whether the dead are present or absent in our daily affairs, whether causation is material or spiritual or both simultaneously. Kierkegaard warned us that a preponderance of outwardness drains the self of meaning.

"On account of our vastly increased knowledge, men had forgotten what it means to exist, and what inwardness signifies…it does not take a parade of millions, or of generations of men; it does not take humanity in the lump, any more than the police arrest humanity at large. The ethical is concerned with particular human beings, and with each and every one of them by himself." [2]

The outward orientation, with its narrow focus on utility, strengthens the technical mind at the expense of the ethical mind – the ethical mind Kierkegaard found only in inwardness. Outward-ness situates us in a linear, expansive, accelerating environment whose major turns are determined by material, scientific and tech-nological changes. Quick turnaround time, rapid job turnover, on-demand inventory management, worldwide instant credit approv-als, faster business exchanges, electronic voting, quicker hand-eye coordination, special effects and rapid cutting in the entertainment media, shape the current zeitgeist. Market share has to increase, markets have to expand, and productivity has to improve. Every-thing has to "grow or die". A steady state suggests stagnation, a kind of mitochondrial decline, a sign of energy death. Speed is the happy essence of things, the savor of life. Corporate public rela-tions firms reframe the maddening pace as a desirable characteris-tic.

By contrast, the ethical mind is slow in its ponderings. It grows stronger with patient practice. It needs time to consider. Rushing impedes it. The ethical mind supports that part of the non-material culture concerned with making choices. Its inwardness exposes the character of conscience to solitary reflection. It upholds the cul-tural standards for shame and guilt. It shows the way dream states are valued, the roles contemplation and meditation play in decision-making. It can be discerned in how we make and keep promises, and how we cope with pleasure and pain. It maintains beliefs about the connection of the soul to the body, including its fate after death.

## 182

Our estrangement intensifies with the pace of technical progress. The combined effects of industrial, electronic, informational and media advances turn times and seasons upside down. Work days, holidays, transportation schedules, shopping hours, professional, legal and financial hours dominate the calendar. All these arbitrary schedules reconceived and digitized, timed by TV programs, smoothed by timed-release medications, authorized by commercially dictated shopping seasons, tracked by the interlinked stock markets of the world, impose new rhythms and new kinds of turbulence on us and in us. In many instances, the events of our lives turn from full off to full on instantaneously. Time doesn't flow.

The stress/aggression thresholds, whose set points were retarded during the third *expansive* epoch and advanced in the fourth, have been thrown into overdrive in the fifth. Aggressivity has become a central pillar of business strategy. Scientific research, art and entertainment show it. Who can dispute that ambition linked to aggressive energy directed rationally has improved the human prospect? However, who would be willing to argue from where we sit now that modern science and technology in the hands of industry has been good for nature?

## 183

Jerry Mander grappled with these issues in *Four Arguments Against Television*. He put his finger on the trance-inducing characteristics of the flicker rate on TV screens. And television just touches one aspect of the syndrome. The broader situation is that in trying to adjust to the digitized jitter of events, we keep trying to redefine what constitutes an interruption. Do the millisecond gaps in digital music recordings interrupt the flow? Do we notice them or not? Do the down-to-the-second starts and endings of TV programs bend our expectations to fit the hour?

We leap over time zones. Nothing and everything become interruptions. The sensory challenges bring on neural changes that reset the conditions under which we feel stress, sometimes torpifying us, sometimes setting us on hair-triggers.

## 184

My fear is that unless we can reduce the hold of vigilant outwardness, we will become permanently glued to the sensory surface of a superficially fascinating civilization lacking in depth. We will use ourselves up dealing with intrusive cultural stimulation, much of it subliminal. Mander expanded his analysis of our dilemma in *The Absence of the Sacred.* He wrote,

"Our entire society has begun to suffer the madness of the astronaut; uprooted, floating in space, encased in our metal worlds, with automated systems neatly at hand, communicating mainly with machines, following machine logic, disconnected from the earth and all organic reality, without contact with a multidimensional, biologically diverse world and with the nuances of world views entirely unlike our own, unable to view ourselves from another perspective, we are alienated to the nth degree." [3]

The human body and the whole biosphere have been increasingly subjected to fundamental frequency, amplitude and phase resettings coming from the industrialization of the sensory surround by the operations of the technical mind. These changes are occurring on a scale not seen since the Agricultural Revolution, but much faster now, and everywhere at once, because our political systems, economies, transportation and communications networks are linked, and our needs and shared crises are shaking all of us up. Right now, our labor/capital/ entrepreneurial choices, distorted by the fundamental biological rhythm resetting problems whipping around us, are precipitating dangerous planetary imbalances.

## 185

These imbalances are bringing the transformational crisis of the fifth epoch to a head. From the deep places where the body meets history, new symptoms of distress and fresh opportunities are rising. They pull us back and forth between the technical and the ethical mindsets. We use the technical mind to devise ways to get what we want. But what we want, what we *choose to value* comes from the ethical mind.

To play variations on the theme:

o The technical mind thinks in terms of ways and means, the ethical mind thinks in terms of hopes, values and responsibilities. The technical mind builds the vehicle that gets us where we are going, the ethical mind decides where we want to go and what to do when we arrive.

o The technical mind asks the question "will it work?" and defines "work" in many ways: personal pleasure, power, recognition, psychic gain, material benefit, enhanced personal safety, strengthened family and community ties, public service, and more. But its work ethic also applies to vengeance, destruction, torture and war.

o As a habit of thought, it lists, categorizes, analyzes, plans, builds, pulls apart, theorizes, tests, improves, accumulates data, organizes facts into categories and makes decisions. The ethical mind, on the other hand, tends not to work in a vast variety of brilliant ways. Instead, it looks at our plans, and it worries whether we are making the best choices regarding the things we care about most.

o Where the technical mind is built for speed, and wants to go faster, the ethical mind resists acceleration. Thoreau saw the contrast in the railroad running through Concord. "It moves too fast. Men think that it is essential that the *Nation* have commerce, and export ice, and talk through a telegraph, and ride thirty miles an hour, without a doubt, whether *they* do or not; but whether we should live like baboons or like men, is a little uncertain."[4]

o Confucius, in the China of 500 B.C., told his students "The superior man understands what is right; the inferior man understands what will sell. The superior man loves his soul; the inferior man loves his property. The superior man always remembers how he was punished for his mistakes; the inferior man always remembers

235

what presents he got." In a more subtle observation, he added, "The superior man is easy to serve, but difficult to please, for he can be pleased by what is right, and he uses men according to their individual abilities. The inferior man is difficult to serve, but easy to please, for you can please him (by catering to his weaknesses) without necessarily being right, and when he comes to using men he demands perfection."[5]

The hard, frequent and abrupt swings between our technical problems and opportunities and our ethical fears and hopes are wearing us down. We may be entering a culture-wide version of the exhaustion stage of Hans Selye's General Adaptation Syndrome (see # **148**).

To use a medical analogy, we are suffering from an essential tremor in our grip on life. We are afflicted with spasmodic movements between crass outwardness and inept inwardness in the frequency bands I identified earlier. These relentless swings challenge our ethical and technical ideations in ways I shall soon explain. So far, we have only recognized the chronobiological face of these problems in a coarse-grained way: as relatively trivial circadian clock disorders coming from shift work and jet lag. But far more serious and insidious problems are occurring in other frequency ranges. They are particularly troubling in the fast ultradian bands on which our nervous systems run, where they interfere with mirror neuron functioning and degrade our capacities for human empathy in a spectrum of disorders extending from atomic individualism to full-blown autism (I explain this below in # **203-204.**)

**186**

I frame the distinctive character of the modern dilemma this way: The world created by the triumphs of the technical mind obstructs love and wisdom by skewing the alternations between inwardness and outwardness that we need to travel the legs of approach/separation and withdrawal/return to our turning points unimpeded.

In the wisdom dynamic, our culture admires the heroic qualities of the return leg while disparaging withdrawal. We support the seeker when he comes back, not when he goes away. We forget that the struggle in the depths of withdrawal shows the real nobility of the human aspiration for truth much more clearly than does the socially acceptable successful return.

In love too we are biased by outwardness. By sentimentalizing intimacy, we lose the dynamic reality of relationships as events in process. We favor continuous false approach over alternating approach and separation. And this makes us identify attachment with relationship as such, as if a relationship depended on a fixed emotional distance between the partners. Moreover, some of us, through an overabundance of saccharine sentimentality, unwittingly tune out empathy and sympathy. We are as likely to feel the same pity for a forlorn puppy we do for a homeless person. And we don't have to act on either. It's enough to savor the feeling.

## 187

Though we can remake the world through the power of outwardness, we can only transform *ourselves* through inwardness. Only in inwardness do the crucibles of metabolic change open to volitional shaping in turning point dramas. On this shaping power we rely for the aptness of our feelings, drives, motivations and ideations. Only in freedom, in our fluid form, can we recast ourselves, reach out and act in new ways.

## 188

Arnold Toynbee insisted that in our times the technical and ethical mindsets had moved decisively into opposition. The battle went on everywhere, inside our offices, schools, courts, deliberative bodies, businesses and laboratories, in our beds, in our marriages, with our children, and inside our heads. Back and forth we bounce between the technical and ethical minds. "... the two incompatible states of mind and standards of conduct are to be seen today, side by side, not merely in the same world, but sometimes in the same country and even in the same soul."[6] It was undermining our civilization: "...Western man's technological control over non-human nature—his stupendous progress in 'know-how... has created enormous problems that have brought our civilization to its 'time of troubles'."[7]

According to him, our propensities for creation and destruction had not changed. We don't build up and tear down with new motivations, we just do it better (or worse) because science and technology, the boldest gifts of the technical mind, have handed us a ferocious power to pursue our inclinations. This power, he noted, is "just what gave our fathers the confidence to delude themselves into imagining that, for them, history was comfortably over." But they were wrong. Instead by "these triumphs of clockwork the Western middle class has... set Juggernaut's car rolling on again with a vengeance."[7]

## 189
### *Our Four-Part Illness Syndrome*

I identify four endemic cultural pathologies. I call them successomania, consumerism, addiction and juvenile arrest. They break the rhythms of nature, get in the way of our turning points and corrupt our values. The double bind is that after blocking the way to real love and wisdom, they palliate for the losses with addictive commercial substitutions. A brief description of the pathologies follows.

**1. *Successomania.*** In the absence of any other ethic that really holds, and lacking strength of character that keeps us sensitive, gentle and intact, we make success the chief aim of life. Success is measured outwardly because we do not have the empathy or imagination to get to each other's inwardness. Because of our outwardness, we rely on appearances. We gauge success by competition. We want to be winners—or at least look like winners. And winning is measured in salaries, property, esteem, access and membership, but not in happiness because happiness is an internal condition and cannot be quantified and scored competitively. Wendell Berry, Kentucky farmer, philosopher and environmentalist wrote in 1990,"The ideal of competition always implies, and in fact requires, that any community must be divided into a class of winners and a class of losers." [8]

Money has become the measure of all things and many of us pursue it devotedly, with religious fervor. We worship money. Money is our God. Money is our lingua franca. With it we work transubstantiation: we exchange it for anything. It is the universal solvent. It makes everything possible. Success with money lets us conjure up the feeling of belonging out of thin air. We can buy it. Our purchase of *memberships* reinforces our sense of place and gives us status. It brings with it more than the *illusion* of access and position. Big spending brings the *reality.* Making money is our real territory, our central dominance contest. Our prestige, our protection against the accelerating intensity of change comes with it, an acceleration supercharged by the electronic money exchanges via satellite uplinks and computer networks, that instantly transfer assets while maintaining the illusion of a closing date two days hence, a legal "kiting" of funds producing as yet unknown effects in the planetary financial crisis.

**2.** *Consumerism.* We base our sense of identity increasingly on the products we buy. Our suppliers and intermediaries see us as "markets" and "demographics". Executives decree the fashion seasons. Fads come and go and our corporate suppliers very deftly create needs where none were before, with the aggressive motive of improving their bottom line. They manage the ritual calendar and drive us into seasonal buying frenzies that falsify all celebrations and turn holidays into twisted, Disneyfied buying opportunities no longer performed to align ourselves with nature or seasons or stages of life. With this aggressive corralling of consumption, we get the impression that our options are always increasing, while what is really increasing is our range of petty choices. The more circumstances overmaster us, the more trivial and more cosmetic our choices become. It is as if we are driven to find our freedom somewhere, no matter in how inconsequential a place. And we're likely to let ourselves be persuaded of anything for a while. The fad itself is an acknowledgment of our existential deadness. So we buy useless products and tantalize ourselves with empty hopes.

But when freedom is pushed out, the real drama of life fades and we piddle away our energy for intensification and reversal. Who we are becomes conflated with what we buy, and when that happens, we lose the transformational power to make our future different from our past—to make life memorable. Instead, we take an inflated view of character. Personality is everything. At first glance, it makes sense to see the capacity for action coming from strong character, until you realize the compensatory grandiosity in it. The conceit is that to live fully you need to live like a character in a soap opera. It makes us 'celebrity conscious'. We want to believe that great deeds come from great people and great people are rich, beautiful, and famous. They're "somebodies" not "nobodies." How long can life on the shopping mall keep its fascination?

At bottom, the consumption game depends on our poor judgment. Corporate management and governmental bureaucracies encourage it. But at the same time, they do not want their board members to be afflicted by it. Nevertheless, they are. The acceleration and grandiosity that go with corporate growth predispose the most ambitious executives to consumption manias of their own, sometimes directed to financial instruments, derivatives and leveraged buyouts. They fly like Icarus on the poor judgment Daedalus decried.

**3. *Addiction*.** Looked at closely, advertisers represent all the products we buy (and not just psychotropic medications) as shortcuts to states of mind. They are promoted in such a way as to encourage dependency, and this is done with clear aggressive intent. A few years ago, advertising campaigns on TV extended the allergy season to all year round. This year they're pushing "new allergies" and new diseases like restless leg syndrome. To make us believe we are sick when we are not has malicious motivation behind it.

We're told we need these products. We are taught to feel that our lives will be deprived, deficient, and empty without them. "Feel empty?" Now you know why, you lack something attainable. Just get the money. Buy it on your credit card. Significant sectors of our economy are held together by the pursuit of substitute gratifications for experiences missing from our lives.

But substitute gratification is a principal root of addiction. When you cannot feel good in real life, you get to feel good from a little pill. The pill substitutes for the real life changes. Behavioral addictions work the same way. Gambling, shopping, excessive aerobic exercise, compulsive pornography aren't substance based, but they can be just as addictive when used to substitute for gratifications no longer coming from the main thrust of real life. Atlantic City, Las Vegas, Indian gaming, the state lotteries are all run with aggressive if not actual criminal intent.

Behavioral addictions sit at the root of consumerism. They do not just create "psychological dependencies," they grab us physiologically. Just as with a substance addiction to alcohol, in behavioral addictions driven by marketing, when you buy the product the body reacts. The pleasure of the substitute provides some of the biochemical reinforcers of the genuine experience; otherwise, they would not work as substitutes. Once we have the substitutes, whose neurochemical components bind to our natural receptor sites, we stop producing the endogenous biochemicals. Downhill skiing provides the endorphins for heroic accomplishment. Sex on Ecstasy gives the body responses of passion and intimacy. Soccer riots give fans a primitive battlefield experience. The stress of workaholic behavior alters our sugar metabolism. Sugar craving is itself an addictive urge perhaps built into the infant's attachment to the sweetness of the mother's milk, standing for nurturance and security and favored by evolutionary selection. So behavioral addictions driven by fashion and consumption cause the craving, withdrawal, dependency and isolation of substance addiction.

For most of us, the body becomes conditioned to the substitutes and this further pulls the rhythmicity out of us. We develop tolerance and need higher doses to get high. We become obese. The induced disruptions to blood sugar regulation eventually make us diabetic. Our pursuit of success itself plunges us into acceleration and acceleration has its own physiological gratifications; it turns the body on. When unchecked by real intimacy and real integrity, the convergence of addictive substitution with the acceleration intrinsic to modern society is drawing us into a manic addictive dependency on the acceleration itself. The "career" is an exercise in momentum; we go "careening" about. The accelerated pace of life is to some extent supported and even amplified by the addictive rush to activity itself, prodded on by secondary reinforcers, things to buy and consume, the focused attention on gain and loss. Buying and selling are our most widely available substitute gratifications. This makes addiction one of the linchpins holding together the economic system. Behavioral and substance addictions substitute for real joy in life.

Of the four pathologies, behavioral and substance addictions exert the most direct visceral pull. Breaking addictions requires one to turn aggression against the self for the sake of the self. It invokes a higher, reformed aggression very much at odds with the aggression encouraged by capitalist competition and greed. One can only learn this from inwardness in the face of suffering.

Getting clean opens the way to a life of action for some people. It redirects efforts from the substitutes to the real meanings. And concern for meaning often leads to a confrontation with consumerism, successomania (and juvenile arrest, we shall see.) One begins to take back responsibility from the experts; one seeks for real satisfaction and rejects the purchasable outer signs of it as a sham. With this resolution, we can get back our transformational powers.

**4. *Juvenile Arrest.*** Despite the emptiness, in fact because of it, we cling tightly to the superficial aspects of personality. Instead of deep penetration, paradox, and uncertainty, we cultivate a cheery indifference. We seek false youth, and crave a life of permanent child's play. But the play is not childlike. It is childish, phony, pre-structured, commercialized, and blocked from its creative possibilities (which probably depend on our access to random generators in the nervous system that open us to the creative depth of play, even in the face of rough perturbations.) It takes rare fortitude to carry real childlike exuberance into adult life while retaining a sense of responsibility for the consequences of our actions. Our long training in false play has taken the power of playfulness away from us. We're not tested or called to account for ourselves. We accept the appearance of boldness without the real experience of courage. Our antics are full of busy make-work that is empty of substance. We jump from this to that. We want a Humvee. Where virtuous play is full of risk and freedom, juvenilized play spends its energy in a busy flight from responsibility. Lewis Mumford described it this way:

"With the successive demands of the outside world so frequent and so imperative, without any respect to their real importance, the inner world becomes progressively meager and formless: instead of active selection there is passive absorption ending in a state happily described by Victor Branford as 'addled subjectivity'."[9]

The addled subjectivity of juvenilized play is doubly regressive. It prevents us from knowing what we are doing, and at the same time encourages us to believe that somewhere out there *real grown-ups* are watching over us (even though, like children, we reserve the right to resist and refuse their parental restrictions.) But where are the grown-ups? Are there any? I've heard this absence of mature judgment called "rule by boys." The main thing to succeed in is success, not in competence, not in recognizing merit in others, not in participatory democracy. The training is in self-aggrandizement smoothed over with reciprocal altruism. We do not want our fears of abandonment awakened. We too are waiting for the real parents to return.

Perhaps that is why so often I see well-meaning people who live with integrity on the long legs fade and fail at turning points. A certain brightness and attunement to the moment is missing in them. The presence and joy in life is clouded over. Some nerve fails, some fear supervenes, some blockage keeps them from dealing with life. They're obtuse. They reject self-transformation except as it is pushed upon them by aging and circumstances. They lose hold of the adventure of life and its turns of fortune, and memorable events elude them and instead they cling to false adventures and photo opportunities and empty meaning. They become Pinteresque characters. Their best intentions come into the world detached from action. They loose their depth of feeling, become sentimental, and live a Soap Opera reality where everything happens but nothing changes, and genuine intimacy is stylized and artificial and love and wisdom fail.

It is a juvenile conceit to think that we grow by accretion, sequentially, in a linear way. Mature people know that achievement does not come from a unidirectional climb up the ladder of success. The desired direction is not always upward, forward and ahead. Sometimes it is backwards and down.

The leading edge of juvenile arrest is a literal quest to end biological aging and extend the period of youthful vigor indefinitely. The juvenile mindset finds the natural stages of life repellent. Baggy eyes, wrinkles, parchment skin, hanging fat no longer reveal character, as if life had no worthwhile tale to tell on the body. The *reductio ad absurdum* of this goes deeper than cosmetic surgery. It tries to fund and redesign the medical establishment so that it affirms that our real quest is not for the meaning of life or for excellence in living it, but for its preservation and extension, for perpetual youth. We have generated a vast industry into which the ill enter in a state of passive patienthood to do battle with death as an adversary, a strategy that focuses on the fact rather than the content of life. The struggle against illness has substituted for the struggle for meaning.

We cling to the statistics that show that our life spans are increasing, but who is the "us"? Have we factored in the whole planetary population and taken account of the scores of millions of people who die by war and hunger and preventable diseases in childhood? How do the statistics read then? What if we include the health and survival of the animal and plant species we are driving to extinction?

The consequence of juvenile arrest is to make our leaders, who are themselves immature twits, treat us like idiots. You buy a ladder. It's plastered over with cautions to protect the manufacturers from lawsuits: "these are the rungs; this is the way up, how not to climb a ladder, never climb higher than the next to last rung." It's an idiocy that produces a life in which the learning period for competence is indefinitely postponed. Competency belongs to experts alone, and only in their specialized field.

Obviously, a meaningful life requires a cultural world that invites rather than inhibits participation, which encourages not passivity and renunciation, but mature commitments and meaningful action. Rather than treating each other as sentient beings whose inner needs, including the needs for competence and responsibility, are puzzling and important, we have let ourselves become statistical entities. We're polled, studied, observed, questioned and manipulated. We build a world out of data and think we live in it somehow as discontinuous, digitized, screen-sized, pixilated info-packets. We even try to live up to our 'demographic'. Moreover, with the mountains of data swirling around us come tribes of specialists to explain us to ourselves, to interpret and manipulate the model and make sense of the statistical tangles they create. But the emphasis on minutiae and detailed quantification reified by expert analysis and interpretation only keeps us from appreciating each other for who we are or who we yearn to become. We hardly know how to judge our deeds or gauge their excellence. The love one can know only by loving we replace with an outward, tendentious understanding of the evolution of reciprocal altruism. And the inwardness of wisdom with its moving drama of withdrawal and return we turn into evolved intelligence based on increased brainpower. And aggression becomes overcrowding, territoriality and dominance. And all of these remain curiously detached from each other in their own specialized niches, guarded over by a priesthood of experts.

# 190
## *Brave New World*

No one has dealt with the illness syndrome more imaginatively that Aldous Huxley in *Brave New World*. If you substitute genetic engineering for the bottled babies in his birthing labs and replace Soma with the full range of mood altering drugs in our pharmacopeia, his forecasts seem prescient.

Huxley's most important insight was that juvenile arrest debases us by banishing real ethical choice from our lives. People living in The Brave New World were so devoid of real choice in action that they lost the capacity to reflect on themselves as doers of meaningful deeds, and without it, they lost the memorability of their lives. In the Brave New World people became insouciantly frivolous, seemingly happy, but in despair without knowing it. Only when Huxley's hero, Bernard, stops taking his Soma does he see the civilization for what it is. However, he cannot establish an alternative path for himself because heroic solitude has no place in the Brave New World. There are no routes for withdrawal and return, and there is no support for genuine approach and separation. Absent of love and wisdom, though yearning for them inchoately, Bernard stops consuming. He starts to suffer and make trouble. He's spotted, watched and eventually weeded out. Even after his arrest and detention, society gives him a fresh start. Mustapha Mond, the World Controller, sends him to his choice of islands where dissidents live according to their own lights.

In a very unusual and strangely kindhearted conversation with Bernard, his friend Helmholtz and the "savage" they brought back from the vacation-land reservation to see civilization, Mustapha explains, "What's the point of truth or beauty when the anthrax bombs are popping all around you? That was when science first began to be controlled—after the Nine Years' War. People were ready to have even their appetites controlled then. Anything for a quiet life. We've gone on controlling ever since. It hasn't been very good for truth, of course. But it's been very good for happiness."[10]

Mustapha Mond did not get to the Feelies much. He stood apart. He had to keep his perspective clear. He accepted his estrangement and sacrificed the pleasure in his own life in order to protect life for the citizenry. In this sense, he was the one mature person in the Brave New World. The juvenile arrest he promoted was for others. Our pouffed and pomaded experts, by contrast, have thoroughly juvenilized themselves.

## 191

We are caught in a destructive, aggression-causing recursive process. The fifth epoch's double bind torments us: We cannot find safety in the current climate of danger until we fully realize the dangers and experience our suffering as suffering and truly care about ourselves, but we can only suffer and feel danger with the ethical mind. And we cannot dwell in the ethical mind until we slow down enough to feel our naked vulnerability from a position of inward self-examination. That our overbearing technical interests will not let us do. It will not slow down. Even cleaning up our mess will require us to move fast. We dare not waste time. The more troubling the threat, the less time we can give it. Inward consideration gets curtailed first. It goes against the grain of the culture. So before we think things through or fully feel our way into them we rush back to the resources of the technical mind to get things done.

We rush briefly to the ethical mind, yearning for a life full of value, only to be swept abruptly back into the technical mind to find solutions, most often to pressing problems brought on by previous technical excesses. Juggernaut's car lurches ahead before the ethical mind can clarify the issues to itself.

## 192

The tectonic plates grind most forcefully where physiology and culture interface, where our inside and outside realities meet. That is the point of highest tension. From this permanent fault-line all great historical movements erupt, though in each epoch people understand the tectonic edge differently. In our period, we feel the physiological/cultural interface wherever the ethical and technical minds fight for dominance. Here the historical world trembles. The tremor is a sign of rhythmic dysfunction. From the turbulent body of nature big upheavals come.

Fertility and birth rate differentials will continue to widen across cultures. Worldwide resource scarcities will produce graver economic problems. Species extinctions will mount. With Shakespearean regularity, pandemic diseases, weather anomalies and massive migrations will appear as the global system breaks apart.

All over the world, communities are being destroyed. Traditional systems of belonging have broken down. Urbanization is gobbling up the countryside. Half the world now lives in cities with populations reaching 20 million or more. Land put into food production reduces species diversity. The biodiversity needed for planetary ecological self-management is failing. Hot zone diseases are spreading.

## 193

We have entered the early stages of a global population shift that will fundamentally change planetary demographics. It will stimulate new rhythms of aggregation and dispersal.

The current demographic changes started in the sixteenth century with the colonization of the New World. The capture and export of millions of African slaves to the Western Hemisphere was part of it, as was the decimation of Native American cultures as millions of European working class settlers pushed west across the continent.

This marked the first wave of modern migrations. Now climate change, economic need, hunger and thirst are driving people from their homes. Two hundred million immigrants push across the world. In China alone an additional two hundred million internal migrants have moved from rural to urban areas. Four million people have been displaced by the Iraq war. Millions more are huddling in refugee camps in Africa. Hundreds of millions exist in precarious situations. They can be cast out at any time.

Since the end of WWII, the African, North African and Turkish migrations into Europe have enlivened and challenged Western civilization. Declining birthrates in an aging European population create needs for workers that soaring populations elsewhere, provoked by drought and famine, disease and impoverishment, supply. The relentless shifts overwhelm social institutions. They put ideological and religious systems into conflict. Xenophobia, nativism, racism and economic exploitation (including slavery and sex trafficking) increase. Economic crisis encourages an uptick in anti-semitism.

# 194

Mass migrations have produced great cultural revolutions before. The Kurgan migrations broke the matriarchal societies of Old Europe. The mass migrations along the Russian river valleys in 2000 BC brought the Aryans into Europe and south Asia. The Hyksos migrations undid Mesopotamian and Babylonian civilization and broke the dynastic succession in Egypt.

In the same epoch, the Indo-European migrations swept across the Old World. Aryans overran the Dravidians, the Dorians subjugated the Achaeans.

Later the pressure of the Mongol hordes on the Asian heartland remade Chinese civilization. The barbarian invasions, pushed west by Asian peoples, undermined the frontiers of the Roman Empire.

Later still the Mongols and Huns crossed Asia into the European peninsula and changed the language patterns of Europe. Add to this the movements of Norse people through Europe in the Middle Ages, the Conquistadors and Puritan refugees who destroyed old civilizations in the Western Hemisphere, and the upheavals in the twentieth century, and you get a sense of the demographic powers that have shifted geopolitics.

Migration is the hammer in the clash of civilizations. Almost every major historical period begins or ends with it.

# 195

Rhythms supporting migration ride on older and deeper pulses than approach/separation and withdrawal/return. They go back to the ur-rhythms of primordial dispersal and aggregation directly responsive to slow-wave environmental oscillations. Recurrent dispersal-aggregation episodes, from this perspective, serve as a primordial fallback mechanism, a regression in the service of self-preservation when more recently evolved capacities fail under sustained environmental and psychosocial assaults.

Settled civilizations may have emerged and receded many times. Mythic resonances for Atlantean type social orders go way back. Perhaps Atlantis represents, mythologically, early civilizations that actually emerged in Paleolithic times, only to recede into primordial dispersal-aggregation rhythms when adverse turns of climate, predation or disease hit them.

These alternations between fundamental nomadic dispersal-aggregation societies and settled urban societies may have been sharpened, honed and selected for during the Ice Ages. Now again climate change may tip the balance.

Ibn Khaldun, the great Medieval Islamic historian, treated the alternation between settled and nomadic cultures as the key to historical revolutions. The capacity to change rhythmic patterns in response to persisting environmental and climate cues must have had biological roots going far back into hominid times.

## 196

The historical changes brought on by global migration in the modern period may again bring on a rhythmic shift in the basic patterns of life. Approach-and-separation and withdrawal-and-return, already under assault across the world in the ambiance of technology, commerce and faltering global finance, may regress to dispersal and aggregation once again, freeing the older limbic system drives in those populations under mounting migratory pressures.

For most of our species history, we have lived nomadic lives. To revert to the dispersal-aggregation pattern, even after settled conditions have prevailed for millennia for great populations, would give us more not less robust species survival skills.

## 197

Within a few generations, hundreds of Diasporas will have spread across the planet. The divided populations will probably maintain ties with each other using electronic media. Looked at this way, we may find decentralization, devolution and globalization advancing together. They may even turn out to be economically beneficial on the large scale.

Confronted with failing national states and splitting polities, non-governmental global institutions of a different sort might develop. Commissions, agencies, corporations, panels, conferences, electronic town meetings, multicultural research universities, staffed by people of the many diasporas may come together to find ways to cooperate.

In this planet of smaller polities, we may be able to develop a mixed, cosmopolitan multi-racial society of city-states arrayed in small bioregional associations. As cross-border transmissions multiply, a new system of cultural values, supported by new passing rules better adapted to function in natural sized communities might help bring the next human nature to birth, inaugurating a sixth human age. From roots struck in renewed turning points in love and wisdom, the interwoven Diasporas will introduce new possibilities for civilization.

In a new kind of cosmopolitanism, a generation of *ad hoc* policy advisors may rise up to advise and guide us on science, health, economics, trade, courts, renewable energy, disaster relief, agricultural production, environmental clean-up, and other issues.

In a decentralized world hospitable to inwardness, love and wisdom would achieve a better balance between techniques and ethics. In such a situation, people would be inspired to solve their problems under the lead of values and virtues, relying on shared ethical rather than proprietary technical innovations in a marketplace of ideas.

## 198

Immanuel Wallerstein, a leftist political sociologist, treats the devolution of nation states as an established historical trend. According to him, four hundred years of global connectedness and economic imperialism are unwinding. Devolution and decentralization started with the dismantling of the Ottoman and British Empires and the end of European colonization. This movement, "by deligitimating the state structures, has undermined an essential pillar of the modern world-system, the states system, a pillar without which the endless accumulation of capital is not possible." We're witnessing the "swan-song of our historical system."[11]

Wallerstein tries to console us. He says that this crisis "though it will be terrible to live through will not go on forever. We know that chaotic realities produce, by themselves, new orderly systems. This may not be much consolation if I add that such a process might take as much as fifty years to complete."[12]

Fifty years? He is optimistic. His schedule will be delayed by migration, nuclear proliferation and climate change. He barely mentions them. Two hundred years is more like it.

## *End Notes*

[1] William Blake. *Four Zoas* vii 168-84

[2] ibid p. 216, p. 284

[3] Jerry Mander. *In the Absence of the Sacred*. Sierra Club Books. 1991 p.188

[4] Thoreau.. *Walden*, where I lived and what I lived for

[5] *The Wisdom of Confucius*, ed Lin Yutang, Modern Library 1938, p.189-190

[6] Toynbee, Civilization on trial, p. 136

[7] Toynbee. *A Study of history*. vol I. D. C. Somervell abridgment. p. 284.

[8] "This division is radically different from other social divisions: that of the more able and the less able, or that of the richer and the poorer, pr even that of the rulers and the ruled. These latter divisions have existed throughout history and at times, at least, have been ameliorated by social and religious ideals that instructed the strong to help the weak. As a purely economic ideal, competition does not contain or imply any such instructions. In fact, the defenders of the ideal of competition have never known what to do with or for the losers. The losers simply accumulate in human dumps..." Wendell Berry, *What are People For?* North Point Press.1990. p.130

[9] Lewis Mumford. *Technics and Civilization*, 1934

[10] When the Controller and the Savage are alone, the Savage comments: "Art, science—you seem to have paid a fairly high price for your happiness...Anything else?"

"Well, religion, of course," replied the Controller. He later explains: "We've now got youth and prosperity right up to the end... we can be independent of God... And why should we go hunting for a substitute for youthful desires, when youthful desires never fail? A substitute for distractions, when we go on enjoying all the old fooleries to the very last? What need have we of repose when our minds and bodies continue to delight in activity? of consolation when we have soma? of something immovable when we have the social order?...There isn't any need for a civilized man to bear anything that's seriously unpleasant. And as for doing things – Ford forbid that he should get the idea into his head. It would upset the whole social order if men started doing things on their own." Aldous Huxley, *Brave New World.* 1932. Bantam Books. P.155-161

[11] Wallerstein. *Futuristics*

[12] ibid p. 63-64

# 11. WHAT CHANGES WHEN PEOPLE CHANGE

## 199
### Challenge and Response

Toynbee based his understanding of history on the concept of challenge and response. But what is challenged and what responds? I say the body is challenged and responds, most sensitively in its clock functions, particularly in the fast ultradian ranges, for they are most susceptible to cultural and environmental influences. And once perturbed, the altered *body functions* drive the person inward, outward, towards or away from others. In my challenge and response schema, many small changes undertaken by many people returning from their depths, or reaching out to others, when they are responsive to common challenges, have historical relevance.

According to Toynbee, when creative persons return from withdrawal they bring with them new inventions, theories, religions, social reforms. These attainments meet the needs of many people who did not make the inward journey. Many more are followers than leaders. The whole panoply of culture, with its periodic outward renewals, in Toynbee's calculus, originates in the growth and inner transformation of those individuals who constitute a creative minority.

In one place he focuses on the lives of great historians and their historical influence. He traced the careers of Thucydides, Xenophon, Josephus, Machiavelli, Clarendon, Ibn Khaldun and Confucius. He found that

"in the 'practical' first part of their careers these future historians have all set themselves to produce an effect upon their fellow men by the obvious and crude and finite 'direct method' of bringing their wills to bear upon the wills of their neighbors. The compulsory withdrawal, which has inhibited the exercise of their activities on this 'practical plane', has compelled them to find a new vent by transferring their action to another plane and transmuting their energies into a new medium. In prison or internment or exile, the energies that can now no longer discharge themselves in the impact of will upon will have been transmuted from willpower into a heightened intensity of perception and thought and imagination and feeling…"[1]

I have studied periods in which many coordinated love and wisdom journeys undertaken by many people across a culture brought on revolutionary changes, generational changes that diffused over scores of years and whole continents. Not great heroes, but the resetting of the passing rules makes history change. In this chapter I will use a few examples to illustrate how the changed passing rules renew the content and scheduling of the signal events of life. These changes, moreover, come from physiological shifts brought on as a result of neural Darwinist competitions in the brain. (See # **200, 215** below.) Motivations, cognitive styles, sensory distance perceptions and variations in love and wisdom develop from them. And historical adventures, economic movements, wars and regime changes follow this.

There is no single winning strategy for advancing our inner-personal or interpersonal lives. Our needs change. When new motivations, attitudes, gestures and actions prove responsive to prevailing human needs by persisting and spreading across social classes, revolutions take hold. Great men and women play parts, but the massed turning points of many mainly anonymous people, and the personal choices that come from them, produce the real changes.

That is why Tolstoy considered the proper subject of history to be "the study of the movements of peoples and of humanity, and not episodes from the lives of individual men…" History was built on "laws common to all the equal and inseparably interconnected, infinitesimal elements of free will."[2] He wrote in War and Peace that

"The movement of peoples is not produced by the exercise of power, nor by intellectual activity, nor even by a combination of the two, as historians have supposed; but by the activity of *all* the men taking part in the event, who are always combined in such a way that those who take the most direct part in the action take the smallest share in responsibility for it, and vice versa."[3]

The "inseparably interconnected, infinitesimal elements of free will" count because they bring life to the things we care about most, love and wisdom. They revitalize them and tune them for the times in new passing rules. These have *ethical* content. They guide our *choices.* We "make history" in the narrative sense a few generations later.

## 200

When people intend new things, they imagine differently. They walk and talk differently. Their heart/breath interactions follow different patterns. Their circadian hormonal schedules subtly advance or retard. Their dreams change. They develop new interests. They fight with each other or cling together differently. They acquire new socioeconomic ties to nature that give them new class-related interests and biases.

These changes introduce new patterns on the legs of love and wisdom. They send people into their turning points differently, though with a shared impetus. And once in them, in their moments of freedom, the responses they come up with have unusual adjacencies. They interlock.

The newness shows early in altered childbearing and childrearing behaviors, in new standards for mate selection encompassing a broader community of interests, and in new feelings of belonging, all of them shaping evolutionary or cultural modes of selection. People discover new affinities or antipathies that dispose them to share meanings never fussed over before. And together these open the way to new political forms.

Love and wisdom, in these crucial moments, become what they have never been before. They take on fundamental political dimensions that show in new loyalties, new senses of belonging or estrangement, new disenchantments and disillusionments, new kinds of aggression, and also in our genius for invention and for self-discovery.

I will try to make clear how, specifically, these changes move from the human soul to the social world. Then I will discuss who is likely to be bearing them into world culture now, and why.

## 201

We must now return to our discussion of late human evolution and introduce three new themes.

1) Emergent love and wisdom and the skills that go with them function at the interface between learning and instinct. They are products of natural selection whose selective advantage is to facilitate learning.

2) They make their appearance in a deep cultural past already transformed by fire, clothing, shelter, cooked food, tools, weapons and more. By means of an interactive evolutionary selection process that theorists call gene-culture co-evolution, they produce adaptive changes built as refinements on already existing physiology.

3) This fine-tuning points to mutations in regulatory genes and their promoters, enhancers and transcription factors. They change the function of existing DNA-RNA-protein expressions by altering their timing or by linking them to different biochemical processes.

In the course of refining our stress responses, for example, natural selection did not dismantle and redesign the limbic system. Instead, in changing the set points between stress and aggression, evolution built onto what came before, following its long established method.

## 202

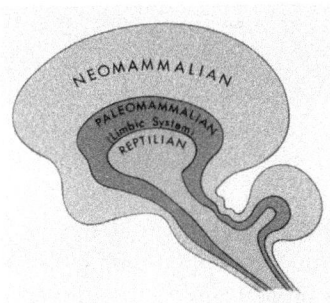

In the mid-twentieth century, Paul MacLean showed that the brain evolved by means of successive overlays. He argued that each major evolutionary development in the animal brain functioned as a cap on a cap. The cerebral cortex surrounded a paleo-mammalian brain, including the midbrain and limbic system, the seat of emotions and drives. The paleo-mammalian cortex, in turn, surrounded the even more primitive reptilian brain, the core of our reflexes and autonomic functions, our brain stem.

All three brains, MacLean called it the triune brain, continue to function in Homo sapiens. The reptilian brain–the "R-complex"–in our brainstem, organizes our basic reflexes and instincts. I conjecture that the drives responsible for territory, dominance and aggression have their seats here. The paleo-mammalian cortex, our limbic system, adds emotion and passion, fear, rage, desire and perhaps personal bonding behaviors that differentiate mammalian from anonymous reptilian social life. And over them both, the neocortex develops capacities for abstraction, prediction and foresight.

According to MacLean, this series of caps on caps gave us the ability to filter out input from lower centers. The old response capacities were still there. Evolution did not do away with the functionality to begin anew. Rather, overlying circuitry inhibited them. Nevertheless, they can be selectively disinhibited under certain circumstances. The higher filters can open their gates and let the impulses through. This switch – the inhibition/disinhibition switch — was seated in a higher more recently evolved center. Of this, he wrote, "It is this new development [the neo-mammalian cortex] that makes possible the insight required to plan for the needs of others as well as the self and to use our knowledge to alleviate suffering everywhere. In creating for the first time a creature with a concern for all living things, nature accomplished a one-hundred-eighty degree turnabout from what had been a reptile-eat-reptile and dog-eat-dog world."[4]

## 203

In the increasingly complex social and cultural settings of Paleolithic times, we evolved a set of connections, an overlay, that let us forestall the bifurcation point between stress and aggression. In a modest way, modern stress reduction techniques tap into this circuitry through relaxation, meditation, exercise, imagery and biofeedback. With voluntary efforts, we can tune down sympathetic innervations and reduce stress.

However, the neocortex, the "thinking cap", could not have achieved "a concern for all living things" simply based on increased computational power. I have already hinted that the con-

trols for stress modification and delay and for wider access to empathy and sympathy, required acquisition of internally structured
inhibitory tracts. Possibly in the physiology of mirror neurons,
as follows .

In all behaviors, the preparations preceding movement stimulate neurons in the frontal cortex. They light up before we go into
action. They even light up when we watch others going into action.
Persons with autistic disorders, who have great difficulty empa-
thizing with others, show specific deficits in the functioning and
number of these mirror neurons. Perhaps linkages between mirror
neurons, facial recognition, and infant/mother bonding circuitry
worked as a governor over the neural and humoral preparations for
stress. It would have allowed empathy to shape and modify our
responses.

However, our highly technologized sensory surrounds now
may impair the working of our mirror neurons in the following
ways:

1) We produce dendritic connections to machine movements. We develop a kind of empathy toward machinery that helps us get around. These new neural
   branchings mingle with and mix up our human-to-
   human contacts.

2) With mirror neurons firing both to mechanical and organic presences, we view human beings as mechanical
   entities, while treating machines as having wills of
   their own.

3) The empathy generated by mirror neurons, partly conditioned by machine movements, refuses to move to
   sympathy. It is less able to travel on hippocampal-
   amygdala-hypothalamus pathways. It stays cool. It
   goes directly to the reasoning mind. Violent video
   games seem to work this way on players. A mechanical "you" hunts down mechanical enemies in an environment of machine rhythms and cries, and the hand-
   eye coordination it takes to get good at video games
   fosters a following response from which empathy
   without sympathy emerges. Now online virtual realities like World of Warfare are doing the same on a
   much broader social scale.

4)  We deal with all kinds of culturally specific "hurry up and wait" situations today in which human social and technologized mechanical mirror neuron calculations are set in play. We race around airports, sit still in planes, and fly across time zones. The body needs to change to keep up. Deficiencies in mirror neuron functioning would make it hard to learn these strange new steps from others. In settings that are more natural the mirror neurons might function well enough. However, under the confusing inputs of technological culture, society makes new demands on them and they may fail to rise to them. Perhaps this shows up in the increased incidence of autistic spectrum disorders (many with mechanical movement tropisms.)

## 204

Consider this possibility: a refinement in the amygdala let it store and distinguish threatening from non-threatening emotional memories. Selection for amygdala modification, along the lines Jerome Kagan suggested, would keep the aggressive components of the stress response from firing unnecessarily.

To support this notion, Kagan identified differences in amygdala sensitivity in the development of childhood temperaments, showing that differences in the relative sensitivity of the dorsal and basal areas of the amygdala involved in emotion incline people to uninhibited or inhibited temperaments.

"Theoretically, the excitability of each area should be inherited independently. Thus, there should be at least four different types of infants. Infants who are highly excitable in both the basolateral and central areas should display frequent flexing and extending of the limbs, spasticity, and arching of the back, together with crying to unfamiliar stimulation. Infants who are minimally excitable in both areas should display low levels of motor activity and minimal irritability. Infants with a highly excitable central area but a minimally excitable basolateral area should show infrequent motor activity but frequent crying. Finally, infants with an excitable basolateral area but a minimally excitable central area should show frequent motor activity but minimal irritability. The first group should be the most inhibited the second group the most uninhibited."[5]

I conjecture that these differences reflect continued evolution in the amygdala into Paleolithic times, and perhaps later. In these, the avoidance behaviors associated with the basolateral areas were refined into more subtle responses through selective processes effecting both incoming and outgoing traffic. Perhaps the evolutionary refinement of the incoming traffic could have been managed most easily by changing the sensitivity of dendrite populations heading toward the amygdala. One way would have been to increase serotonin secretions in order to produce more prolactin, a testosterone inhibitor. Kagan favors modifications to the outgoing traffic.[6]

## 205

In the period we have been discussing, love, wisdom and aggression grew free from their embeddness. Mutations to regulatory genes must have preceded the emergence. But which? Earlier we rejected the standard that attributes all late human advancements to "increased cortical capacity." We decided that human behaviors have more complexity than that. Other neural assemblies control them. For one thing, more resilient access to empathy and sympathy cannot rest solely or even mainly on increased cognitive powers. We need highly developed capabilities for identification and imitation. They had already been evolved in parent/child and mating relationships. Did it take foresight and calculation to extend their holding power to broader social circles? I don't think so.

Enhanced powers of imagination must have played a part. They provided the grounds for empathy and sympathy. Imaginative identification involving mirror neurons may have provided the means for deducing the internal reality of others from our own states of mind.

Certainly putting yourself in another's shoes is primarily an imaginative not an intellectual act. But imagination needs *feeling tones* to give it the emotional understanding it needs to *intuit* what was going on in others from indistinct and distant data. You have to feel your way inside from outside before you can even frame a notion about it.

To make reasoned judgments based on imagination requires something even more basic: an internal observer, a modular entity who, from a stance of inwardness could watch the self watching the other.

These capacities have little to do with Calvin's notion of "general purpose" processing powers (described in # **165**.) They depended, first of all, on subcortical changes that modified the stress response and shifted the bifurcation point between stress and aggression. With that, you got the time, space and leisure to imagine and feel.

On the foundation of an altered stress response, we acquired wider responses to novelty, threat, uncertainty and social complexity. But I don't think we got them primarily through social learning, or reasoned argument built on increased cortical processing power, since the fight and flight response notoriously resists reasoned argument.

**206**

The experience of trust itself – to believe in another person's good intentions – does not rely on logical proofs. It may surface first in infantile feeding behaviors, reinforced by sucking rhythms, facial following responses and other rhythmic interactions with caregivers. These presuppose empathetic connections at least on the part of the mother. And the flow of love, or at least attachment, in these settings, not the increase of knowledge per se, would have fortified the trust. Later the trust went into promise–keeping behaviors in support of deeper socialization.

The new circuitry would not have added another layer of automatic responses to the underlying fixed action patterns but, through frontal-limbic connections, awash with the quantum-coherent cellular ground substance of protein in water that Loewenstein projected, would have let them come under intermittent voluntary control. (Review # **118-119.**)

The supplanting of autonomic by voluntary behaviors could only have happened if natural selection utilized the quantum indeterminacy in biological systems. Perhaps the fine-tuning mutations built on randomness, and by linking to chaotic signaling events in membrane processes, bridged the gap between fixed and voluntary responses.

## 207

Consider a mother responding to an infant biting her breast. If the pain passes a certain threshold, the mother will push the child away. There's aggression on both sides here. The child may get angry at being pushed away and bite back harder. Tantrums, opposition, childhood destructiveness of all sorts stir aggression in both infant and mother. They awaken one-sided ethical challenges. Mothers rarely abandon aggressive children. The rejection of abandonment is a main pillar on which fourfold love stands, and it is tested and tempered by aggressive interactions early on. That which *becomes love* draws the infant and mother back from separation to approach.

Over long periods of evolutionary selection, the set point between stress and aggression moves back, enabling broader and deeper mother-infant bonding. A kind of moral restraint toward the child in the face of its infantile aggression proves adaptive. (Observe the antics primate mothers put up with from their youngsters.) The longer the period of childhood dependency, the more involved the educational process, the more adaptive it becomes.

Moreover, the forbearance of the mother must have been communicated to the child. It would travel directly in the gift of sugar, in the warmth and sweetness of the mother's milk, and in the giving and taking of the breast at a time when the immature brain was still adding circuitry at great speed, much of it genetically scheduled, and most of it completed by the end of nursing. These *developmental limits to aggressivity* would involve both nature and nurture. They would express gene-culture co-evolution. The child would be prepared for its own moral emergence.

## 208

I am suggesting that the qualities of character that to some extent control the linkages between reason and aggression – and allow or forestall aggressive encounters – might depend in subtle and significant ways on the close-up senses, on taste, touch and smell, and on the intimate visual eye contact and facial recognition that occurs during person-to-person contact within certain safe and close sensory distance ranges. These bonds of trust would not depend on evolutionary changes to the cognitive processing areas of the brain. Instead, the evolved restraints that kept stress from escalating into fight or flight may have been *reflexive products of gene-culture co-evolution, learned autonomic responses* stimulated by facial expressions and body attitudes sent and received at certain distances.

In our increasingly anonymous, electronically mediated social world, those sensory experiences have been altered or overridden. Faced with new civilized stresses, the instinctual restraints on triggered aggression falter and fail. Violence increases. Reason, keyed up by the need for heightened vigilance, could lead us to design aggressive responses that let us maintain our advantages in unprecedented ways, for instance, from 60,000 feet in the air, or hundreds of miles away at missile launch sites. From those distances, we call killing "collateral damage."

## 209

Rudolpho Llinas suggests that the linkages between prefrontal and limbic areas evolved on pathways passing through the thalamus, a routing that would have given us "the ability to override a liberated FAP [fixed action pattern]...an ability born out of an increasingly elaborate thalamocortical system."[7]

Combining Kagan with Llinas, we come up with the reasonable conjecture that the new circuitry evolved to filter the FAPs *connected with the flight and fight response.* New cortico-thalamic circuitry could have put a brief delay in motor output, a pause for consideration, during which time the brain could poll relevant neural, hormonal, and neuropeptide inputs. That would be consistent with known thalamic functions. It seems to work as a coordinating center, linking both to the hypothalamus (through which the brain orchestrates major components of the stress response) and to emotional memory centers in the amygdala and hippocampus.

Llinas then suggests the 40 Hz carrier wave implicated in consciousness may have roots in the thalamus.

"Studies indicate that the 40-Hz coherent neuronal activity large enough to be detected from the scalp is generated during cognitive tasks. Furthermore, some propose that this 40-Hz activity reflects the resonant properties of the thalamocortical system, *which is itself endowed with intrinsic 40-Hz oscillatory activity.*[my emphasis]... What does it mean? We are confronted with a system that addresses the external world not as a slumbering machine to be awoken by the entry of sensory information, but rather as a continuously humming brain. This active brain is willing to internalize and incorporate into its intimate activity an image of the external world, but always within the context of its own existence and its own intrinsic electrical activity."

Coherence matters to Llinas because "temporal coherence is believed to be the neurological mechanism that underlies perceptual unity, the binding together or conjunction of independently derived sensory components, called 'cognitive binding'."[8]

### 210

New routes for two-way communication between the limbic system and the forebrain could have been etched in the nervous system during embryological development. Relatively simple mutations to regulatory genes could orchestrate cell migration patterns and control the 'selective pruning' back of neural branching processes through programmed cell death. Since these changes affect soft tissue deep in the brain, and are completed by the end of gestation, they would not show up in the fossil record of skulls. However, we may someday find evidence for them in fossil DNA studies.

### 211

With links between neocortical expansion and thalamocortical pathways established, we can now draw together a great deal of material of relevance to human nature, culture and history.

1) Once the cortico-thalamic-limbic circuitry had been put in place, our vastly improved language skills would have given us the power to put the musings of the inside world out into the world of others *in words*. In charged moments involving intimacy, trust or dire necessity, we could have used these words to make things happen. Why would stress modification allow this? It's like when kindergarten teachers say to their pupils "use your words, not your fists."

2) Along with deepened dialectical exchanges, and the bargains, deals and contracts that came with them, new

kinds of enmities must have bubbled up, new griev-
ances, suspicions of injustice, an ever more exacting
set of standards for differentiating "us" groups, generating special hatreds and aversions. Novel expressions
of aggression must have sprung up, with violent out-
growths, not least moral outrage based on absolute
convictions of rectitude.

3) Even with the paradoxical results brought about by the
liberation of aggression from its original conservative
functions, a wider range of responses to novelty had a
clear survival value. As we developed a more advanced material culture, change became endemic.

4) With modifications to the stress response in place in an
increasingly acculturated environment, with stronger
empathy and sympathy, and deeper imaginative faculties, we would have developed abilities to not only
frame clearer intentions for future projects, but also to
hold them firmly in the face of jarring reversals. To
hold vivid imagination steady, and observe it without
flying into action, requires precise skills for timing action, knowing when to act and when to hold back. This
*discretion* gave us conscious fortitude in the midst of
turmoil.

Taken together, these abilities helped us move into the historical period, into worlds of our own making.

## 212
### *The Neutral Traits*

In my story of origins, gene-culture skills drift in the neutral
traits. They're heritable though not selected for. Should conditions
change, however, they can give their possessors a survival advantage and become adaptive traits.

First, the neutral traits get into the DNA by random variation.
They surface in the behaviors of individuals and families, and perhaps in communities whose gene pools have been isolated by geographical and climatic barriers. Second, they persist as long as they
don't prove disadvantageous, or until, on the contrary, they are
called forward by environmental changes that render them adaptive, (or make them positively detrimental.) Third, changed criteria

of environmental or sexual selection favor or disfavor them, particularly sexual selection insofar as choice of mates has come to rely increasingly on qualities of character.

The workings of the neutral traits are always on display. They are expressed prominently in the skills of musicians, artists, athletes, dancers, creative geniuses, new thinkers, and savants of all kinds, even acrobats, jugglers and sleight of hand artists. Their physical, cognitive, emotional and spiritual abilities are partly heritable. When times change, they can move into cultural consciousness at an accelerating rate. They shine out in our individual differences, in quirks of temperament, in all kinds of aesthetic, ethical and emotional sensitivities. Parents teach them to their children, but at the same time, they pass along the genetic equipment to learn, practice and enjoy them.

I am suggesting that access to neutral traits at significant moments may provide the "tiny cause" that brings on big changes. Because the neutral traits persist in behavior in unobtrusive ways, they become potential sites for evolutionary selection in periods of rapid geological and climatic change, nowhere more so than in cultural venues favoring gene-culture co-evolution. Here the traits, while being circulating in a milieu of genetic drift as part of our biodiversity, are tried and tested in action. It is as if we have been rehearsing behind the scenes, and now wait to make our entrances.

What summons us to the main stage? Three factors:

1) New environments present us with unusual gene releasers. These stimulate quiescent tendencies in our DNA, some hidden in the junk DNA. Other persons stimulate the releasers, as do our own new hopes and fears, environmental stimuli, metabolic needs, disease vectors, or the stimuli of culture.

2) We explore new pathways through play. Play can call us onstage by discovering untried possibilities. We played with fire before we used it. So too other technologies, weapons, idols, art, magical objects, sounds, words, thoughts and stories. In play, we explore new possibilities in family life and intimate relations.

3) Unfolding patterns of self-organizing complexity keep us, on molecular levels, close to the sites where indeterminacy influences reactions. Stuart Kauffman imagined a topology in which these poised moments

formed "adaptive peaks." On them, we are always poised and ready to change.

Following Per Bak, Kauffmann likens these small changes to avalanches in the dynamics of sand piles. He writes, "The size of the avalanche is unrelated to the grain of sand that triggers it. The same tiny grain of sand may unleash a tiny avalanche or the largest avalanche of the century. Big and little events can be triggered by the same kind of tiny cause. Poised systems need no massive mover to move massively."[9]

The neutral traits provide a reservoir of possibilities for evolutionary fitness widely distributed across populations, perhaps randomly so. They travel *sub rosa*. Human biodiversity, in other words, maintains a cache of behavioral possibilities that we carry with us in our historical peregrinations.

### 213

Our desire to sing and dance, to compose music, make art and perform dramas, to play together as children, to ponder the beauty of nature and observe the heavens with wonder and to spend a lifetime learning to do it come from the neutral traits. How a stranger plays with a child, the changes in child's play itself from one generation to the next, how a happy person affects a neighborhood, the capacity to imitate through observation, the curiosity to get to know each other, to remember and report dreams, to render experience in writing and pictures – all of these have genetic components that travel in the neutral traits. They give us richer and more interesting lives, not limited by reproductive success, economic rivalry or low-level selfish gene strategies.

Take a world without music, what a terrible loss that would be, but not one that has a meaning for an evolutionary thinker in his role as a chronicler of selfish genes. Of course, researchers can tie any quality, including musical abilities, to reputed survival instincts and explain them as emergent properties of the big brain, etc.

By not taking account of the future adaptive potentials in the genetic drift, many evolutionary theorists overlook the importance of the neutral traits. In doing this, they ignore some of our most interesting and promising characteristics. Richard Dawkins follows this path when he insists, perhaps too didactically, that "...if we happen to be focusing our interest on adaptation...a neutral mutation might as well not exist because neither we, nor natural selection, can see it."[10] It makes no difference to "legs and arms and wings and eyes and behavior", by which he means adaptive behavior. He writes,

"Darwinism is a theory of cumulative processes so slow that they take between thousands and millions of decades to complete. All our intuitive judgments of what is probable turn out to be wrong by many orders of magnitude. Our well-tuned apparatus of skepticism and subjective probability-theory misfires by huge margins, because it is tuned—ironically, by evolution itself—to work within a lifetime of a few decades." [11]

Dawkins' framework rules out Darwin's concept of pre-adaptation, the notion that random mutations may *later* prove adaptive under changed circumstances.

Yet when I try to understand what sustains the life of people, I find that most of what I think of as essentially human is carried in the neutral traits. The crucial contributors to human nature in its huge central areas of content-filled drift, it seems to me, are precisely those unknown persons who carry the pre-adaptive traits with stalwart strength, who make a difference to the world because in the fullness of their natures they shine as beacon lights in ways not measurable by selfish gene theories.

These adaptive possibilities, we must remember, are not abstract entities. They inhere in people, the same people who, hardly noticed by cultural historians, war sweeps away, who repeatedly suffer the slaughter of the innocents. We cannot measure the greatness of the waste by what was left undestroyed or came next, or built upon the bones. We cannot know what might have been but never was. But we cannot summarily dismiss its possibilities either. I cannot help feeling that the monuments left by the martial spirit might look far less appealing to us if the lovers and poets who died along the way had been left alive to make their own different contributions.

Every day nature forwards new powers from the neutral traits into the world through the agency of individuals who serve as our Johnny Appleseeds. They work their creative influences into life through the investments they make in their children, mates and friends. These skills constitute a body of practical knowledge, an art of life really, that serves to tune us to circumstances with great refinement and creative range. They draw on far more than our rational problem solving powers. They sit in the gate where our nuanced emotional join our intellectual qualities to vitalize judgment. They bring us new virtues and mental abilities. They make us rich with possibilities.

However, evolutionary researchers can explain anything with selfish gene theory. Once they find evidence for the expansion and complexification of the musical areas of the brain in cranial endocasts, they will no doubt study selfish musical genes and uncover reciprocal altruism in musical enthusiasm, and promote the sexual selection of troubadours on the analogy with bird song. They will view musical instruments as expressions of human tool making skills.

## 214

When the right conditions prevail, groups of people, sometimes significant portions of whole populations, make coordinated changes along the pathways opened by environmental change, childsplay and the inherent powers of self-organizing complexity present in tipping point moments. These coordinated changes actually show the shared operation of regulatory genes expressing a subset of skills already present in the neutral traits. Moreover, when many people undergo turning points responsive to commonly suffered problems these right conditions become a fulcrum for leveraged change. To say it another way, in highly charged times, powerful leaps come from the neutral traits.

How, specifically, do personal discoveries get into history? Do great figures, culture-heroes, bring them into the mainstream? Are inspired leaders less innovators than apt representatives of the times? Or do social revolutions move from person to person anonymously? All three.

But the ground is prepared anonymously. The neutral traits mostly move into readiness without claims of authorship. Often they pass in subtle gestures and expressions. They move across populations by inflections of voice, by gestural inflections, by looks in the eye, by current jokes, irony, and by the infectious qualities in laughter.

Both action and inflection are necessary for communication. They are the message kernels with which we construct the signals we send and receive. Sometimes we carry the message in the meaning of words; sometimes the medium is the message; sometimes the meaning is carried in deeds that contradict the words. Sometimes communications are carried in changes in the tremor rhythms, in touch, in thermal agitations, in the scanning rhythms of the eyes, in pheromones, in subliminal facial expressions, in conversational rhythms and pauses, and in subtle alterations of distance regulation rules that revise the boundaries between the four social distances.

Every social and cultural change, every innovation, every technological revolution, travels personally in real-time, between living individuals, with the "wink of the eye" or the "query in the voice" traveling as modulations on carrier waves – and first of all between parents and children. However, their contents vary with the historical moment; they push forward with different degrees of frontality; they occupy different niches in the social strata.

They spread as memes.

## 215
### *Memes, Imitation and Introjection*

Researchers describe memes as communications that pass widely through populations. Some pass quickly and spread far. Others die out. Memes are refined through competition and selection, as are biological traits. Some thrive, some fade. Cultural provenance, not biological reproduction, favors their selection.

Fashions and styles are memes; expressions, idioms and turns of phrase; habits of thought, melodies, melodic phrases that play in the mind; urban legends, gossip, jokes and stories are memes; but so are planted ideas, propaganda, *obiter dicta* and genuine and spurious words of the wise.

Many memes travel in the exchanges between parents and children, or they pass in play among siblings and friends. Some basic cultural and spiritual memes (and family truth) pass down through the generations; they travel as customs, received ideas, family legends, etc.

Beyond their rational content, *catchiness* favors neural Darwinist competitions among memes. Jingles travel well. Rhythmic phrasing in duple and triple meter generally favors mimetic transfer. Nuanced language use, and the fine expostulations of orators, can ride as much on rhythms as on meanings.

Some memes only advance through brutal competition in periods of political turmoil and revolution. They fight their way across town. A meme can cost you your life, as in the Wars of Religion in Europe in the 16th and 17th centuries, in geopolitical movements and ethnic cleansings.

Both mimetic competition and cooperation can move through whole societies. Memes that travel cooperatively sometimes have beneficial effects on the culture, but sometimes have destructive outcomes. Some competitive memes have destroyed the people who adopted and transmitted them. Others have won space for freedom, as in the American Revolution. Even those fueling wars can have long-term beneficial effects. Other truths and lies of war travelling as memes have ruined societies past recovery.

The mimetic transmissions that spread most widely and last longest stir up the most multifarious action possibilities. They can change habits of thought and foster new ways of seeing. They can even modify the sensory filters that determine what we *can be seen*. Historians of science following Thomas Kuhn call them paradigm shifts. Following Copernicus, the spreading meme of a sun centered solar system made us see the sun, the planets, and the whole cosmos differently. We can actually feel the earth turning when we see a full horizon line, for instance at sunrise – or not yet but maybe someday.

The connection between neutral traits and historical turning points will become clearer to us if we remember that ongoing withdrawal/return and approach/separation processes serve as principal carriers and spreaders of memes. Those memes, drawn into and through turning points in love and wisdom, become agents in gene-culture co-evolution.

Memes circulate first within individuals. The parliament of the brain ponders them in its modular centers. Certain ideations win internal neural Darwinist competitions.

Gerald Edelman develops a "neural Darwinist hypothesis" to explain modular brain functioning in which neuronal groups act as units of selection. Neural groups ally and send out feelers. Groups build territories. Their neural nets expand or contract over the surface of the brain. They form distributed maps.

That is to say, the modules of the personality do not need to be located near each other in discrete areas of the brain; they can be disseminated, linked by strengthened synaptic connections sending fibers over far dispersed regions.

"When maps are connected by reentrant fibers, the individual fibers generally extend their arbors over many locally linked neurons. When secondary repertoires are formed, the strengthening of synapses within these arbors may then select groups of neighboring neurons, changing borders over smaller dimensions than those of the arbors." [12]

John McCone's gloss on Edelman's theory of neural Darwinism makes the physiology clearer: "Every moment would begin with a battle in which some networks of activity would blossom, gaining the neural territory needed to become conscious-level percepts, while other, weaker nerve ensembles withered away... If a sensation was anticipated, or deemed important in some other way, positive feedback from higher areas would help swell the mapping activity, elevating it above the general clamour."[13]

Parts of the self, competing with each other in neural Darwinist networks, their qualia travelling as internal memes, extending their territories, can come upon skills residing in local areas as state-dependent memories. The local areas sit there ready for action. Some of these not only "swell the mapping activity" and enter consciousness, they do so repeatedly, and at important times, and they break into the world as deeds – some with a great deal of energy behind them. Once expressed in action, others perceive them. They start their journey as memes in a Tolstoyan world of people with "equal and inseparably interconnected, infinitesimal elements of free will."

They press their claims by crossing mind/body frontiers. They ride upon three basic culturally learned ways of being in the world: selective sensory stimulation based on the events, objects and movements predominant in the cultural environment; internal division of the self supported by the culture's biasing of the passing rules; and cultural restrictions on movement and action, more significantly, on what it is even possible to conceive as action. I will describe these routes in some detail:

1) The culturally learned deployment of sensation between inwardness and outwardness becomes a major shaper of human culture. The culturally specific

shapes of love and wisdom themselves come from the inwardness and outwardness we are trained to undergo. This is so because approach-separation and withdrawal-return, the basic meaning bestowing rhythms of life, are built on successive stages of outwardness and inwardness. Depending on the historical moment in which they live, people develop different understandings of what it means to love and grow wise and how to go about it. We lead fundamentally different lives in different times. We become who we are by what the times allow.

Every civilization makes rules about the place and function and worth of inwardness and outwardness. The rules tell us what we ought to favor and what we ought to suppress. The conditioning of the sensory mix, in turn, defines the four social distances and creates cultural boundaries at them. We develop distinct notions of where intimacy occurs and what senses are used in what proportions in it. We learn where personal space begins and ends. We acquire its sensory mix. So too for social and public distances. Each civilization sets out the four social distances differently. The map of social distances shapes the character of a civilization. In a similar way, cultures establish internal boundaries in mental life. They distinguish between permitted and prohibited fantasies. They set the rigor or lassitude of the barriers between the compartments of the mind.

Our established order of sensation is changing now. Under the impact of electronic media and other technologies, a great struggle is being fought over the borders between inward and outward ways of knowing.

2) In our current cultural milieu, we have stiffened up the passing rules to protect us against the loss of the I-sense, a loss brought about by the all too frequent and sudden interruptions, shifts of attention and fast cuts that characterize our outwardness. People cling to their passing rules as long as they can, usually long after their adaptive utility has passed. They patch them up; throw a patch here and a patch there, with always shoddier stitching and more crooked seams. Finally the patchwork tears. Boundaries break down; we lose

our sense of place. The cultural biases cease carrying conviction. Internal confusion overwhelms the passing rules. We oscillate between extremes. The internal play of part-personalities becomes a dramatic ferment. The *characters in search of their author* advocate for change. The divided self cries out for order. Self-division becomes an agent for change.

3) Though in every culture we eat, digest food, sleep, rise, laugh, and cry, and share the physiology behind these behaviors, the particular way we do these things, the times we choose to do them, the places we do them in, vary widely. The designs of houses, streets, and meeting places influence our capacities for action. The cultural momentum sometimes lasts for generations past its useful life. Those who benefit most from the status quo protect it. Those who suffer from it struggle against the prevailing order – at least those capable of recognizing their own real interests.

At all three interfaces events shape the conditions of belonging. They make an "us and them" unique to the culture. They set the balance between inwardness and outwardness. They distribute happiness and misery differentially. As the self/world interface shifts, the character of self-regard changes and with it the standards of belonging and the deployments and purposes of aggression change. Simultaneous tectonic strains at the internal self-division and sensory stimulation frontiers speed up the historical change process.

## 216

Successful memes persist in action for as long as they are favored in the reassembly of the passing rules. Revisions to the passing rules – and hence to the models of reality we carry – occur most forcefully during turning points in love and wisdom. With every turning in love and wisdom, therefore, the opportunity comes to review and transform our essential virtues and hence our historical roles.

## 217

However, no civilization fully employs the resources of its citizenry. Each civilization differently encourages or suppresses us. In different settings, some parts of human nature will inevitably be called forward while others will be neglected or suppressed. The culture plays favorites. It nurtures certain personalities at the expense of others.

In some epochs, a person's strengths are demeaned while his or her weaknesses are encouraged. Sometimes it is the other way around. Some personal failures are really successes that the culture cannot recognize. Moreover, while some of us will be weak where the culture would have us strong, and will suffer for it, others will be strong where the culture would have us weak and will have to pay the price for that too. Still others will be strong where the culture wants strength and weak where they wants us weak and they will be sought after and schooled for success and, within the boundaries of class and status, will become our leaders.

The unlucky people have their own contributions to make. Creative opportunities denied to the well-adjusted open to them. In crisis times, when the power structure petrifies, its victims become the creative minority who plant the seeds for the next epoch. Who suffers now? Whose suffering contains the strongest, sanest energy for cultural change?

## 218
### *Aptness and Readiness*

Who are these pioneer people today? How do they differ from you and me? Do their infinitesimal doses of free-will spin differently in the stream of life? Why do their small deeds resonate and spread?

First, they are apt and ready. They possess a certain kind of sensitivity that, in estranging settings, entices them to withdraw to seek answers, to find wisdom, and perhaps to love in new ways. The best of the meme-senders bring the resonant energy of their own turning points into their messages. These turnings are transformational: they draw out new skills from the neutral traits. They demonstrate their relevancy. They open to ethical inwardness. They bring moral sentiments into the world. The Jurist, Richard

Posner calls them moral entrepreneurs. (I refer to his work in # **281**.)

**219**

Though Toynbee found examples of culture-changing withdrawal and return patterns in the lives of famous people, his researches led him to locate it more often in the underclasses.

He pointed out that when a civilization experiences its "Times of Trouble" help may come to it, if help comes at all, from a disenfranchised internal proletariat.

"This deep desire for changes and the strong resolve to bring them about by one means or another were not, after all, surprising in the underdog, as represented by underprivileged classes and defeated or unliberated peoples."[14]

Taking a very long view of civilization, Toynbee anticipated that the creative springs of our next epoch would surface in the underclasses:

"The historians of AD 3047 will, I believe, be chiefly interested in the tremendous counter-effects which, by that time, the victims will have produced in the life of the aggressor. By AD 3047, our Western civilization, as we have known it... may have been transformed out of all recognition by a counter-radiation of influences from the foreign worlds which we, in our day, are in the act of engulfing into ours..."[15]

This counter radiation may happen sooner than he supposed. Current migratory, technological, economic and climatic pressures are driving it to the surface now. And I can identify its carriers.

***End Notes***

[1] Toynbee. *A Study of History* vol. 3 p.288
[2] Tolstoy. *War and Peace.* Pt II. Chap 7. Modern Library Edition. p.1134
[3] ibid p. 1121
[4] MacLean quoted in Restak. *The Brain.* Bantam. 1984 p. 136-7

[5] Jerome Kagan. *Galen's Prophecy*. Basic Books. 1994. p.172

[6] "Robert Adamec has suggested that local chemistry within the amygdala influences the probability that an event will evoke greater activity in one or the other of the two major fiber bundles that leave the amygdala...These two bundles and their targets mediate different behaviors in response to novelty." Ibid p. 104

[7] Rudolfo Llinas. *The I of the Vortex*. MIT. 2001. p. 242

200 ibid. p.124, p. 121

[9] Kauffman. ibid p. 236)

[10] *The Blind Watchmaker* p. 304

[11] ibid p. xii

[12] Gerald Edelman. *Bright Air, Brilliant Fire*. Basic Books. 1992. p. 87.

[13] John McCone. *Going Inside*. Faber and Faber. 1999. p. 167.

[14] Toynbee. The Present Point in History. p.29

[15] Toynbee. Encounters Between Civilizations. p.190

# 12. NEW HISTORY

## 220

Some day historians will be able to show how the energies of approach-and-separation, withdrawal-and-return and dispersal-and-aggregation underpin large-scale historical movements. They will demonstrate that the perturbations of outward events on those endogenous rhythms, especially as they near their turning moments, make life changeful. By doing so, they will join historical narrative with evolutionary psychology and give them both a richer more realistic ethological, physiological, phenomenological, environmental and sociological basis. They will be able to link the phenomenology of inwardness, including the experience of moral discovery itself, to brain science. They will explain how the Golden Rule, and the tried and true values, refresh themselves and are made relevant for changing historical circumstances by neural revisions to the modular structure of personality. They will explain how these enter the new passing rules and how the new passing rules bring different sets of the neutral traits to the fore, revising and expanding our human possibilities.

Historians will be able to show that the starting place of a civilization's influence on human nature is always in the inputs and impacts it makes on the human body. They will demonstrate that the strongest waves of influence are assimilated in the early years of life when, given the plasticity of the brain, fundamental circuits are made and remade. They will illustrate how these have been conditioned by cultural influences at different times in different ways (much like the capacity for language that, though universal, always provides the children with the language of their parents.)

With this new conceptual framework, we will be able to monitor the sensitive zones where the body touches culture, starting with the three main interfaces described earlier: selective sensory stimulation, internal self division based on the culture's passing rules and cultural inhibitions on action (see # **215.**) We will come to understand how cultural constructs influence our physiological functions by reshaping public spaces and reconfiguring social and economic exchanges between people, between man and woman, woman and child, man and his children, in sibling and kin relations, between compatriots, between strangers.

Historians will demonstrate that the strongest waves of influence come from the institutions that organize space and time, surfacing in the layout of streets and neighborhoods, in the sequence of holidays, in the setting of work hours, in the EM fields surrounding the body, in the technological rhythms influencing physiology.

We will understand then how every civilization shapes love and wisdom differently, encouraging, inhibiting, conditioning and in other ways influencing our biological tendencies by perturbations made to the basic rhythms, disturbances that get to us through the presence of persons, objects and events in complex combinations. We will be able to trace how these shape our upbringings, the developmental crises we go through, the contents of our consciousness, and our take on the meaning of life itself.


## 221

Every major historical approach/separation or withdrawal/return pattern plays out through many mini-turnings. Possible revisions to the passing rules open at every turning point. In the periods of consolidation between turning points, play itself opens us to the neutral traits. In historically charged moments the mini-turnings, though they may come too fast and not last long enough, become the training ground for the development of new options. George Leonard described mastery as the skills we develop and elaborate between the peaks. These skills we carry into our subsequent turning dramas.

The neutral traits that find a place in the newly organized passing rules influence our transits along the legs of love and wisdom. This incorporation of neutral traits, via memes, into the passing rules brings the genius of ordinary people into the culture. Over time, these contributions transform the opportunities of people by supporting new criteria for success, belonging and meaning.

## 222

We will learn that the strongest culture shocks produce the most dramatic re-orderings of rhythms. Moreover, from them we make our steepest adjustments to the passing rules. Our new historians will be able to trace how they pass into human nature during significant life stage turning points. We will understand, then, how the newly instituted passing rules change the rhythms of life specifically. We will quantify these reset rhythms, showing how they alter their phase relationships with each other, and how they differently fuse with aggression, producing our violence and torpor, passionate engagement or indifference, our sense of what is true and false, of what we can and cannot do, our loyalties and rebelliousness.

We will appreciate then how thoroughly the byplay between the ethical and the technical mind, as taught by the culture from our baby days, bends character and influences temperament. We will understand how their apportionment uses the incoming and outgoing paths in the central nervous system differently (in our civilization the technical drawing more on exteroceptive, the ethical on interoceptive sensation.) We will be able to follow the routes by which each enters our belief systems and actions in some blend that colors the outward face of ambition and the inward sense of self-regard. Moreover, we will realize that there have been other alignments and distributions between inwardness and outwardness in other times and there will be new ones still to come.

The study of history will show that the deployment of interest between our technical and ethical interests goes a long way toward defining our ways of belonging in culture. It will demonstrate how our need for belonging with its visceral dangers of alienation, ostracism and abandonment in turn changes the way the body triggers stress and aggression.

## 223
### *Reason and Aggression; the Technical and Ethical Mindsets*

By blocking our access to the natural clocks that once entrained the organism to the pulse of nature, and by filtering so much of our social lives through electronic media, our current technological ambiance fosters an easier, earlier alliance between reason and aggression. The evolved fronto-limbic modifications to the stress response now operate under critical strains; too often, they actively *prevent* us from overriding the fight or flight response.

The liberated aggression with which we built civilization has by now so altered the signaling sources in nature, and so accelerated the rate of change, that we can find neither the time nor the means to restore perturbed rhythms before hurtling ahead to our next appointment with destiny. Instead, we employ the evolved neocortical-limbic links to join reason to aggression to counteract what we take to be the immediate threat. We soldier on as best we can. But we do it in culturally learned ways: by seeking technical solutions to what are essentially ethical problems. Even the capacity for empathy cuts both ways. Amalgamated with sympathy, it gives us real intimacy and continuity in the four loves and wisdoms. However, without sympathy, the same empathy delivers enormous powers of manipulation into our hands.

Once we defuse the aggressive triggers from their primordial roles as preservers of the rhythms of approach-separation and withdrawal-return, our hold on love fades and our devotion to wisdom flags. We are alienated from our inwardness and have only the resources of our technical intelligence to cope with our social problems, unpinned from our ethical intelligence, the seat of caring.

We have big problems to deal with. How can we survive in an overpoweringly technical world without an unshakable connection to the ethical mind? To be technical in all of our relationships without countervailing ethical inclinations is to be sociopathic. But how can we live with *only* a connection to ethical action because, despite our good will, our ignorance and incompetence will make us unwittingly destructive? How can we find a balance?

## 224

Knowing more about how love, wisdom and aggression intertwine with territory and dominance, the new historians will understand better what makes a person care or be indifferent to the happiness or misery of another, what makes one person laugh and another cry, one hopeful and another despairing, and for how long and with what evocative qualities. In every polity, and inside ourselves, we will study two spectra of aggression: one sparked by frustration over territorial and dominance arrangements, the other, equally insistent, coming directly from frustrations to love and wisdom, and we will recognize that the openness of cultural institutions to alteration by love and wisdom will vary with the prevailing kinds of aggression. And we will learn to keep a balance between them, understanding that technical predominance over ethical sensibilities will lead to pathologies of lovelessness while ethical dominance over technical interests will forestall research and enterprise.

We will see aggression broadly, even compassionately, then. Its 'aggredi' and 'agonal' aspects will be respected, the nuances of its creative energies cultivated, the "themness" in its militant enthusiasm tempered and diminished, reason not so exclusively fastened to aggressive projects as to lose its primal bonds to love and wisdom – you might call it a more philosophical attitude.

There will still be heroes in this kind of history – the exemplars of love and wisdom who rise into prominence as great creators and doers of deeds – how they handle aggression, how they live in and shape their communities, the memes they generate – the way they typify their times.

Armies will still march in this new way of writing history and dynasties will rise and fall and the price of bread and the changes of climate will be basic data, but discernible beneath the surface of what Fernand Braudel called the "froth of history" will be the deeper swells that undergird change in the mesocosm, the historical middle zone, which is the human body itself.

The great periods in human history will be seen to be great because great shifts in love and wisdom and aggression occur in them.

## 225
### *Fan Shaped Destiny*

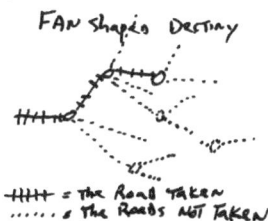

FAN shaped Destiny

-HHt = The Road Taken
....... = The Roads Not Taken

**Bifurcating pathways form
a fan shape**

With these historical techniques in hand, we can now build a metahistorical model that will account for the collapsing probability waves of alternate pathways. It will represent all that did not happen of what could have happened within what actually happened along all the bifurcating rays of our historical turning points.

281

Our historical development conceived this way would take on the shape of a fan. It will trace out what I call our fan-shaped destiny.

By this, I mean that the complexity of life is such that from each turning an indefinitely large number of possible pathways diverge, each leading to new turnings generating more alternative pathways. Therefore, we can picture our destiny lines like slats on a fan radiating out from our turning point moments.

At each turning the fan snaps open, its radials of possibility push out in many directions. Each slat of the fan represents an alternate route through life. But only one road can be taken. The first step down one path of approach, separation, withdrawal or return forecloses on the others. The causal connections to the past legs are attenuated, weakened, and then broken by what you do. Whatever you do next sets the seal on the last phase, closes it out, whether for good or bad, and starts you on the new path that leads to the next phase of life. A physiological resetting accompanies this as new passing rules congeal.

Though you can take only one road, each radial has a different likelihood of being taken (and in every present moment, we can plug in different numbers to try out possible scenarios). As lived, it is our bearing in the moment, our starting gesture, the firmness or trepidation in our first steps, and the texture of the immediate surroundings, that set us on the paths leading to our outcomes and their likelihoods. Once we take that step, the wave function collapses and the "roads not taken" disappear.

### 226

From all of our turnings in love and wisdom, our fan of possibilities extends into the future.

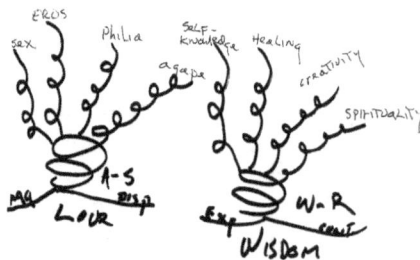

From time to time, we change paths because every turning is a junction point where each kind of love and each kind of wisdom can transform into another, and love and wisdom can become each other.

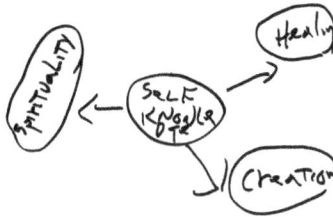

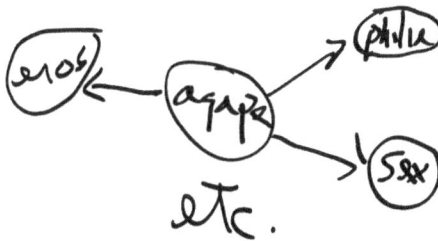

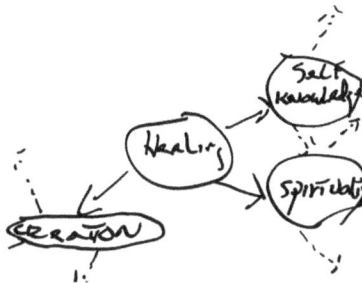

## 227

The better we learn to move along the radials from one turning to another, the more complete our understanding of life becomes and the clearer our intentions get. As a result, we achieve more of what we want in our moments of freedom. Moreover, the fuller we grow the wiser and more loving we become. As our experience of life deepens, the goals we choose tend to favor the convergence of love and wisdom.

The best fanning is neither the widest nor narrowest but the one that leads most favorably toward the next full turning. And the vectors most consequential for future turnings appear during the signal events in life.

In matters of greatest concern, the best fanning means the one that makes the fullest use of freedom when we have it. Living this way, our destiny unfolds not as a defensive bulwark to protect what we already know and value, but as a tendril growing toward what we do not yet know and have not yet been and what the world may not yet value. In pursuit of new meaning and value, we live creative lives.

## 228

One cannot fully know where the fan of destiny will lead. No one can read the future. Its possibilities radiate out unpredictably from its hubs. Yet in retrospect, the paths we actually take seem to have been necessary, even inevitable. Why is that? Because we only know the paths we have taken. The many bifurcations along the way leave hardly a trace in memory (though with special efforts we can recover them.) In the big turnings, we focus too sharply on outcomes to register the collapsing probability wave of all we will not do. In retrospect, the major bifurcations in the pathways of choice may look like inconsequential events. That's because the conditions in the turning region are particularly responsive to small inputs. They only offer themselves in brief moments that quickly pass.

All we know is that the deeds leading to and from the turning moment have changed us. What we did and what we suffered changed us at a given moment. Looking backward the path to the hub of change makes perfect sense, as if it could not have gone otherwise. Nevertheless, try to look ahead to where there are yet no hubs. The way seems utterly mysterious and unknowable.

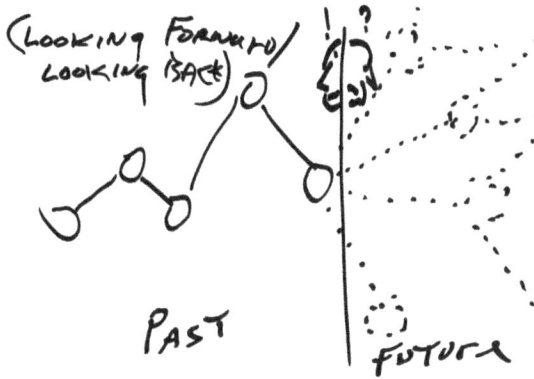

**229**

Unfortunately, this already complicated picture turns out to be too simple. The fan we have drawn posits an atomic individualism never found in nature. All our social behaviors are transactional. So we must conceive of the fan-shaped destiny that happens for each person as happening for everyone simultaneously. That is to say, our turnings influence each other and interpenetrate. They are parts of a larger pattern. Our destiny lines cross. We spin our lives out on a billion interleaving fans of love and wisdom (and hate and folly, their mirror sides) and we generate our destiny from the os-cillations and perturbations of innumerable spots of freedom along the way and their unpredictable outcomes.

When viewed in the frequency domain, we can model human reality as a vast array of intersecting fans: your fan, my fan, every-one's fan, and fans coming from fans, growing, changing, our des-tiny lines collapsing, crossing, joining and parting with great complexity.

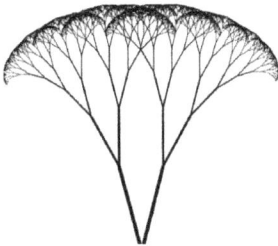

We can envision the overlapping fans taking the shape of a tree, the grow-ing, living and indeed phototrophic tree of life, turning toward the sun, seeking the energies that built it. Bifurcations, branchings out, are arguably the basic form of change in life in all the orders of nature.

**Bifurcation Tree**

285

Too simple still. there's not just one tree. We are all trees. The vision of life is less a tree than an orchard, a *Pardes* in the Cabalistic rendering. The same earth, water, air and sun nourishes all trees.

**230**

Complexity theorists could formulate a mathematical description of the growth and decay of the bifurcating trees of possibility in this orchard by treating the rhythms of nature as *oscillating* systems rather than instances of *expansive* events brought to their "tipping points" by the explosive consequences of initial conditions. When Stuart Kauffman writes, "the flash of the Big Bang 15 billion years ago has yielded a universe said to be expanding, perhaps never to fall together into the Big Crunch. It is a non equilibrium universe filled with too many hydrogen and helium atoms compared with the most stable form, iron,"[1] he reflects the consensus view that the laws of nature are compounded out of a few perturbable but unidirectional starting elements: 1) the initial conditions in the early universe in the Big Bang, 2) the "simple" laws of physics, 3) the strong forward bias of time's arrow in this stage of the cosmic unfolding. However, many tipping points are actually turning points in oscillating systems that our intellectually captivating, expansive, accelerative, unidirectional sensibility hides from us. Perhaps Hiroshima and Nagasaki seared the explosive quality into the sensibilities of the scientists of that generation who then passed it on to their students.

By shifting the emphasis to mathematics of bifurcations within a field of nonlinear oscillating systems, complexity theorists might develop a chaos mathematics to the base two that uses data on ultradian, circadian and other biological rhythms in the frequency bands we identified earlier to describe the processes by which events succeed each other and life and death embrace. They would study the edge between chaos and order in terms of recursions, renewals, reversals and returns in oscillating systems.

## 231

With a certain frivolous fancy, we can even hypothesize that reality comes from an interference pattern, a kind of standing wave cast into time by the generating sources of love and of wisdom. These two basic oscillating systems, in this fancy, constitute the *branes* at the fundament of reality. Like the Buddhist universe dissolving and recreating itself many times a second, the two great wave fronts of love and wisdom would oscillate, make waves and interfere. The interference would create a hologram that we experience as our universe. There would be epochs of relative harmony following others more dissonant, from the perfect octave, to the fifth to the dissonant tri-tone. On occasion, the reversal points would line up with each other bringing on a universal collapse. Then a new universe would be born. In some universes, there would be life.

In some universes with life, there would be consciousness of different grades and in those consciousnesses always some kind of balance or imbalance between love and wisdom would manifest. The way they "interfere" would constitute the active life of those beings. Humans on earth, (possibly other animal species here,) evince love and wisdom. There may be others in our own or other galaxies. If so, they would also be created as holograms cast up from vibrating fields of approach/separation and withdrawal/return. In this model universe, sentient beings, once they mastered the meta-historical patterns of fan-shaped destiny, would be able to understand each other. They might even love each other. Or come to hate each other. Or misunderstand and misconstrue each other. Or grow wise together.

\* \* \*

### End Notes

[1] At Home in the Universe. p. 19

# 13. MY GOLEM

## 232
### *Techno-Enthusiasm*

The confusion between ethical and technical thinking reaches a peak of insensitivity in the writings of our techno-enthusiast futurists who overlook the role of aggression in human affairs altogether. The strivings of the heart are matters of indifference to them. Love and wisdom have no place in their paradigms. Consequently, they gravely misapprehend the human potential. Take Ray Kurzweil as a case in point. His best selling books portray a future based on accelerating technological advances in an entrepreneurial, democratic culture, thriving on neural implants, bionics, virtual reality, nanotechnology, biotechnology, life extension, and thinking computers. He states his basic position unequivocally:

"Stopping computer technology, or any fruitful technology, would mean repealing basic realities of economic competition, not to mention our quest for knowledge. It's not going to happen. Furthermore, the road we're going down is a road paved with gold. It's full of benefits that we're never going to resist..." [1]

According to Kurzweil, no limits to growth exist. If we run out of resources, we'll shuffle available atoms around and make new supplies. Progress is more than inevitable; the rate we rush toward it inevitably accelerates. Even acceleration accelerates. The graph of "accelerating returns", which looks for all the world

288

like an erect penis, is about to get steeper, stiffer and longer (naturally).

"The answer we can predict from the Law of Accelerating Returns is that the ladder will nonetheless continue to reach ever higher, implying that humans need to become more capable by other means. Education can only accomplish so much."[2]

Darwin had a *theory* of evolution. Kurzweil has a *law*. If he were standing on the peak of his graph, betting on the next leg up, he would, in a collapsing economy in a world full of violence, be headed for disappointment if not outright ruin.

According to Kurzweil's law, we are approaching a "Singularity", a sudden, unprecedented change of state that makes the future scarcely imaginable. The old paradigms will no longer apply. Conveniently for the lapses in his reasoning powers, the Singularity draws a curtain over all speculation concerning the outcomes of the trends leading to it. Nonetheless, "the Singularity will allow us to transcend these limitations of our biological bodies and brains. We will gain power over our fates...There will be no distinction, post-Singularity, between human and machine or between physical and virtual reality."[3]

Because the brain's wetware cannot keep up with the rate of change, "the only way for the species to keep pace will be for humans to gain greater competence from the computational technology we have created, that is, for the species to merge with its technology."[4] He tells us "these technologies will create extraordinary wealth, thereby overcoming poverty and enabling us to provide for all of our material needs by transforming inexpensive raw materials and information into any kind of product. [5]

There are perils but the promises far exceed them. A few decades of economic contraction? No problem, a miniscule tick on the graph. A hundred million civilian deaths in war? A blip not worth mentioning. Hundreds of millions starving? An unfortunate holdover from an outmoded system.

Kurzweil's vision of accelerating acceleration cannot conceive of reversal itself as desirable, or more significantly, as inherent in moral and intellectual growth. The material culture pulls us along. The forward momentum is irresistible. Progress has a technological foundation (now conflated with self-organizing complexity.) Reversal for the techno-enthusiast is a setback, a fall into atavism. No turning point dramas open to the new. The psychology behind the notion that the real interests of humanity cannot progress outside of material advancements is widely shared by popular intellectuals. But it makes numerous culturally relative fashionable but spurious assumptions: 1) that we are goal-seeking animals, 2) that material advancement is our goal, 3) that intelligence is the way to it.

The world seen through this lens is cold. Happy and peaceful it may someday get to be, but cold. Intellectual delight stands at its pinnacle – delight coming from the beauty and excitement of scientific discovery and its applications in a life of creative play. It is not essentially motivated by love, wisdom, aggression or self-examination. In fact, the techno-enthusiasts exclude them from their palettes. Wisdom they subsume under technical intelligence. On love, they keep silent.


## 233
### *My Golem Theory*

In my student days, I devised an equally misguided but much more imaginative technological future vision. To me, history unfolded in a series of externalizations of human capacities.

We were replacing our physiology with technology and in the process losing our primary relationship to nature. The historical momentum, as I understood it then, came out of three successive, overlapping and ongoing revolutions: the machine, the computer and the sensor. In the first, we gave our machines our movements; in the second, we gave our calculators our reasoning. In the third, we gave our monitoring instruments our senses.

Our technologies resemble our body systems in very significant ways. The simple tool increases leverage, gives us mechanical advantage. We're manually attached to a tool. Its movements follow closely on ours. A sensitive sculptor, for instance, can feel the stone through his chisel. He can work on it with augmented muscle power. Machines, however, go by themselves. They no longer work as extensions of muscle and joint but as replacements for them. When labor saving devices replace the simple hand tools, we lose our visceral knowledge of mechanical advantage.

In the Industrial Revolution, accordingly, our emphasis changes from manual to rational control of our productive efforts. The relationship between the brain and the hand changes under the historical pressure of industrialization. The factory with its production line of machines and men exemplifies this.

And what happens to our musculature? Of course we continue to use it, but for fewer people (or so I thought) would laborious muscular movement be of survival value. I believed then that our essential sense of ourselves would have to change. The part of it that we externalized in industrial machinery would come back to us in a different way: muscle would be aestheticized and go to sport. I foresaw a great increase in the importance of sport. Behind it stood the need to relocate our essential humanity to a place where it was still free and under our own control. I thought that physical playfulness, amped up, would create entrepreneurial opportunities that would carry professional sports and social dancing and outdoor activities to new heights in a mix of technology, media, fitness, fame, wealth and heroic activity.

The serious work left to be done I conceived as brainwork, because powerful machines that move by themselves had to be *governed.* We have to be able to turn them on and off, speed and slow them, adjust their activities, change their functions, and the essence of governance was human choice – and choice came from brainwork. We controlled technology by hand, as we used our tools before, but now we did it by turning dials and making various adjustments. The hand movements were easy. Sequential thinking mattered more.

In the course of the second revolution, as the industrial age matured, the machines became too complex and fast for hand governance and we had to develop automated control devices, servomechanisms, governors, and we gave the governors on our machines our reasoning powers and autonomy to carry out their tasks. Steam engine governors and weaving machines show it. The character of the second revolution was that, just as we needed to externalize our muscles before, we needed to externalize our brains to control machine technology now. About this time, 1821, out on the cusp of technology, in advance of the need, though a harbinger of it, Charles Babbage and Ada Lovelace designed the first programmable computer.

When the computer revolution came in the fifties, I wondered what would happen to the brain inside the human head. What would keep it from becoming vestigial? Just as the body went to sport, the mind I figured would go to intellectual play. It would maintain its vigor in the absence of serious applications in a kind of rarified research and speculation. The disenfranchised brain would pride itself on abstruseness and obscurity. The intellectual innovations would be difficult to follow and so would become a sport of the mind for an effete academic and technical elite, something like Castalia in Hermann Hesse's novel, *The Glass Bead Game.* This may be happening now.

With muscles and brain externalized, what defining human characteristics were left to us? I thought it might be art. Artistic expression would become the cutting edge of human endeavor. However, by my theory a third revolution would overcome the arts and make then trivial too. The third revolution was required by the computers and the robots. They had to have ready access to information. The computer needed to open its own sensory pathways. The army of programmers and data entry clerks would be too slow. The electronic brains would be starved for data unless we could provide them with direct access to the world. We would have to externalize our sensory capacities in a technology that would feed the computerized culture without human intermediation. The third *sensorial* revolution, as it gained momentum, would challenge our humanity in unforeseen ways. Our eyes and ears would not be needed; they would be too slow and coarse. The world we woke and slept in and the culture that fed and clothed us would be incommensurate. All the real exchanges of goods and knowledge would be invisibly fast.

Without muscles, brains or senses, everything about us would become vestigial. Once again, we'd have to wage a struggle for identity. We'd have to find our freedom. We'd have to learn to use our equipment differently, and for new purposes. I figured we'd move to exotic tastes and smells. We'd have special sensory chauvinists just as we had brain chauvinists and muscle chauvinists. Just as the body went to sports and the mind to higher speculations, the senses would commandeer the free parts of human life by centering our interests not on content but on sensory impact, on fast cutting, on extreme blaring blasts of colors and tones, on sensationalism as a compensation for the lost power of our personal presence in the world, eventually making all traditional arts irrelevant.

There I was, a twenty-year old young man feeling superannuated, outmoded, if not vestigial. Where could I go, what could I do? In my lighter moods, I fancied I would lead the world into a new epoch based on play. We would play in order to discover in the unknown something unique about us, a higher purpose, and then through play develop new human capacities to explore it. I saw the crucial challenge in each technological revolution as the loss of freedom. I thought that whenever we were threatened with self-obsolescence (threatened by ourselves ultimately) we would grow restless, challenged by the loss of a unique property, restive, like children exploring, and by nature or instinct we would propel ourselves farther into the unknown.

This impulsion to freedom I considered basic and "inalienable," inherent in human nature, a tendency closely related to child's play. It put us into "search mode" driven by a biological imperative that surfaced in play. All mammals do it to explore the full capacities of the genome, possibly to overcome restriction. My maxim became to live to justify the tonic sympathetic drive on the body, to develop a level of activity that fully utilized the sympathetic arousal we generated in the stress of modern life. We would discharge surplus arousal in aerobics, sports, feats of endurance, high-risk activities, fast cars, sexual adventures, etc. To live well was to strike a balance between the parasympathetic and sympathetic nervous systems and between the slowdown and speedup rhythms of the body generally, and that would give us a well-toned, empowered physical vehicle, a sheath, a cloak, for a destiny quest hidden in play romping ahead under the banner of insouciance.

But what could we quest after by playing our way into the unknown? What could we discover in the last zone of freedom? That's where I got stuck. Considering the great deeds people had done, a life of play would hardly be worthy of me. It wouldn't lead to real accomplishments. The only great deed would be to overthrow the system that made us trivial but I was too scared (and realistic) to try anything like that. Besides, I didn't see a way. I couldn't figure out what was missing in my logic and I did not realize how negligent and cruel my theory was in the face of all the suffering it ignored.

When I presented my work to my best Brandeis professor, Frank E. Manuel, the historian of ideas, he asked me a question that made me go blank: "What do you do with aggression?" I didn't know. I hadn't considered it. Not my own or the world's, or the aggression seething below the surface of the theory itself. What *do* you do with aggression? I've been trying to figure it out ever since.

* * *

## End Notes

[1] Ray Kurzweil. *The Age of Spiritual Machines.* Penguin. 1999. p. 130

[2] ibid p. 81-2

[3] Kurzweil. *The Singularity is Near.* Viking. 2005. p. 9

[4] *The Age of Spiritual Machines.* p. 81-2

[5] He continues: "We will spend increasing portions of our time in virtual environments and will be able to have any type of desired experience with anyone, real or simulated. Nanotechnology will bring a similar ability to morph the physical world to our needs and desires. Lingering problems from our industrial age will be overcome...Most significant will be the merger of biological and nonbiological intelligence, although nonbiological intelligence will quickly come to predominate. There will be a vast expansion of the concept of what it means to be human." *The Singularity is Near.* p. 397

# MORAL DISCOVERY

# 14. FULL ENGAGEMENT OF OUR POWERS

## 234
### *Unlimited Freedom of Inquiry*

The techno-enthusiasts forget that freedom of inquiry advances our non-material interests as well as our material ones. Indeed, our greatest discoveries have always been non-material and non-instrumental and these have come into the world with very simple technological mediation, for instance, in language use, in learning to write on soft clay with a stick, in devising the alphabet, in sitting down to reason together, in organizing group activities, and in caring, sharing and moral discovery.

Given these accomplishments, the mitigation of aggression and the restoration of its function as a conservator of love and wisdom (the necessary conditions for species survival) are likely to come, if they come at all, from nonmaterial rather than material advances, and from ethical rather than technical thinking. Self-preservation itself is a non-material concern, despite our prejudice for thick armor and powerful weapons. Spinoza understood this perfectly well. He insisted that self-preservation had an ethical core, because the mind too endeavors to exist. Thoughts that increase our acting power, he argued, survive better than thoughts that stymie them. We'd do well to heed him before we rush precipitously into the future.

"Happiness consists in man's power of preserving his own being; secondly, that virtue is to be desired for its own sake, and that there is nothing more excellent or more useful to us... Further... that we can never do without all external things for the preservation of our being or living, so as to have no relations with things which are outside ourselves." And the most important supplies we need from outside ourselves come from other persons. We can't thrive without help from others. "Therefore, to man there is nothing more useful than man – nothing, I repeat, more excellent for preserving their being can be wished for by men, than that all should so in all points agree... Hence, men who are governed by reason – that is, who seek what is useful to them in accordance with reason, – desire for themselves nothing, which they do not also desire for the rest of mankind, and, consequently, are just, faithful, and honorable in their conduct." [1]

He reasoned his way to this position. But you need more than reasoning powers to make sound judgments. You need to *care* who you are and be motivated by what you are doing. Spinoza may have built his system entirely on reason, but his "ratio" meant more than logic. It had emotion in it. Arne Naess, a contemporary Danish philosopher whose work inspired the deep ecology movement, argued that our contemporary take on reason, by its narrow focus on technical logic, completely distorts Spinoza's understanding of the mind and thwarts the fuller use of ourselves.

"The conventional translation of *ratio* is unfortunately 'reason'", he wrote, "But what we call reason today is enormously different from what the philosophers of the seventeenth century called *ratio*. When we are confronted with a vital choice of action, *ratio* indicates which choice is in accord with human nature or essence... We must imagine it more in the nature of what we call the voice of conscience, an inner voice that most often communicates through the emotions... a choice based on active emotions rather than passive ones." [2]

In the philosophy of full engagement I am espousing here, these desire-filled understandings are what give us the strength to go through our turning points in love and wisdom. They are the active emotions that provide the motive power behind our deeds. They *turn the turnings*. By Spinoza's definition the power of the ratio rested in the "emotions that are related to the mind in so far as it acts." Arne Naess quotes Spinoza as saying: "in a state that is not marked by emotion one makes no progress in anything that is essential to mankind".

## 235

The path to judgment that goes through wisdom I consider the growing tip of human progress. It stirs in each of us and seeks expression in every generation. We get to it through full turning points, most directly when love is convergent with wisdom (see # **97, 98** above.)

The kind of reason Naess and Spinoza advocate honors the fullness of human nature but takes a skeptical view of our eager fascination with quick fixes. Quick fixes don't go deep enough. They leap over inner contemplation of goals. Even so, I do not think Spinoza, who was an eager follower of developments in the sciences, and a high-tech lens grinder himself (I believe he built a microscope for Huygens,) would ask us to curb the pursuit of knowledge and invention wherever they led.

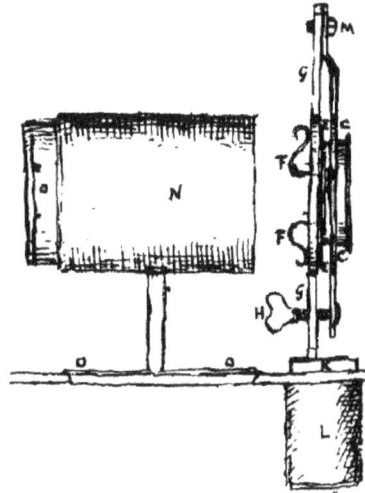

Spinoza's microscope lens designs

## 236

We need to find new ways of valuing life and communicating those values. The learning and teaching that Spinoza described, and Wittgenstein readmitted through his analysis of real-life language usage (# **168**) generate the most consequential of all discoveries: the discovery of new ways of being and belonging in the world. With new belonging, we will discover in each other's intimacy the nascent abilities carried in the neutral traits (# **212-214**.)

With our turning points restored and our legs of approach, separation, withdrawal and return strengthened, we will get the courage we need to employ the creative energy of aggression trustworthily.

William James thought we might be able to bind up aggressive energy in a "moral equivalent of war". By focusing on worthy cultural goals, he hoped that "the manliness to which the military mind so faithfully clings... intrepidity, contempt of softness, sur-render of private interest, obedience to command..." would have a role, possibly inaugurating a "reign of peace and in the gradual advent of some sort of socialistic equilibrium." In his private cor-respondence, he was not optimistic about the chances. Writing to H.G. Wells, (perhaps reflecting on his long experience as a teacher of the privileged youth of America at Harvard) he deplored "the moral flabbiness born of the exclusive worship of the bitch-goddess success. That–with the squalid cash interpretation put on the word success-is our national disease."

Our safest, best and most adventurous path into the future would use the technical mind to reject the current model of accel-eration and linear progress as outmoded and worn through. It would treat it as one among many conceptual systems, the resid-uum of a particular intellectual, political and economic order. It would pursue instead a model of intensification and reversal be-cause it fits the evidence of physics and biology more succinctly. The technical mind, on this pathway into the future, would design strategies for a world built of rhythms and turnings. In this setting, (a field of endeavor I call our fan-shaped destiny, # **226, 227,**) the core virtues would reflect the rhythmic substratum of nature and human nature. With a changed model in a world of nonlinear oscillating biosystems, we would be much more likely to recognize the centrality of love and wisdom to the conduct of life.

* * *

### End Notes

[1] Spinoza.(Ethics. IV.xviii. Note)
[2] Arne Naess. *Life's Philosophy*. University of Georgia Press. 2002. p. 10

# 15. ON LIVING BRAVELY

## *Subtle Sensation and Non-Reactivity*

To navigate well along the routes of love and wisdom you need to acquire subtle sensory awareness skills to help you clearly discriminate internal body sensations (heat, pressure, pleasure pains, tingles, energy runs, etc.) from outer world experiences, particularly in the presence of others onto whom we project our internal realities and from whom we introject our interpretations of theirs.

To parse sensation takes bright steady attentiveness. But to *stay* in the stream of sensation takes more than sensory awareness. It takes non-reactivity. Acute sensory awareness without non-reactivity puts you on a hair trigger. Non-reactivity without sensory awareness makes you dull and indifferent.

With non-reactivity, you can rest in the moment with quiet readiness. Every moment has a bosom to rest in if only you can find it, though sometimes it's a scraggly one. Resting in the bosom of the moment diminishes defensiveness. Calm open awareness reveals possibilities for action. In this resting presence, we can learn to tolerate ambiguity and embrace paradox. We stay in touch with the deeper systole and diastole of life.

I am counseling comfort under duress. It opens you to each moment's terrible beauty. Even in the hardest times, you can find beauty. Goethe believed that beauty was "the manifestation of secret laws of nature, which, were it not for their being revealed through beauty, would have remained unknown for ever." (see # **313, 314.**)

My clinical experience in biofeedback shows me that with training most people can become far more capable of subtle discrimination than they suppose. And when this couples with non-reactivity, they acquire abilities to make surprising changes in previously inaccessible autonomic body functions.

They can not only attune themselves to the resonations of the body, they can *pace and lead* them.

With directed use of imagination, when it is coupled with subtle sensory awareness, they can change the texture of the interoceptive sensory field (which changes the body) and then allow the changed body to change the sensations so that a positive feedback loop develops.

With these skills, people change heart rhythms, selectively alter blood flow and skin temperature, recruit immune functions, change brainwave rhythms, and from a broader perspective bring oscillating body systems into phase with each other.

With interoceptive sensation, people can attend to the ebbs and flows of the limbic system as they pass, not unlike attending to a kind of emotionally labile music that expresses dozens or even hundreds of emotions in the course of a few minutes. With subtle sensory awareness, they can experience the interplay of bio-oscillating systems as if they were audio tonalities, feeling their frequencies moving into and out of resonance with each other.

With subtle sensory awareness, we can tune into more of ourselves and brighten our presence before others. In social situations we can pick up each other's delicate changes in complexion, thermal fields, pheromone flows. We can sense the tremor of intent in another person and observe it changing from instant to instant. We can acquire knowledge of the character of another through eye contact.

We can even get a direct take on the brain's functioning by experiencing the phosphenes and scotomas generated by the play of impulses from the optic nerve on the retina with eyes closed.

## 238
### *Intention and Imagery*

With subtle sensory awareness, non-reactivity and the strength to know our wounds, we can learn to frame strong intentions when the turning moments come.

Future actions depend on our intentions, not least the moral character with which we face choices in love and wisdom.

A broad look at intentions reveals that:

o   We establish our most consequential intentions on the long legs of love or wisdom. Some intentions hold steady, some change, and some fleet by at the edge of consciousness. Some go and then return.

o   Intentions are more than ideas, more than intellectual acts. Intentions along their legs of love and wisdom and at their turning points closely correspond to Spinoza's positive emotions, "emotions that are related to the mind in so far as it acts." These he contrasts with passive emotions. Harry Wolfson describes these as "characterized chiefly by the fact that there is always an external cause which produces them, and that man himself is therefore only their inadequate or partial cause."[1]

o   With its rich mix of yearnings and hopes, we compound our intentions from a mood palette of interoceptive sensations and state-dependent learnings. We often incorporate bits of memories and fragments of remembered dreams into them.

o   Because they are future oriented, intentions want to be fulfilled. They *wait* to be fulfilled. This makes them forward seeking. Hence, they carry within them imaginative constructions of the time-line. They are acts of imagination that include a fable of origins, a tale of changes and a vision of outcomes.

o   Intentions not only *refer* to the future, they *go into the world as timed events* set to be released at a particular moment in that future.

o   We try to send intentions into the upward communicating channels of freedom when freedom of thought briefly opens to the freedom of action in the turbulence of the turning. Then and only then do yearnings metamorphose into the kinds of choices that cast out seeds. These seed-forms first manifest as oscillations in the central nervous system and then, through the body's own rhythmic connections, influence the rhythms of the world.

o   To know what you are intending you need to have a vividly pictured goal. Intention and imagery work together. The intention/imagery complex comes from the body and speaks to the body. It produces visceral

303

rushes, rumbles, frissons, distinctive heart rate and breathing patterns. We can observe these in ourselves with subtle sensory awareness and non-reactivity. By these means, we can distinguish intention and imagery from fantasy.

o A fantasy is an imaginative production in which the "you" having the fantasy is swept away. Time passes and you don't notice it. You react to the inner adventure with all its perils, triumphs and losses as if you were there. The fantasist breathes to the story with no more consciousness of breath than a dreamer has of rapid eye movement. By contrast, the silent observer stays present in imagery work and guards it. The observer maintains perspective. The breath stays conscious.

o A desired outcome, experienced in fantasy, no longer needs to go into the world. Fantasy discharges the energy of hope, while intention potentiates it.

o We are naturally endowed to make imagery. It wells up spontaneously as we approach sleep as *hypnogogic* and before we wake as *hypnopompic* imagery. It blossoms during interoceptive states in love and wisdom, inspiring art, poetry and invention. Imagery is the natural language of the body to the mind. As you ponder your intentions, images form. Imagery comes from the body and works its way into consciousness under the influence of the intention. It gets its vocabulary (rebus-like, in hieroglyphic form) from the body itself and speaks back to it in a kind of baby talk, which sometimes gives an intention/imagery complex the quality of a jingle whose rhythms go round and round and whose images play like film loops. Joseph Campbell, who pondered over the origins of myth for more than sixty years, came to the view that "From the outer world the senses carry images to the mind, which do not become myth, however, until there transformed by fusion with accordant insights, awakened as imagination from the inner world of the body."[2]

**239**

Now for some details.

Propitious Moments for Framing Intentions: Love

Intentions, though they can be planned, thought out in advance, rehearsed, troubled over, still have to be exercised. They must be seeded into the turbulence of life at a specific moment. The best opportunities for framing intentions come during solitude. Times of inwardness occur in both love and wisdom, though wisdom solicits much more of it. In both dynamics, the solitary times blossom up at the ends of the long legs as we head into the turnings. In love, we find them in the transitions from furthest separation to new approach and in wisdom from deepest withdrawal to return. Intentions framed then and sent into the world take hold best.

In these inward periods, exteroceptive sensation diminishes and interoception increases. The shift frees the imagination. It builds its own landscape and inhabits its own fabled world.

At first imagery functions as a backdrop to other cognitive functions, a penumbra, unobtrusive, hardly noticed, a stage-set, like a decorated proscenium arch in a theatre, framing pragmatic thinking. However, if the interoceptive state continues long enough, as it does during a sig-

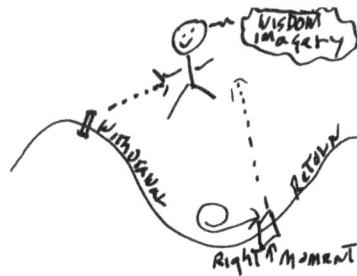

Propitious Moments for Framing Intentions: Wisdom

nificant withdrawal or separation, the images jump off the proscenium and come to life; its boundaries soften and dissolve; imaginative productions grow stronger, more colorful, more redolent, vivid and multi-sensory. At some point, the imagery crosses a threshold of relevancy that captures our attention. Such captivation can be terrifying or tempting, coming when the naked self has trouble holding steady. The intention, now buoyed up on imagery, takes over and provokes physiological responses in the same way a sexual fantasy does, or a lust for power or the premonition of disaster.

## 240

The imagination's empathetic images, its night dreams, its fantasy arousals, its conscious efforts, worm their way into the body. The body receives the image the same way the thought of biting into a lemon stirs the salivary glands. But it stimulates other glands, produces other signaling substances, sets specific ionic transport mechanisms going that open membrane bound protein gates. The imagery/intention complex uses body organs to express its energy.

Whatever else it may become, the imagery/intention complex resides in the body as a resonant biological system. Its message content travels in signals between organ systems along pathways starting in the brain. These pathways form or break as the result of neural competitions. (See discussion at # **215.**)

This means that imagination itself is able to provoke genomic impulsions to action. Some such impulses, sparked by regulatory gene combinations may call out new affinities, attitudes, feelings and behaviors, new territorial and dominance interests, new sexuality, new sexual selection standards, new belonging. These enormous changes do not come easily – and they do not manifest frequently – but when they do happen in a significant minority of the population in a relatively brief time-span they can well up into moral discoveries.

## 241

How can we best use our intentions?

o We use intentions to organize future opportunities by imagining them vividly, usually, as if the desired results have already occurred. Imagery teachers call this "goal rehearsal imagery."

o The more we learn from the body by receiving and interpreting its spontaneous imagery, the more effectively we can frame intentions that talk back to the body through deliberate imagery.

o Intentions that lead to freedom have imagery in their core.
o An effective imagery/intention complex leads to freedom of action when it has been timed to the turning moment (see # **107, 108, 112-117** on indeterminacy.)

**242**
*Skills of Timing*

We fortify ourselves for action in a world of pulses, waves and oscillations by engaging our intentions with the rhythms of nature. We make day and night, gravity and mass, balance and agility, breath and heartbeat important guides to the timing of our actions. With small, subtly timed nudges, we try to bring the fullness of ourselves, including our intentions, into alignment with natural and cultural patterns in the turning drama.

Timing is a wisdom concept. The sensibility that there is "a time for every purpose" sits at the core of Solomonic wisdom. To know the right times you must obviously experience the flow of events in the world, and know the proper intention (in Hebrew *kavannah*) to hold in different circumstances. With Solomonic wisdom, one learns to penetrate the forces of the time, not to fight them, to learn their secrets, to work with them, to master the subtle qualities of intention that operate on the membrane between self and other, in the frequency domain, on the world's vibrating drum head.

King Solomon knew that reality had a variable structure, sometimes determinate and sometimes indeterminate, and that "everything has an appointed season, and there is a time for every matter under heaven."

These spots in time can be apprehended. Solomon represents them as dualisms:

*A time to give birth and a time to die: a time to plant and a time to uproot that which is planted.*
*A time to kill and a time to heal; a time to break and a time to build.*
*A time to weep and a time to laugh; a time of wailing and a time of dancing...*

But alternations don't only fall into patterns of two. Life comes at you in threes and fives and other numbers too. The phase relationships between them are complex, their turning moments overlap, even merge. The math behind it is nonlinear and difficult.

If to everything there is a season, there must even be seasons for the rare and recondite; though our knowledge of these times is limited, (some know more than others do.) Likely, some things people knew we have forgotten, and some we never knew we will soon learn.

**243**

With subtle sensory awareness and non-reactivity, and in the purview of the observing presence – *that* I call the *Watcher* - you can learn to construct a workshop for seeding intentions in love and wisdom. You can devise a special field of endeavor in which you bring time, opportunity and intention together. The Taoists who wrote *The Book of Balance and Harmony* speak of this when they say:

"To master change, nothing is more important than to know the time; to know the time, nothing is more important than to understand inner design; to understand inner design, nothing is more important than open calm."[3]

In my Western way of rendering this state of mind, I'd say that one can bring volition into play as the turning comes with right timing acquired through properly directed sensory awareness. With skills of timing and with immersion in natural rhythms, people can learn to steer imagery and intention into the turning moments in real-life situations, and in those moments invoke the powers of imagination as an embodied neural event that works as a lever for change.[4]

We all do it. Yearning for an absent lover recreates them in imagination. The longing can generate clear sexual arousal or feelings of love and devotion. The person active in your imagination, buoyed up by internal sensation, you hold present in an image amalgamated with an intention. We ponder the image, react and respond to it, enjoy it, suffer over it, plan about it, make moves. Our imaginings to some considerable extent steer the course of our relationships in all four kinds of love, particularly when the lovers experience correspondent imaginings on both sides.

In the same way, sensation, imagination and intention shape the cognitive contents of wisdom processes. The turning from withdrawal to return, for example, encourages a shift from inner to outer sensation and this, when it happens, awakens us to worldly possibilities.

## 244
### *Who is The Watcher?*

You need courage to study each moment for its readiness. The moment for seeding intentions may come only after brutal contests between rival intentions in opposing self-states. It takes practice and training to set your interventions into the slipstream of time with minimal disruption during the disintegration of the personality. To accomplish these tasks requires the presence of a least-operative self, the Watcher. Only the Watcher can consider all possibilities and yet be able to select one at the right moment.

To find the Watcher is an acquired not an inborn skill. You must seek out Who is left of me when I am lost in the turning drama. You must ask, What is it in me that feels the deconstructed scintilla of subtle sensations? Who does the intending? Who has perspective on a sub-personality while in it? Who recognizes a fantasy while having it? Who is the "me" that is there when "I" am gone? The answer: the Watcher.

I experience it as a bare center of awareness left over in the depatterning of the passing rules (See # **76-81**.) Do not confuse this still-point for the part of you that makes the ego's choices. It is smaller than that. I know it as a part of me that sits quietly by, observing other parts of me, small because it makes no claims and goes along with all changes, a kind of eye of calm in every storm, quiet but keen and alert, a way of being in the world characterized by equanimity, intelligence, dispassion, steadiness, imperturbability – a self but not too much of a self, a relatively selfless self, the least operative self still capable of attending, accompanying a decision and joining an action.

I first experienced the denuded I-sense in my teenage years. I discovered that when I was in a state of low arousal some part of me inside was always brightly awake, alert, energized, dancing with vigor and watching. Moreover, with this, I came to believe I had encountered the core, the inner eye, the place of the observer in me, and I called it the Watcher. I noticed that when we focus on something and we are aware we are focusing on something, neither the thing we focus on, nor our focusing is the Watcher. But the part of us that is aware we are focusing on something is the Watcher. It watches caring, but it doesn't care. No joy, no grief touches the Watcher. But the Watcher watches the joy and grief. However, the Watcher is not remote or indifferent; it is interested. Curiosity it has, and mild amusement. It is emotionally thin, uses few words, and mainly keeps quiet, but it can issue verbal commands and can intervene in action. In this least operative self, we have the clearest vision to recognize the shifting phases of the causal texture of our surroundings while having the least susceptibility to being swept away by it. It takes this small self to navigate the turbulence of the turnings and to seed intentions with proper timing.

When it finds the Watcher, the I-sense links to presentness itself. But presentness is not a simple state. It too passes. The Watcher functions with the same state dependency as our other identity states. While it holds steady, if we are quiet in the turning drama, the Watcher can focus on the intention without panic or clinging, without presumption or resignation to mystery and authority, enduring its uncertainty. The Watcher can plant intention in the turning moment by electing it, and with slight amusement, even in trepidation, can empower the deed that comes next. The non-reactivity of the relatively selfless self, however, does not make it passionless, because passion may be necessary for choice. The choice is fueled by passion but aimed by non-reactivity. As the surgeon's knife is steady but empowered by the fire of effort, so in our turning moments we are both passionate by virtue of our caring and yet, by allowing and observing this passion to flow, we are non-reactive.

Emily Dickinson was the American poet par excellence of the Watcher and the choices it could make. She celebrated it in many moods.

*Of all the souls that stand create*
*I have elected one.*
*When sense from spirit files away,*
*and subterfuge is done;*

*When that which is and that which was*
*Apart, intrinsic, stand,*
*And this brief tragedy of flesh*
*Is shifted like a sand;*

*When figures show their royal front*
*And mists are carved away, –*
*Behold the atom I preferred*
*To all the lists of clay!*

## 245
### *The Prime Doer*

I know now that what I called the Watcher when I was young has no spiritual status. It is one among many of our modular personalities, though not one easily accessed. With skillful efforts, however, as part of a meditative practice, one can reach it in quiet moments. It surfaces on its own in extenuating circumstances.

In my youth, I didn't understand the ontological status of the Watcher, or its special virtues. As I learned more about Eastern thought, I came to dismiss my experience as trivial, thinking that if Patanjali was right the Watcher would soon dissolve into a greater emptiness. But mine didn't, so I figured I was stuck at a lesser place. For decades, I looked for the deeper ground and though I experienced many unusual states of consciousness that fit the descriptions I came upon in Eastern books none of them seemed necessarily deeper.

Eventually I came to understand that the Watcher was less a witness than an actor. Its depth and virtue appeared in action. It held the stage at the instant the intended action began. It was less the Watcher than the Prime Doer for whom life was an adventure in onceness. (I only got a sidelong glance at it from the meditative state when I was young, so it looked like a "Watcher" to me. However, when I was meditating, I was just "practicing"; I was not at the crux of change. So I did not comprehend its power as a doer.)

The denuded I-sense, I discovered, held the all-important spark of common sense in it. I found that when the chaos came and my will, passion, epiphanies, and intuitions were scattered abroad, my common sense remained intact. Common sense is the *mind* of the Watcher/Prime Doer. However, that mind is only good for the simplest observations and the smallest choices. Yet when I look over my life those small choices have guided me best. The decisions I made, though softly enunciated, became the springboards for all subsequent action.

## 246

Having now covered the basics, we can devise a strategy for morally based worldly action on the following principles:

o  Intention, imagination and imagery held in a stabilized internal sensory field, under the aegis of the *Watcher*, provide the materials the diminished I-sense needs to steer a readiness into worldly action.

o  In this process, the imagination becomes the link between the soul and nature. Responding to the products of our own subtle sensory awareness, it decodes the chronobiological carrier waves generated on the mind/body edge.

o  By the strength of its presence, imagination can tip the turning process in a desired direction at the right moment. But only then.

o  The incandescent imagination, flaring with life in those charged moments, becomes the mightiest of the tiny neural agents that we can bring to bear to open the fan of our destiny.

But the crucial question remains: what makes these actions moral?

## 247

The Romantic Poets understood the power of imagination.

The literary critic Frederick Pottle, though he acknowledged that "it is hard for us nowadays to understand why Blake, Wordsworth, Coleridge, and Shelley made such a fuss about the imagination" goes on to explain that "Wordsworth and Coleridge were convinced that imagination was creative; and they wished to make imagination not merely creative but a power for apprehending truth."[5]

In Book VI of The Prelude, William Wordsworth focused directly on the power of the imagery/intention complex:

*in such strength*
*Of usurpation, when the light of sense*
*Goes out, but with a flash that has revealed*
*The invisible world, doth greatness make abode,*
*There harbors; whether we be young or old,*
*Our destiny...*[6]

He is saying that there is a 'flash' moment in the turning process when the light of outer sense goes out, and in this moment, we get a sudden glimpse of a reality to come. Some of what we construe then, decoded from carrier waves by physiological processes I dealt with earlier, tells us about ourselves, about our potentialities, but some tells us about the possibilities in human nature more broadly, or about the historical moment and its promises and perils. Taken together, these revelations become auguries of the future.

In the flash moment, much happens. One's model changes. One briefly experiences or rehearses what it would be like to live in the world in a new way. Nevertheless, we inhabit a joint field, filled with other living beings, so we do not have easy access to these elevated moments. To act and create in new ways in rare "spots in time"[7] depends on subtle alignments of inner and outer rhythms.

The "spots in time", because they are transformational moments, manifest in the hubs of turning points. In Wordsworth's accounts, they hit when reversals impended; they popped up in scary and ominous times, particularly in events that touched on mortality, as in his youthful premonition of his father's death.

Wordsworth believed that these experiences, if recollected appropriately on later occasions, would have the force of continuing epiphanies.

*So feeling comes in aid*
*Of feeling, and diversity of strength*
*Attends us, if but once we have been strong.*
*Oh! mystery of man, from what a depth*
*Proceed thy honours.*

In the concluding book of *The Prelude,* he makes clear that in our creative universe, imagination gives us our fullest link to the whole.

*This spiritual Love acts not nor can exist*
*Without Imagination, which, in truth,*
*Is but another name for absolute power*
*And clearest insight, amplitude of mind,*
*And reason in her most exalted mood.*
*This faculty hath been the feeding source*
*Of our long labour...*[8]

## 248
### *Planting the Imagery/Intention Complex in the Turning Moment*

When the consensus world crumbles and the ties to the past break and the model dissolves, you enter a state where body and mind grow more intimate with each other (I do not say *comfortable*) by virtue of these downfalls. Though fearful and desperate, one feels put upon, extremely exercised, at a limit, down to a last extremity. Physical presence, intensified by internal sensations, commands the mind.

In the near-chaos conditions of the turning moment, our attending, our conscious awareness itself, when attached to an intention made vivid by imagery, *generates* a phase resetting. In the turning, your own intentions perturb you. The instant you rise to the perturbation, a choice point comes, a temporal pivot from which the rays of your potentialities extend in alternate directions.

With the right training and motivation, people can learn to etch volitional efforts into the autonomic circuitry of the central nervous system, utilizing their own neuroplasticity as an ethical and aesthetic medium for change.

The raw ingredients for change come up from the neutral traits. (See # **213-214**) We develop them using the joined powers of the technical and ethical minds.

Without new tools for self-examination, however, we cannot walk the technical/ethical boundary zones, so we will not get far with our changes. Nevertheless, the upwelling from the neutral traits does push us toward self-examination and from it, we begin to build bridges between phenomenology and the world detected by instruments using the methods I have described here.

They produce effective action with the least expenditure of energy. They keep us from hitting dead-ends or running in circles. To push the levers of change we do best by employing low-mediated intentions.

## 249

In our highly charged, high power technological civilization, the whole notion of low-mediated intention with its architectonics of imagination must strike the reader as strange, self-deceiving, and even delusory.

What can we do with low-mediated intentions that we cannot do with brute force? The question hints at its answer. However, let's push the objections further. Can we, for instance,

1) use low-mediated intentions to shift the bifurcation point between stress and aggression to bring more peace into the world?

2) Can we rebuild the interfaces between voluntary and autonomic functioning to change both how we perceive and act?

3) Will any of these alterations, even assuming we can learn and teach them to others, produce more than personal meanings and intimate satisfactions? Doesn't cultural transformation require fundamental shifts that go far beyond the placement of low-mediated intentions?

4) Can we, in certain special moments at turning points in love and wisdom, send the power of self-directed change into the world on impeccable starting gestures that fire forth like Blake's arrows of desire? Can we shape worldly events with intentions made vivid in imagination? Can we release them, precisely timed to enter the world by looping through two turning moments simultaneously, one inside us, the other in the surrounding environment? What access do we have to turnings in the world?

I answer Yes to all four questions. We can learn to leap the gap between phenomenology and physiology.[9] In fact, only low-mediated intentions can accomplish 1) and 2). 3) takes knowing each other in the fourfoldness of love, and *engaging* each other, and the tenderness and respect for this are greatly helped by low mediated intentions. As for 4), impeccable starting gestures, precisely timed, move more effectively as the deeds arising from low-mediated intentions than as grand heroic exploits.

## 250

Empowered intentions make an impact on the world when big changes come from small shifts in initial conditions. Turning points, even local, personal, intensely private ones, reset initial conditions in some small part of the world. They evoke indeterminacy.

By holding our intentions steady and making them apt, we play parts in resetting those initial conditions in venues inside and adjacent to ourselves. By swinging into action at the right moment with the right starting gesture, we open a fan of likelihoods in the world. Each radial of the fan points to different possible outcomes, indeed, to different future encounters, meetings, engagements, to different possible lives.

Though the vectors on the last long leg determine the way into the turning point, the way out is free. To use that freedom and to achieve the dramatic transformations promised in a turning crisis, takes a precisely timed starting gesture energized by the active emotions held in the imagery/intention complex. The precision of the timing necessitates the smallness of the gesture. Undertaken at the right moment, the starting gesture, only a baby step in the right direction, stands high as an affirmation, a "yes" that has a "no" in it because the hard choices in love and wisdom always require us to give something up. We not only do, we don't do. What we don't do becomes the road not taken. What we relinquish matters excruciatingly. In turnings of love and wisdom what we care about most is at stake. But the relinquishment brings liberation into action, a lightening of the load finally, and a streamlining of the delivery system by which values pass from the self to the world and from one generation to the next.

## 251

To get the right starting gesture you must inhabit the temporal world. When time is your principal abode, you can rest on the crest of the moment and watch space go by. When you do you see eddies in the flow, standing waves, rhythmic reversal moments, Wordsworthian spots in time that portend new beginnings.

## 252

The deeds we undertake in the turning moment count most when they fall *closest to hand*. Inevitably, they start in the moment and place we occupy. Their hold on the world comes first from their personal relevance. They go on from there, from that spot in time (# **247**), by establishing rhythmic resonances with nature. These resonances link us with others. In certain social settings, our starting gestures generate entrainments that travel as memes. I explained the basics of mimetic transfer in # **215-219**.

## 253

Since our turning dramas intrinsically concern love and wisdom, our creative deeds, when we time them well, ride into the temporal flow on the active emotions of empathy and sympathy. They may even spread far from their points of origin, though the great moments for far-reaching change come infrequently. But when they do, they always start small in the hearts of individuals. Of these times, turnings and persons, Lewis Mumford wrote:

"At rare intervals, the most significant factors in determining the future occur in infinitesimal quantities on unique occasions. Such behavior is too erratic and infrequent to lend itself to repeated observations and statistical order... This doctrine allows for the direct impact of the human personality in history, not only by mass movements, but by individuals and small groups... At a moment of ripeness, the unseen will become visible, the unthinkable thought, the unactable enacted; and by the same token, obstacles that seem insurmountable will crumble away. This experience has many parallels in human history. Suddenly, at what seems the peak of their efficiency and power, dominant institutions lose their hold on their most devout supporters or their most favored beneficiaries; while at the same time, millions of people who seemingly conformed with docility to these institutions throw them off, like a dirty garment."[10]

By being present to our own turnings, we increase our power and range of choices in them, and that amplifies our influence on events. If we conduct ourselves well, we can invigorate the life around us in wider ways than we ever supposed. This maximizes our usefulness in the evolving freedom of the whole and is a source of happiness to us, as Spinoza expected.

### 254

I will say more about the starting gesture. It provides the jolt that makes the turning turn. The mildest of forces then works to reset initial conditions, producing a bifurcation, a forking in the road. Imagine that from this 'Y' two paths of potential diverge, though you have taken neither yet. You have never had to make a choice between them before.

After the turning, there's no going back. The 'Y' has vanished. The spot in time it occupied is gone. Alfred North Whitehead described this situation well when he wrote, "Locke's notion of time hits the mark better: time is 'perpetually perishing.' In the organic philosophy an actual entity has 'perished' when it is complete."[11]

Robert Frost deals with this circumstance in his poem *The Road Not Taken.*

*Two roads diverged in a yellow wood,*
*And sorry I could not travel both*
*And be one traveler, long I stood*
*And looked down one as far as I could*
*To where it bent in the undergrowth;*

*Then took the other, as just as fair,*
*And having perhaps the better claim,*
*Because it was grassy and wanted wear;*
*Though as for that the passing there*
*Had worn them really about the same,*

*And both that morning equally lay*
*In leaves no step had trodden black.*
*Oh, I kept the first for another day!*
*Yet knowing how way leads on to way,*

*I doubted if I should ever come back.*

*I shall be telling this with a sigh*
*Somewhere ages and ages hence:*
*Two roads diverged in a wood, and I –*
*I took the one less traveled by,*
*And that has made all the difference.*

At the 'Y' in the road, the smallest step down one path negates the other. The road not taken effectively disappears. Frost in his laconic way intentionally understates the uncertainties of that moment when he says, *"Oh, I kept the first for another day!"* He immediately adds *"Yet knowing how way leads on to way, / I doubted if I should ever come back."*

He meant to take the less traveled way, but whether the road he followed was actually less traveled, who knows? Frost admits this when he says "Though as for that the passing there/ Had worn them really about the same."

To my way of thinking, all roads are new roads. The road not taken and the road taken both only exist as tokens of possibility, figments of the imagination rising in the moment of choice. The forking in the pathway is itself small, insignificant, inconsequential as viewed from anywhere but the path itself. The Taoist maxim that the journey of a thousand miles begins with a single step catches the feeling. However, the hidden truth is that the special conditions created in the region of the turning do not happen every day. The big thing is to make the choice and take the step when they do, and to know you have done so. That's what Frost celebrates.

To make a real choice, to have the alternatives vanish, to move from the hub of a fan to its radials, you must achieve a full alignment of the hierarchy of freedoms. Imagining, intending, willing and acting all must come together in the moment. Otherwise, the indeterminacy in the bifurcation will not rise to freedom. When one or more freedom is missing, you make illusory choices.

## 255

In real life, the real transformational opportunities don't appear on command. They're not conveniently scheduled to meet your hopes. Nor does the alignment of freedoms present itself through an act of will. You seize the moment because the moment is ready to be seized. Tremulous and ripe, the seizing is a gentle not a rough act.

To some extent, an authentic previous turning sets you up well for the next. But past success is not a prerequisite or a guarantee. Life throws strange obstacles in our way. We are fallible, so we often fail.

## 256

It is all about doing (See # **1**.) Behind the intention and the colorings and textures of its imagery, we are *managing a biophysical event happening inside us*. We manage it by bringing the legs of love and wisdom, approach, separation, withdrawal and return, into new momentary relationships. We try to line them up with happenings in the world. We want them to be apt. To get this alignment to happen, we alter the phase, frequency and amplitude relationships in our own bio-oscillatory systems (See # **19-22**.) We bring them under partial voluntary control. (Illustrated in detail for self-directed healing in Chapter 16.) We use intention, imagery, timing, gesture and deed as *tools*, as techniques. On the mind-body edge, we employ the technical mind to ethical purposes. There, on that edge, at the vanishing 'Y' in the moment of choice, physics and phenomenology may touch. Intention – couched in imagery, observed by the Watcher, and animated by an impeccably timed starting gesture – creates the spark that flies the gap.

## 257

It takes a special *courage of gentleness* to find the right intervention points and then perform the right starting gestures in them from a place of immersion in the field of endeavor. To implement these tactics you need daily practice and a training that sharpens the skills and explores the neutral traits so that you will be ready in the real flow of events.

Sometimes starting gestures as slight as a conscious breath or the making of significant eye contact or the lifting of the telephone, or saying "yes", especially that, can generate powerful Aikido-like low mediated follow-up actions. They move smoothly because in their unheroic rightness they use the energy of the world to move the world. Sometimes the actions emanating from the simplest low-mediated intentions seem more like the passing of volition from the world to the world – as if they don't pass through a person at all. But that's not so. No intention works without a follow-up gesture "peculiar" to the person, an idiosyncrasy, a temperament, a *bent* that adds a spin, a "body English" that makes intentions effective in the world.

## 258
### *Follow Up: The Template for Recognition*

After the turning the game changes. The imagery in which the intention was contained provides an "almost complete gestalt". It becomes a template for recognition. You hold it up to the world looking for a match. It's like a puzzle with some missing pieces. The missing pieces draw us into the world. We're searching for a corresponding reality. We know the shape of the missing pieces so we know what we're looking for. We shine the template around like a spotlight. We seek confirming instances. (The template is not a static but a moving image, more like a film loop than a photo, a vivid imagined depiction of an event or process whose main emotion is an expectation of "something evermore about to be," as Wordsworth put it.)

Every confirming instance strengthens the neural net that stores the intention. The template is refined by confirming instances; it gets better, its spotlight brighter.

The anticipation we carry in the template sticks with us as if it was memory, a future memory, the memory of a sensibility, a feeling, a wash of possibilities conjured in the imagination, a facultative memory of what it feels like to move into non-habitual mode, to overcome automaticity and get free.

Sometimes a person watches for obvious matching events and recognizes them by the 'rightness' of the moment- by the "ah, hah!" you experience when a puzzle piece fits.

The "ah, hah!" experience rivets our attention to events in the outer world. The confirming instances strike us as special spots in time, sometimes as mystical occurrences, as synchronicities or acts of divine providence. Love at first sight does that to us.

It's not easy to see why a longed for event manifests in the moment it does. It's not that the world presents us with startlingly different options. The opportunities that follow a turning point come mostly as chance events, or crossing points of divergent causal lines. They are part of the usual event sequences in the habitual arrangement of our lives. Nevertheless, they become opportunities for change because we rise to them; we see them as triggers for the doing of deeds. My general sense is that a new attunement lets you see what has been there all along. But the new attunement comes forth only from turnings in love or wisdom.

The template matching experience increases hope and diminishes fear. Moreover, when the biology of hope blossoms, its spreading roots firm up one's connection with life. The neural coalitions cooperate better, the template gets clearer and the 'Yes' of life keeps one looking forward.

I can make it clearer: the world pricks the image/intention complex into wakefulness by presenting clues that seem to say "Pay attention now, I may be a puzzle piece." On the occasions when a match comes, the "ah, hah!" experience affirms it. The template tells your body that you are pleased and you want more. It says you can solve the whole puzzle. It urges you to find the inner/outer resonance again. It poses this not abstractly but with a visceral rush of interest and excitement that feels like a prevision, an elation, and a memory of the future. Often empathy for another wells up in it. The *mu wave* in the sensory motor cortex is suppressed and the mirror neurons are activated.

The charge you feel has a biophysical character. It comes from rhythmic resonances based on frequency matching between an expectation and an event. It feels good because it intimates belonging. When your internal readiness meets a readiness in the world, it affirms your role in the cosmic whole. If these are mystical experiences, we all have them. We use them to guide ourselves along the bifurcating ways of love and wisdom. That's how writing this book is working out for me.

## 259

We cannot know which among our starting deeds will turn out to be the most relevant when they first enter the world. We cannot really plan our moves ahead of time, only in time. We cannot know for sure the exact intention we will frame in the heat of the turning until we do it. It may be different from the one you thought you would frame. Nor can you know in advance the exact instant you will put it into action. The uncertainty here is not regrettable. On the contrary, it's the strong bulwark of our freedom. In it we receive our call to adventure.

Historical penetration may depend more on simple vivid intentions than on big plans. Perhaps the smaller gesture opens the way for the larger deeds by setting a pathway among the interconnections of the "infinitesimal elements of free will." On the analogy with relay switches, the smaller will trip the larger.

Gandhi said, "There comes a time when an individual becomes irresistible and his action becomes all-pervasive in its effect. This comes when he reduces himself to zero." One would like to think that even with a fractional self, a least intrusive presence, not zero but not all pervasive either, we can frame clear and sane intentions, human intentions, bright and caring. But we live in a culture that highly values technological prowess, economic might and celebrity over delicate timing, refined intentions and unobtrusive presence. How can they have any influence?

## 260

King Solomon told a story that illustrates these leveraged possibilities. "There was a small city, with few people in it, and a great king came upon it and surrounded it and built over it great bulwarks. And there was found therein a poor wise man, and he extricated the city through his wisdom, but no man remembered that poor man."
(Ecclesiastes. 9:13-15)

What does he mean? Is Solomon condemning the poor man's lack of recognition? Or is he telling us something about the workings of wisdom here, in which, as a technical offshoot, anonymity plays a part? The aphorism is really about aptness, courage, timing and intention. First, the city is saved on the initiative of an individual. Second, it happens in a small city where ethical memes fly quickly. Third, the wise person was poor and inconspicuous, not heralded, not celebrated, and not sought out. Fourth, the wise person saves the city by unspecified means that were in some way not memorable, maybe not even noticeable. Then he returns to his former life. He does not wait around for a lifetime achievement award. The deed matters, not the doer of the deed. Naturally, no one remembered his name, for few knew what he had done.

But that still begs the question. What did that wise person *do*? Does Solomon tell us something about effective action here? How did the poor wise man bring his intentions into the world? What was the nature of the "extrication" that saved the city surrounded by great ramps and towers? If no man remembered that wise man, did he do nothing memorable, nothing stupendous, nothing involving weapons, armies, violence or treachery? The word "extricated" suggests something different. But what? Did he "reduce himself to zero?" That doesn't happen.

## 261

Nobody attains the limiting case of pure selflessness. Not in the East, not in the West. Only a living, breathing person can seize the moment. The act asserts a personal presence. It gives model change arms and legs. Its starting gesture sets up the sensorimotor pathway that connects the emerging new model to the world. What did the wise man do? He walked around, he talked, and he lent a hand here and there. Nothing conspicuous, nothing to make him stand out. Nevertheless, he saved the city. How we cannot know. To know, we would have either to be there with him, or be him.

### *End Notes*

[1] Harry Wolfson. *The Philosophy of Spinoza*. Vol ii p. 218)
[2] Joseph Campbell. *The Inner Reaches of Outer Space*. Harper and Row. 1986. p. 31
[3] *The Book of Balance and Harmony*. Thomas Cleary, translator. North Point Press. 1989. p. 14.
[4] Arnold Mindell in *The Dream Body* and later books explored some of these possibilities.
[5] Frederick Pottle in *Modern Critical Views. Wordsworth*. Ed. Harold Bloom.Chelsea House Publishers. 1985. p.18)
[6] William Wordsworth. *The Prelude*. Book VI 599-608
[7] *There are in our existence spots of time*
  *That with distinct pre-eminence retain*
  *A renovating virtue, whence, depressed*
  *By false opinion and contentious thought,*
  *Or ought of heavier or more deadly weight,*
  *In trivial occupations, and the round*
  *Of ordinary intercourse, our minds*

*Are nourished and invisibly repaired;*
*A virtue, by which pleasure is enhanced,*
*That penetrates, enables us to mount,*
*When high, more high, and lifts us up when fallen.*
*This efficacious spirit chiefly lurks*
*Among those passages of life that give*
*Profoundest knowledge to what point, and how,*
*The mind is lord and master – outward sense*
*The obedient servant of her will.*
 ibid Book XII 208-223
[8] *Imagination having been our theme,*
*So also hath that intellectual Love,*
*For they are each in each, and cannot stand*
*Dividually. –Here must thou be, O Man!*
*Power to thyself; no Helper hast thou here;*
*Here keepest thou in singleness they state:*
*No other can divide with thee this work...*

*Oh, joy to him who hath here sown, hath laid*
*Here, the foundation of his future years!*

[9] But this is a power that can be abused, because you don't need to be "good" to attain it. The fictional Don Juan had it. Perhaps the actual Casanova did too. And Giordano Bruno, the Renaissance magician, included it in his techniques as a way to make people fall in love with you instantly so you could do magic with them (at least in their eyes). You can think of it as acquired charisma, and charisma does have power.

[10] Lewis Mumford. *The Transformations of Man.* 1956. Collier, p. 136-7

[11] Whitehead. *Process and Reality.* p. 81

# 16. SELF DIRECTED HEALING

### 262
### *Healing Intention*

Low-mediated intentions play important roles in self-directed healing. By learning to heal (if you will allow me to explore that possibility, without deeming it absurd in advance,) you will gain access to the powers inhering in time-sensitive freedom that, with further practice, will help you move events in the world with special efficacy in ways I will explain. In this chapter I take on healing, in the next the world.

### 263

Earlier I showed that shamanic wisdom worked through a dramatic withdrawal-return process that employed the inward skills of imagery, intention, sensory awareness, rhythm and timing (see # **66, 67.**) Shamans are called wounded healers precisely because they discovered their skills by healing their own dangerous illnesses. With their self-overcoming disciplines in place, the shamans begin their vocation first as healers of others, and then as interveners in the patternings of nature on a larger scale. (With what actual results, and through which lines of causation, we will have to judge.)

Let us start with the simple, incontrovertible observation that healing intentions are rarely absent from the minds of sick persons. Ill people usually try to direct the body's healing powers to the site of illness using sensation, imagination and hope, though with varying results, including no results at all. But self-directed healing can be done well or poorly. People do learn to get voluntary control of autonomic functions. How?

o Accurate sensory awareness when combined with non-reactivity delivers new information about the illness.
o We receive this information about ourselves in images. We describe the pain as blasting, as pressing in on us. My throat is on fire. Later we put words to the images, but the words refer back, they do not replace the imagery.
o We use the imagery to help us change the body by directing it back to the sites or imagined causes of the illness. As imagery, it can gain better access to deeper-lying neural processes.

We know very little about how this healing (or killing imagery) works, but it does work. Hundreds of controlled studies have established this.

**264**

The best-researched manifestation is the placebo effect. A wide variety of illnesses, from viral warts to coronary insufficiencies, to inoperable cancers, respond favorably to the placebo effect. Everything from the laying on of hands, to sugar pills, to machines purporting to generate wave energies, have promoted actual healing. Some machines work effectively without ever being switched on. Antoine Mesmer's animal magnetism séances, illustrated to the left, were renowned for their successes.

Double blind studies show that patients receiving placebo pills for mild pain relief experienced mild pain relief, while other patients, given the same pills, but told they were strong pain killers, experienced greater pain relief. Evidence shows that active place-bos, pills that fizz or surgeries that open and close the body with-out doing anything else, work better than passive placebos. For similar reasons, one can speculate that new medications, them-selves the product of huge investments and advertising campaigns, work better in their first year of release than afterward –and they do! Evidently, the glamour of scientific authority itself has a heal-ing effect. It may even be the case that in all medicine the authority of the doctor has a placebo effect, and that all medications, what-ever else their virtues, get part of their power by directing attention in imagery-rich ways to the site of the illness.

### 265

People may "spontaneously heal" much more often than we know. Their stories do not get into the medical literature. Health care workers may be reluctant to report them. Some disbelieve what they are witnessing. The greatest number of self-directed cures may happen in people who never seek treatment.

Moreover, these healings are not spontaneous, if by that you mean instantaneous and uncaused. To describe them that way trivi-alizes the body's subtle healing efforts, many still unexplored.

### 266

That healing employs the creative powers of imagination should not surprise us. Creative powers, focused by the imagina-tion, regularly bring human endeavors into the world. Why not in healing? Why can't healing imagery support the intention to heal the same way the architect's sketches move a building project ahead?

A blueprint gets to contractors and carpenters, but how do the imaginative influences on healing find their way to the sites of in-jury? The same way the imagery of slicing a lemon gets to your salivary glands. Perhaps the intervention arrives through longer sequences of "hand-shakes" following along the alignment of free-doms I described earlier: the freedom to imagine shakes hands with the freedom to wish, the freedom to wish with the freedom to will, the freedom to will with the freedom to act, the freedom to act with the receptive readiness of the world to respond to the action. All at the right moment, all in a temporal setting.

Research shows that the imagery supporting the intention doesn't have to be accurate to have a positive effect. However, it has to be suggestive and *apt* in some way. And it can work for or against healing.

The Holmes-Rahe research demonstrated that the onset of illness correlates with preceding life events (see # **106**.) A wife dies. Her widower languishes and dies soon after. The life-event, the loss, is terrible. That's true. However, most of the horror is in the mind. Negative images make the losses harder to bear. The grief, the churning memories it provokes, the intense dreams, the recurring potent images, fearful anxieties, the haunting presence of the departed break us down. These examples demonstrate the power of the imagination to grab hold of our bodies in braced, contorted, constricted, anxious ways producing illness. Voodoo death would be an extreme instance of the power of negative imagery.

The same powers of imagination that can exacerbate illness, however, can sometimes heal it. On the extreme end, there are documented cases of cancer remissions following IV normal saline solution. Patients believing they were receiving new experimental drugs had their tumors "melt away like snowballs." The healings at Lourdes have been well documented.

### 267

When self-directed healing works, how does it work? The salivary response to a lemon is automatic. We're talking about voluntary healing interventions here. They work no differently from what happens when you say to yourself "I'm going to raise my right arm" and you do it. You don't know how you do it or why it works, but you make it happen. You'd be flabbergasted and greatly troubled if it didn't happen. Why? Because your volition travels a common route that is exercised daily. But what if you say to yourself "I'm going to slow my heart"? That would send traffic down an uncommon route. Most people can't do it. But some can. How do they manage it? They build a path of handshakes to it. They practice sending intentions along the path. The intention-imagery complexes are timed events. The more frequently they succeed in relieving symptoms and building healthy tissues the stronger the uncommon pathway becomes.

Like yogis, self-directed healers train themselves to get access to the body's autonomic functions. A person may not know how he does it. But somehow the transmission of intention through the hierarchy of freedoms crosses from the volitional to the autonomic control systems and for a moment at least the autonomic becomes voluntary.

## 268

The pathways transmit instructions to the body. I hope to show they work only to the extent that their rhythmic elements entrain each other by the timing of their handshakes in the hierarchy of freedoms. At those moments resonances occur, vulnerabilities move into phase with each other, field effects occur.

Sir John Eccles supported this view decades ago, maintaining in 1986 that quantum uncertainty in neurotransmitter releases allowed free will. Jeffrey Schwartz and Sharon Begley marshaled a number of these arguments in *The Mind and the Brain.* [1] Schwartz draws on Henry Stapp's work in *Mind, Matter and Quantum Mechanics* (1993) to explain his successes using cognitive retraining techniques with OCD patients.

Quantum effects, insofar as they operate on atomic and ionic scales are always present in these reactions. But plain old secular timing on molecular and cellular scales play the most important parts in steering indeterminacies into freedom volitionally. These connections use the evolved circadian, ultradian and infradian frequency bands we have described and make rhythmic alignments between them.

In healing work, these biological events travel pathways less commonly used than those we use to raise an arm or keep balance on a bicycle. But they're no less real. They send volitional signals along routes in the autonomic nervous system, but they also create new routes as needed. Neuroplasticity blazes trails along new dendritic arbors that when practiced repeatedly and reinforced with positive results remain in place.

It is not a far leap to go from neuroplasticity in neural traffic to biochemical plasticity in neuropeptide flows, as studied by Solomon and Pert, to volitional influences on enzymes involved in cell signaling processes, to selective voluntary DNA expression. In other words, through biological alignments not yet well understood, we can bring our intentions into the communications that catalyze biochemical reactions.

## 269

Whatever substances or processes they use, whatever organs they start in, all self-directed healing processes, physical and emotional, share important features.

1) They unfold in time.
2) By virtue of their transit through the nervous system, they travel in pulses, pulsed packets or waves with periods and amplitudes that can be measured.
3) The pulses and waves play parts in larger processes of approach-separation or withdrawal-return, hence involving social and solitary behaviors and at the highest levels love and wisdom.
4) The physiological transmission move in the evolved frequency ranges we have already identified.

From my observations of rhythmic functioning, I get the impression that self-directed healing using low-mediated intentions always involves approach/separation and withdrawal/return rhythms and they usually have love or wisdom in them to a greater or lesser extent. By paying attention to body time and to sensations of change built on a strong interoceptive self-connection, the sick person may in some cases initiate a turning point in an illness, and by doing so raise self-directed healing into a wisdom process of its own.

## 270
### *Medical Chronobiology*

Self-directed healing strategies often build on the rhythm patterns of remitting/recurring illnesses. By small well-timed efforts, they fortify the remitting side of the temporal pattern. This they do by shifting the turning points, or by strengthening the remitting while weakening the recurring leg with timed interventions of a sort we will soon consider in detail.

Perhaps more illnesses follow the remitting/recurring pattern than we know. By better understanding the biophysics of their vulnerable phases and by exploiting their perturbability more effectively, we may be able to influence remitting/recurring illnesses at their most sensitive moments.

The healing effects of these interventions would manifest not all at once, but in stages, as circumstances permit, as the mind/body connection develops. In this effort, intentional healing would help recruit the body's own healing powers that themselves might wax and wane on schedules correlated with the frequency bands in our physiology.

Indeed, many illnesses are tied to circadian oscillators. Disorders of circadian rhythm have been well researched in hyperthyroidism and adrenal insufficiency syndromes. Multiple sclerosis shows remitting/recurring patterns on an overall downward course. On a briefer time scale, perturbations in the heartbeat (the R on T pathology) make the heart susceptible to lethal arrhythmias. Respiration, sympathetic vascular tone, even a stressful thought can suddenly withdraw vagal tone and produce perturbations leading to the unfortunately timed arrhythmias associated with sudden cardiac death.[2] The breakdown in homeostatic function that accompanies illness, therefore, has, besides its material aspects, temporal qualities. In serious illnesses, in illness syndromes, whole congeries of rhythms may be thrown out of order or entrained to each other dysfunctionally, or all of them may be entrained to a single deranged oscillator, or perturbed to the point where their rhythmicity is extinguished altogether. Franz Halberg did early research in this area. He was one of the first to apply circadian physiology to illness patterns.[3]

Even illnesses without known remitting/recurring patterns have oscillatory characteristics, whose periods, amplitude and phase relations to other oscillators in the body can be identified. Heart arrhythmias are the most widely studied of these, along with seizure and motor disorders. However, as our knowledge of other body pacemakers improves – particularly the pacemakers promoting regulatory gene expression – the list is bound to grow.

## 271

The conclusion I draw from this is that self-directed healing interventions built on low-mediated intention work better at some moments than others. My clinical observations in biofeedback, at least as they apply to the fast ultradian rhythms in the brain wave, heartbeat, respiration and vasomotor ranges, show that the healing efforts are best promoted when a number of component oscillators induce resonances as their peaks and valleys come into phase. I have worked with and documented voluntary alignments of breath, heart rate variability, vasomotor and theta brain wave rhythms leading to resolutions of blood pressure, respiratory and heart rhythm problems.

Attention to the time qualities of an illness, its momentum, acceleration, its patterns of remission and recurrence, its intensifications and subsidences, are likely to become essential tools in self-directed healing as our knowledge of the body/mind improves. Not all of the knowledge will belong to doctors or be approved by them. As patients, we will do our own self-directed healing work using our own strategies. I describe some that I've seen or used below.

## 272
### *Pacing and Leading*

To ride the stages of an illness, to be in touch with its rhythms, to attend to its rises and falls, is to pace it. Self-healers pace the illness until its contours and its temporal variability patterns become familiar and even predictable almost on a second-to-second basis.

We generally time changes to the breath, but sometimes we use the pulse as a timekeeper. It makes a good pacing tool for more precise inner monitoring. I have taught myself and my biofeedback clients to meditate on breath and heartbeat patterns together. Breath based meditations go way back, but to integrate both is new, particularly to time the start of the in and out breaths to the heartbeat patterns in the RSA, (the respiratory sinus arrhythmia; the tendency for heart rate to increase on the in-breath and decrease on the out breath.) With this technique, we align two oscillators.

In a similar manner, we can time our intervention to other body oscillators. We can tense and relax muscle groups, close and open eyes, eat and refrain from eating. We can learn to follow the changes of the illness closely enough to time the breath, heart rate variability or other slower wave oscillators to them.

When events confirm your anticipations, you have achieved pacing. That's the first step. Next you take the lead. Just by a second or so, you intervene before the mini-turning in the illness (or on a larger scale, in the overall remitting/recurring pattern.) By pacing and then leading, you nudge the illness. You make the smallest move at the best moment to put one or more of the oscillators in the illness under momentary voluntary control.

It's a basic hypnotherapeutic technique. The hypnotist, for example, watches the subject, notices his eyes blinking and says, "Your eyes are blinking." That's pacing. Then the hypnotist says, "You're getting tired." That's leading.

Pacing an illness requires subtle sensory awareness and non-reactivity. Leading involves intention, imagery and subtly timed actions that may include but are not limited to vigorous exercise, meditation, diet, breathing, heating, cooling, sweating, laughing, sleeping, love making, etc.

In our conceptual scheme, leading is most effective when a person is nearer to the reversal moments in the oscillatory components of the illness. That's when small causes, properly timed, can have big effects. Nonlinear bio-oscillatory systems have moments of special vulnerability in them when they can be phase-shifted or reset or made to bifurcate. Self-directed healing strategies work with these moments.

We often use the power of imagination, intention and small gestures, rightly deployed, to change the body. It happens best in special windows of opportunity near mini-turnings. We rarely control or direct our strategies to these moments. We use a much less efficient scatter-shot approach.

With training one can pace and lead many ultradian oscillators, including the 90-120 minute rest/activation cycle that Ernest Rossi employs in his hypnotherapeutic work, the hunger/satiation, and sleep/waking cycles. The sleep cycle, including periods of lucid dreaming, can be partially trained. Metabolic rhythms, blood components in the immune system, rhythms within the presentation of pain and maybe even components of the cell cycle itself can be brought under voluntary control. Social/asocial alternations of hormonal rhythms, for instance, sexual readiness or avoidance, stress and relaxation, can be entrained to the extent that the intention/image complexes used in self-directed change can be shaped through tempi, pulses and melody.

## 273

To mount an effective healing strategy directed to the vulnerable times in the illness, you must develop skills in recognizing the frequency, phase and amplitude characteristics of events in the body. Usually you start with breathing because it is the one body system essential to survival that can run either under voluntary or autonomic control; it slips back and forth between them easily. That's one way in. In the state of consciousness associated with good pacing, you move from one oscillator to another. To do this you have to, first, be present to the illness with subtle sensory awareness and non-reactivity, second, generate a robust rhythmic continuity between the illness and the healing effort with pacing and leading strategies and, third, make an apt healing intervention at the right moment. But what is an apt healing intervention, and what favors its taking hold?

## 274

When you bring oscillators in the body under voluntary control, they can influence other oscillators in that frequency and phase range not yet under control. The influence of one oscillatory system on another has been demonstrated in many studies of biological entrainment (with particular force by Kitney, 1984.) You can breathe to influence your heart rate, swallow to influence peristalsis, and blink to drive brain wave rhythms. There are many frequency ranges accessible to volition, and within those ranges, many oscillators can be put into or out of phase with each other. With breath, you can recruit heart rate variability, vasomotor activity and gastric motility. Even faster frequencies, as in the EEG response to auditory and photic driving strategies, can pull large areas of the cortex into synchrony briefly.

In my biofeedback practice, I developed training methods to help clients move an oscillator under voluntary control into a resonant relationship with a dysfunctional oscillator not yet under voluntary control. My experience comes mainly from using the breath to influence heartbeat patterns in biofeedback training of the RSA. The protocol here is to use a trained oscillator to control or "capture" a target oscillator, in this case the heartbeat pattern. The larger the amplitude of the trained oscillator, consistent with a good quality wave form, the stronger its influence on the target oscillator. The influences come from anatomical and physiological connections, mainly these: oscillators provide or remove ingredients that other oscillators need; they intrude on one another's space of operation, adding or subtracting energy, agitating the medium in which another oscillator functions.

Altering phase relationships between the trained and target oscillators can shift the target oscillator with respect to frequency, amplitude or both. The change, for as long as it lasts, helps break down habitual physiological bracing patterns; it opens the pathway for more homeostatic responses. The longer the homeostatic rhythms persist the stronger its pathway becomes.

Once conditioned habit patterns have been shifted in the autonomic systems of the body, a target oscillator can in turn become a trained oscillator. The newly trained oscillator can then be sent out to capture another target oscillator deeper down, further in, closer to the core of the illness. A restored RSA pattern will entrain the fast ultradian oscillators controlling vasomotor activity and restore rhythmicity to dysfunctionally diminished baroreceptor sensitivity, helping to normalize short-term blood pressure. Normalized pressure reduces the fluid shift burden on the kidneys.

The trained oscillator can also be used to move the target oscillator increasingly out of phase with another oscillator, until a different oscillator grabs it.

You can also bring in rhythms from outside, pulsed sounds, video animations, music, pulsed heat, radiation or ultrasonic energies. I designed and built an auditory biofeedback system that transforms body changes into accurate contrapuntal modal music - music that seeks to resolve to a key center associated with homeostatic physiological functioning.

Someday we may be able to combine mind/body with medical technology to intervene in the phases of the cell cycle. Inner-directed sensory techniques may be used to pace and lead the pulsing rates of therapeutic radiation in cancer treatment. Early research by Kirson, Gurvich *et al* suggest that the properly pulsed electrical charges can interrupt the cell cycle.[4]

## 275
### *Turning Point Healing*

Many illnesses present with a distinct, central turning crisis. That's what wisdom and love dramas do too, which suggests, at least on the surface, that it may pay to look for connections between the broader patterns of meaning in life and the healing process.

I believe that our most serious illnesses _are_ wisdom processes. The illness in pursuing its clinical course follows a withdrawal-return dynamic. Like wisdom, the return from illness can change our lives. Sometimes, the healing does more than mend the body. It changes a person's character. It influences what we choose to do with ourselves. Sometimes the withdrawal/return dynamic awakens an inward quest in the course of an illness that searches for its original wound, not so much the source of the infection, or the moment the virus entered the body, but the insult to meaning that made one susceptible to it, the wound to action on the leg of love or wisdom. Sometimes events experienced long ago become life-long sources of stress, confusion and loss. They sow the seed for future illnesses.

The illness, as a wisdom process, sparks a withdrawal larger than the immediate physical pathology, because it evokes questions and doubts and restores memories of hopes and regrets. The illness covers more than flesh. It brings with it deeper attitudinal changes that can transform one kind of wisdom into another. People experience spiritual growth with physical healing. New capacities for love or service may fall into place. Our ties of affection shift when we have dealt with our mortality in a new, more vivid way.

In the central crisis of an illness many rhythms may converge, some amplifying, supporting, mutually entailing, others annihilating, pulling on each other's frequencies, causing clashes and dissonances. Into this cacophony, the low-mediated self-directed healing intervention becomes a factor. It becomes part of the mix. If it is strong and we insert it at the right moment in the right way, it may draw other rhythms to it and help bring the errant oscillators into resonance.

Seen this way, the healing event promotes an unexpected turning point, or makes a potential turning point actual. Just as the thought of the beloved can bring on a turning from separation to approach or the realization of a truth can spark the turning from withdrawal to return, the intention to live until a child marries, for example, can spur a turning. But there are limits to our resilience. In the interpenetrating fields of multiple vibrations that constitute inner and outer life, dissonances may overwhelm us.

By bringing one or more trained oscillators into resonance with the deranged oscillators in the illness itself, and repeating this day after day, over-learning it, one can build the momentum for a turning in an illness. Perhaps it takes a specific mind/body state to reset, delay, or bring a rhythmic pathology temporarily back to normal functioning.

The intervention can be oscillatory itself, but it doesn't have to be (a shout can cure hiccups), as long as it has the energy and timing to catch one or more of the underlying oscillators at a vulnerable moment. These healing interventions induce mini-turnings. You may have to do them many times until the body learns the route to change. You bring a rhythm back, it slips away. You keep it up day after day, session after session, until one mini-turning actually becomes the maxi turning, the turning of turnings, though you may not realize it when it happens. To be present with consciousness, to be immersed in sensation at that moment - to prepare for that moment with medications and standard medical care, to choose the standard treatments consciously, to intend to bring them all into play - all become part of your healing strategy.

Shamans use timing in highly dramatic and effective ways. They work in infradian, circadian and ultradian frequencies, using phases of the moon, time of day, rattles and drumming. They wait for moments of accessibility. They pace and lead the illness and drive the oscillators into resonance with a variety of pharmaceutical, ritualistic, hypnotic, dramatic and musical techniques.

It takes a certain prowess to develop these techniques for yourself. However, on certain paths of possibility we may learn how to apply low mediated intentions to healing, perhaps with the help of monitoring instruments and therapists who teach us how to use them. Eventually, you take off the training wheels. You don't need feedback instruments any more.

## 276

Perhaps there's a rule of thumb here: the greater the number of oscillatory process that can be coordinated to the healing work and brought into resonant relationships, the more effective the healing process. The greater the illness, the deeper the turning needed to heal it. The more profound the healing crisis, the stronger the alignment of multiple oscillators required.

Among these rhythms, as rhythms themselves, we will find the neural and humoral faces of emotional and cognitive understandings. Some will seek the origins of the illness and its meaning. Others will yearn for healing for specific purposes that get their energy from what you want to do with your life thenceforth. All of these together – with the people and issues you care most about – will feed the approach/separation and withdrawal/return dramas as they shape the flow of love and wisdom in your life.

## 277
### *Healing as Living*

What self-directed healing cannot do, you may argue, is change the world outside the body directly. You must perform intervening actions to bring your healthier functioning into the world. But with healing surprising pathways do open up from person to person. With the mapping, sensory awareness, non-reactivity, imagery, intention and timing used for broader healing purposes, transformational memes can go from person to person along the rhythmic channels of approach/separation. Small healing changes in the conduct of love and wisdom can influence outer events with surprising efficacy. To this phenomenon we will now turn our attention.

### *End Notes*

[1] Good recent discussions of neuroplasticity can be found in Sharon Begley. *Train Your Mind, Change Your Brain,* and Jeffrey M. Schwartz and Sharon Begley. *The Mind and the Brain.* Regan Books. 2002.

[2] Cardiologist Ary Goldenberger has done good research in this area

[3] According to Jeremy Campbell in *Winston Churchill's Afternoon Nap*, Halberg "emphasizes that the body responds to events in a spectrum of natural frequencies that are in a harmonic relation to one another... As in the case of the harmonics of a musical sound, the component frequencies of a spectrum of biological rhythms may vary in amplitude; some may be 'louder' than others. Often, but not always, the circadian, or about-24-hour, component has a larger amplitude than the infradian, or longer than 24-hour, components... a circadian rhythm in, say, body temperature might be open to influences from an infradian rhythm whose frequency is a harmonic of the circadian frequency, but be unreceptive to one that is not in a harmonic relationship." (Jeremy Campbell, *Winston Churchill's Afternoon Nap*. Simon and Schuster. 1986. p.128)

[4] Eilon D. Kirson[1], Zoya Gurvich[2], Rosa Schneiderman[2], Erez Dekel[3], Aviran Itzhaki[4], Yoram Wasserman[1,4], Rachel Schatzberger[2] and Yoram Palti[2]. "Disruption of Cancer Cell Replication by Alternating Electric Fields". [*Cancer Research* 64, 3288-3295, May 1, 2004

# 17. POLITICAL LIFE AND MORAL DISCOVERY

Self-directed change techniques do reach into the world. They touch the wheels of events by rhythmic entrainment and work best in small communities.

As larger polities break down under chronically stressful social and economic conditions, however, we have a harder time finding real communities.

But we can take steps to overcome alienation to strengthen belonging even in the absence of genuine communities.

o   Loosening the grip of successomania, consumerism, addiction and juvenile arrest – at least sufficiently to reveal their emptiness and irrelevancy – helps. It draws you into natural-sized groupings where you have a better chance of experiencing genuine belonging. When we clear back the seductive palliations of our larger polities, we listen to each better and respond more honestly. As real communities form, we will find ourselves more inclined to meet in person. Gestures and inflections, and all the carrier waves that bear the freight of real human connection, will gain impact. The empathy now locked away in enchantment to media stimulation will break free. And this will create better conditions for the reception of new memes on the local scale.

o   With training in inwardness, we will restore our access to the ethical mind. We will find its concerns interesting again, because they will be germane and interactional. We will want to consider them carefully. We will take the time to do so. We will know caring is crucial to happiness.

o   More of what we believe, create and imagine will pass through the generations. Memes will move more harmoniously, transmitted by living example from parents to children, and the reverse.

o   In small settings we stand the best chance of getting belonging back. Our heroes will become local again. We will know them and they will know us.[1]

## 279
### *Natural Community*

Social healing restores belonging while mitigating aggression. In circles of affiliation and friendship– and with sensory stimulation made artful by a creative life – many of the frustrations that push stress across the line to aggression will be relieved. Assured of belonging, we will let ourselves be ourselves and as ourselves renew kinship ties in extended family and second family relationships. Democracy would become deliberative then, more open to discussion, more participatory, and politics would be more inclined deal with important ethical issues for which it was worth struggling.

## 280

We are never indifferent to natural communities. They are constantly forming and dissolving within larger civic institutions. The stimulating signals for natural sociability exist in large polities simply by the physical proximity of others, particularly people who share interests and passions. Whenever the sensory stimuli present themselves, including the archaic distance regulation rules, we revisit our social behaviors. Band-sized societies take shape within larger polities as families, secondary institutions, factions, friendships, voluntary associations, and other groupings form.

The population of a great city is intrinsically a mosaic of superimposed and overlapping natural communities. A single neighborhood, a single block, can have numerous overlapping communities in it. Nor do the members of a community have to live next door to each other. In a big city, thousands of independent natural communities can aggregate and disperse in the course of a day to celebrate significant events. The same technologies that isolate us while seeming to draw us together, the phone and the Internet, can *actually* bring us together. In New York City thousands, perhaps hundreds of thousands, of intentional communities overlap. These intersecting webs of natural comity stretch across neighborhood boundaries and reach across all five boroughs and beyond. They give us our real world of Philia.

With their rich opportunities, the likelihood is that cities that do not "wait for their states" in states that "do not wait for the federal government" will become the venues for the rebirth of community in times to come in the USA, precisely because their diversity and complexity create so many possibilities.

This shifting mosaic of communities would meet and overlap at permeable boundaries. Across these boundaries two-way movements would flow. All kinds of information, embodied intentions and actions would travel between natural sized communities. Some with strong mimetic energy would inspire imitation and repetition. These horizontal transfers would bind people into larger groupings. And these larger units, for as long as they persevere within still larger polities, would exert upward pressure for change. In some of these *ad hoc* alliances of communities, love and wisdom would begin to exert a modest pressure on dominance and territory as they have done in the past.

Political and ethical movements travel on these routes.

## 281

For ethical changes to take hold, these movements would have to:

1) Arise from the discovery of correlated unused potentials by many people simultaneously.
2) We would communicate the memes expressing these potentials face to face, by example, *in person*. We would carry them on approach-and-separation and withdrawal-and-return pathways.
3) These ethical discoveries (ethical because they stir choiceful action) would affirm "This I like, this I want,

this I abhor," and these visceral reactions would provide the real transfer material to which imitation, entrainment and introjection for whole communities
could be linked.

Our ethics have no fixed content or priorities. Nature has
equipped us to play many and varied roles and it is our great good
fortune that they come forward in different settings, even when
they prove contradictory. I agree with Richard Posner's affirmation
of moral pluralism. "Given the variety of necessary roles in a complex society," he writes, "it is not a safe idea to have a morally uniform population…We need gentle, kind and sensitive people, but
we also need people who are willing to employ force, to lie. To
posture, to break rules… Failing it, we are better off with moral
variety, and this places the entire project of moral education in
question."[2]

## 282

You may object that most people most of the time get their
moral strength by following the rules, customs or religious laws
handed down to them. They defend what society teaches them to
believe. Few are willing to think for themselves. John Stuart Mill
wrote that for these people it was "as if accepting it [belief] on
trust dispensed with the necessity of realizing it in consciousness,
or testing it by personal experience, until it almost ceases to connect itself at all with the inner life of the human being."[3]

Nevertheless, even transmitted ethics have a creative side.
Reasoned assent or sincere belief in a received or recovered ethical
directive affirms moral creativity because a moral act is moral only
in the moment we choose and enact it. In that moment, the act is
new. Of course, the feeling of newness does not constitute an act of
discovery. But it doesn't rule it out either. Ralph Waldo Emerson
pointed to intuition as the "source, at once the essence of genius,
virtue, and of life, which we call Spontaneity or Instinct. We de-
note this primary wisdom as Intuition, whilst all later teachings are
tuitions."[4]

Emerson suggested that the open-endedness and creative energy in our emergent powers comes more from imagination, play,
caring and daring than from the strict cognitive learning. He insisted, "The way, the thought, the good, shall be wholly strange

and new."[5] These competencies are always "new in nature, and none but he knows what that is which he can do, nor does he know until he has tried."[6]

For some, tuition rules, for others, intuition. The distribution within and between people varies with the spirit of the times. But even assent to tuition has an intuitive character.

## 283

We make our best moral discoveries when love and wisdom converge in tough times. These are the times when we find it particularly hard to comprehend the two root ethical dilemmas: "What do I owe others?" and "What do I owe myself?" (The first gets its essential energy from the approach/separation rhythm; the second builds from withdrawal /return).

For love and wisdom to draw together in troubled, turbulent times, they have to overcome destructive derhythmization before they can even start to find each other. Rhythmic restoration moves consciousness in subtle ways that free up the powers of choice at turning point moments. And freed choices with convergent love and wisdom in them, when their turning points merge into one greater turning event, lead to creative solutions of fundamental human dilemmas. Certain people get good at this in troubled times.

Creative people come upon their moral energies in the interregnum between the old and new passing rules. The longer the interregnum lasts the better the chances for value transformation. Most people fail to generate effective turnings during this time. However, the imagination of creative people, if it can be stirred by empathetic impulses traveling as social memes, can devise intentions that satisfy the pulls of love and wisdom both then. I can make this less abstract with every day examples.

## 284
### *Four Current Moral Discoveries*

Four planet-wide ethical movements meet these strict but fair criteria. Their influence is spreading now. I list them as nonviolent activism, feminism, environmentalism and new corporate governance. Each responds to the altered conditions of belonging brought on by the fifth epoch revolutions (see # **191, 192, 201-203**.) Each tries to restructure belonging to overcome the kinds of alienation that mark the era. Each seeks reengagement with life by making belonging real again. Each serves to move us toward the sixth human age.

Each moral discovery requires decisions and actions. Each recognizes the inevitable presence of aggression in social and political life and tries to transform it by giving it new content. While seeking to assert itself in the world, each uses the language of struggle and the imagery of victory and defeat in its public pronouncements. Nevertheless, at the same time each moral discovery tries to resist the "us and them" mentality.

All four share certain fundamental traits.

o   They are essentially secular – spiritual but secular.
o   They are inclusive. Everybody's invited to join. They promote diversity. They are all global movements that enlarge the "us" without creating a "them".
o   To some extent, they have been held in trust for the dispossessed by the déclassé among the prosperous people. But the catalyzing moment that makes them real comes when the underclasses take on their own ethical discoveries and put them out into the world. This is happening now with nonviolent activism and feminism and will soon happen in popular uprisings dealing with environmental issues.
o   The four moral discoveries rely on inwardness because they require individual acts that bear witness.
o   The movements, despite their call to confront pressing needs, show a *joie de vivre* in action that makes them seem "media savvy", theatrical and sometimes even comic. They respond to situations imaginatively, in ways the state cannot easily predict or thwart. They are playful.

However, ethical movements that produce fundamental and lasting revolutionary changes have to maintain their integrity against enormous sociopolitical and economic resistance. Trying to keep balance while the world goes mad is hard.

## 285
### *1. Nonviolent Activism*

The ethics of non-violence as practiced by Gandhi and Martin Luther King has influenced mass movements everywhere. When nonviolent activism reached the underclasses in America in the Civil Rights movement, a global circle in the emergence of ethical activism was completed. The story, when it is told in the fullness of time, will show how the descendents of slaves created the Civil Rights movement and changed the course of American history.

The Civil Rights movement, the peace movement, the anti-nuclear campaigns, the opposition to the Vietnam war and the farm worker movement used these same methods of non-violent activism. They in turn inspired the strategies for the Velvet Revolutions across Eastern Europe that shook off Soviet domination; these movements along with the internal activism of the Refusniks, to some extent hastened the peaceful fall of the Soviet Union. Interestingly, Tom Stoppard in his recent play *Rock and Roll* attributes to rhythmic dance music – and the jailing of the *Plastics,* a Czech rock band – the mimetic force to inspire political action, and in fact to trigger the demonstrations that led to the fall of the Communist regime.

Movements of non-violent activism have had extraordinarily good track records in the twentieth century. They have generated a momentum of ethical engagement that touches human nature at a pivotal point: the realignment of the "us and them." Moreover, the core value of all of the nonviolent activist movements has been an expansion of "usness" across political and ethnic divides.

As Gandhi wrote:

"It is the acid test of nonviolence that in a nonviolent encounter there is no rancour left behind and, in the end, enemies are converted into friends. That was my experience in South Africa with General Smuts. He started with being my bitterest opponent and critic. Today he is my warmest friend…"

The chemistry of non-violent personal and political action has catalyzed change on a planetary scale.

"Nonviolence is like radium in its action," Gandhi wrote. "An infinitesimal quantity of it embedded in a malignant growth acts continuously, silently, and ceaselessly till it has transformed the whole mass of the diseased tissue into a healthy one. Similarly, even a little of true nonviolence acts in a silent, subtle, unseen way and leavens the whole society."

People have practiced nonviolence before, but never across the whole planet almost simultaneously and with knowledge and support of each other.

## 286
### *2. Feminism*

Here the powers of women have engaged cultural and political realities in a new way. The engagement is transforming economic relations, families and electoral politics around the world. Gender relations, child rearing, artistic expressions and legal systems are being shaped by it. Possibly the feminist movements, as they alter the size and structure of family life, will shift the character of sexual and romantic love sufficiently to bring new standards into sexual selection, changing the stages of approach and separation and influencing the balances between inwardness and outwardness in favor of the ethical mind.

And the moral power of feminine nurture may yet bring more warmth and softness to civil society, perhaps even changing the character of competition in the workplace. Feminism too has had earlier manifestations, but never so broadly based across so many different cultures. Its current historical role is unique.

In feminism as in the civil rights movement in America, success brings to the table new people with fresh outlooks and attitudes. And the newness in both ethical movements promotes a broadening of the "us", reaching out to challenged people, to children, the disabled, the elderly and the mentally ill. And this ties the feminist to the environmental movement on one side and to nonviolent activism on the other.

## 287
### *3. Environmentalism*

The struggle to protect and heal the planet now involves millions of people across national borders. The motivations are compelling; the climate is changing more rapidly than even the most pessimistic of the scientific studies predicted. Many people follow the science. They accept a level of complexity that earlier mass movements ignored. They take an integrative, systems of view of the biosphere. Like the other moral endeavors, the movement rests on principles of non-violence, and seeks an all-inclusive "usness", one that even transcends species. The broadest kind of love infuses it, a direct caring for the biosphere, for other species, for future generations, for climate and geology, an inclusive Agape, a love of nature for its own sake. Like the other ethical movements, environmentalism has had earlier manifestations but never before with full consciousness of the fate of the whole planet at stake.

## 288
### 4. *New Corporate Governance*

One can hardly call this a movement yet. Its ethical implications have just begun to stir. But economic crisis and decline will bring them into sharper focus among larger numbers of citizens. You can see it in shareholder efforts to assure corporate accountability, in scandals over corporate theft, in taxpayer resistance to corporate bailouts, in outrage over reneged pension plans and cancelled health insurance. As things go badly for the economy, the obscene gap between executive compensation and wage labor will become a pressing issue.

Legal efforts to repeal "corporate personhood" will focus future judicial and legislative actions on responsibility and accountability, but only after great failures have occurred. New corporate profit sharing and employee ownership plans may surface. We will get synergistic effects from the other three movements. For instance, the revulsion against corporate greed and incompetence will grow as environmental concerns intensify; and some of it will come from feminist and civil-rights efforts to fight discrimination.

Very little has changed so far. Powerful interests oppose it. If they prevail during the coming economic crises, we are in for steep decline in standards of living and authoritarian government. The preferred dystopian model will switch from *Brave New World* to *1984*.

## 289
### *"Money is the Root of All Evil"*

When I was a child, my father taught me that money was the root of all evil. I believed him because I saw him suffering from the stresses of his business life. He died of heart disease at 53. I no longer take the evil at face value, but I do think money causes us much grief and confusion, whether we have it or not. Deep psychic resonances go along with it. Our most confounding moral problems come from money. It is a pervasive source of suffering. Economic concerns trigger the stress response. The rhythms of love and wisdom break down under chronic stress. The aggression set loose in the wake of the breakdown hurts us in central ways.

You could say that money is the medium of exchange for suffering. We transfer more than goods, services, debts, equities, wealth, budgets, gifts, legacies, priorities and choices through money exchange. We transfer want, injustice, and oppression too. Marriages are broken, family friendships destroyed, crimes committed, even murders perpetrated for money. Whole neighborhoods disappear, cities are torn down and rebuilt for money, with little regard for what was there before or what will be needed next.

And you cannot ignore money or retreat far from it. Money makes constant demands on us. Jacob Needleman insisted that our relationship to money is the principal value challenge of our time. "The outward expenditure of mankind's energy," he wrote, "now takes place in and through money... Therefore, if one wishes to understand life, one must understand money—in this present phase of history and civilization."[7]

There's no pathway into the near or middle future that will circumvent monetary value as the basis of material exchange. Whatever else we do to ameliorate our social problems, money anxieties will still be there to haunt us.

We would be hopelessly naïve to think that we could make deep social changes without bringing in new attitudes to money.

But if the ethical mind *can* overcome its subordination to the technical mind in other areas, why not here? Why not new culturally learned ways of valuing, using and understanding money? What would have to change to allow it? What would change afterward? Would the aggressive energies tied to money flow into other purposes? Could the *motivations* behind aggressive expression shift away from zero sum games? Could we do without winners and losers? Maybe the systems by which goods and services move could "shift itself in ways amenable to new purposes. Perhaps we would have to devise new forms of corporate governance to let it happen. Or perhaps they would come along as the actual exchanges developed.

Needleman noted that "in other times and places, not everyone has wanted money above all else; people have desired salvation, beauty, power, strength, pleasure, propriety, explanations, food, adventure, conquest, comfort."[8] Our present forms of money exchange may be the most effective and flexible we have devised so far, but there have been others and there will be others in the many scores of millennia left to us. There has always been considerable variability in what has been exchanged and why. The Northwest Indian Potlatch giveaways were as much a product of human nature as predatory capitalism. And competition drove both of them. Can we do it without competition? Or can the nature and cultural meaning of competition itself change?

## 290

My own coming to terms with money has been long and mostly peaceful. I understand my relationship to money as one part of all that I do to "make a living". My living includes not only the cash transactions I get and give, but all my labor and all the goods I exchange and all of the ways I exercise my values to give and receive in the world. Money is a part, but not the whole of it. From my vantage point it counts for less than half (whatever your income.)

We must factor every kind of giving and receiving into our livelihoods. The actual balance between overall giving and receiving, with its monetary, non-monetary, bartered, and traded and freely given components, always has been the basis of our livelihood. Giving/receiving shows how we live. The real meaning of "making a living" then is the balance between all that you give and all that you receive. It's a Cabalistic concept. If you are receiving more than you are giving, you are "eating the bread of shame."

The poor and oppressed always give more than they receive, sometimes even in conditions of outright beggary. To be underpaid and under-served is to give more than you receive. Only the criminals among the poor eat the bread of shame. The rest, though they may be "shamed" by the rich for their neediness, and come to shame themselves and consider themselves uncouth and dangerous, eat what bread they can as civilly as they are able. To my observation, they share what they have more generously than the rich do. On a deep level, they maintain their self-regard; they are saner. Mainly the prosperous eat the bread of shame while keeping the shame hidden from themselves.

## 291

Though the fullest suffering belongs to the poor, prosperous people suffer too, from emptiness. To overcome the shackles of emptiness they need a stronger hold on inwardness. Inwardness puts money in its place. It helps us balance giving and receiving. And new ideas about exchange might come from it, opening new areas for moral discovery.

We are living in a time when the forcing conditions for historical change are more likely to come from failure than from success. And this pushes the dominant classes into an ironic posture, because though they have taken success as their highest virtue, failure is pushing them through changes. They've been in a chronic emergency for decades.

## 292

Not that human life is ever without suffering; there is too much that is intrinsically problematical in our mortality to allow that. But poor and rich people don't suffer in the same way. They have different needs and dissatisfactions.

## 293
### *Empty Suffering*

Prosperous people experience empty suffering. Its distinctive pains come from the frustration of higher needs by competition, greed, money-lust, power, prestige, and position. The emptiness they suffer is endured in the presence of surfeit. It's a tribulation on nothingness in the absence of meaning. Successomania is its Mephistopheles. The prospect of failure haunts it.

The suffering of emptiness does not call for massive deliveries of food or shelter or medicine. Instead, it aims for absorbing distractions. It tries to hold on to the outward measures of success by blaming others for every failure. High-class empty suffering doses itself with a subtle armamentarium of psychiatric drugs, cosmetic surgery, life extension therapies and discretely hushed-up crimes. It prefers to whisper to the judge instead of crying out for justice. Its emptiness is not alleviated by food and shelter, but by diet spas and addiction recovery clinics.

The best-situated people, those afflicted most thoroughly with the emptiness of suffering, try hardest to hold on to their power and privileges as proof of their *worth*. In doing so, in clinging hard to greed or fear, they lose their aptness for the times. They think too much about themselves and too little about the world. And because they see themselves mainly from outside, they lose their creative vigor. They no longer lead, they follow. But they hide this from themselves by hiring experts who toady and deceive them (and the experts, as they rise, increasingly fool themselves.) These advisors stay employed by seeming to have "good track records." "Seeming" is the operative concept here. They maintain their edge by blurring the transactions by which others can judge their performance. Acceleration is their sleight of hand. A short-term investment lasts a few seconds, a long-term investment a few days to a few months. The financial managers who "bought and held" are gone.

The top people coast along on the seemingly reasonable assumption that wealth equals power and power equals leadership, an equation that works well enough in good times. And even when things go bad for the very rich, they can buy their way out. They can afford to make mistakes and pay for more mistakes to cover the old ones up, unlike the rest of us. However, their shortsightedness has thrown the larger economy – the economy of goods and services, giving and receiving – into turmoil.

In this historical moment, many people realize that the equation "wealth plus power equals leadership" is a self-soothing fiction of the moneyed classes, a form of masturbation, an antidepressant or tranquillizer.

Like other substance abusers, (they truly are quintessential abusers of the substance of the earth,) the sufferers of emptiness build up a tolerance to their own remedies. They need to feel and feed their power in ever more extravagant ways. A psychosocial analysis of monetary inflation would probably show that inflation-ary fiscal policies have been significantly driven in hidden ways by the need for psychological compensation. We went from multi-millionaires to billionaires in the wake of the malaise of the Vietnam War, the Nixon presidency, the oil shock, Reaganomics, loose money and inflation. I would not be surprised if we come out of our present debt crisis with inflation-created trillionaires who affirm their self-regard with fiat scrip.

## 294
### *Full Suffering*

Poor and oppressed people endure the fullness of suffering. In full suffering, the losses come from deprivation of basic needs: for food, shelter and safety, for respect and self-respect, and for meaningful action. Full suffering is endured in the midst of tangible pains. Through the fullness of suffering, hundreds of millions of people lose the chance to become themselves.

Full suffering creates wounds to love in the breaking of the family, in the forced relocation of populations, in the threats to intimacy, in poverty and despair, in oppression, in the degrading relation of the slave to the master. The wound to wisdom, in full suffering, suppurates not over the loss of trivial choices, but over the crushing of the capability for a self-directed life.

The fullness of suffering has an authentic cry lacking in empty suffering. Writing in the 1850's, Frederick Douglass, the escaped slave, remembered hearing the cry in actual songs.

"I did not, when a slave, understand the deep meaning of those rude and apparently incoherent songs. I was myself within the circle; so that I neither saw nor heard as those without might see or hear. They told a tale of woe which was then altogether beyond my feeble comprehension; they were tones loud, long and deep; they breathed the prayer and complaint of souls boiling over with the bitterest anguish."[9]

More recently, James Baldwin wrote:

"When a white man faces a black man, especially if the black man is helpless, terrible things are revealed. I know. I have been carried into precinct basements often enough, and I have seen and heard and endured the secrets of desperate white men and women, which they knew were safe with me, because even if I should speak, no one would believe me."[10]

The same authentic cry of full suffering is heard from Native Americans. Chief Joseph, the leader of the Nez Perce, said in his surrender speech:

"I am tired of fighting. Our chiefs are killed. Looking Glass is dead... It is cold and we have no blankets. The little children are freezing to death. My people, some of them have run away to the hills and have no blankets, no food; no one knows where they are – perhaps freezing to death. I want to have time to look for my children and see how many I can find. Maybe I shall find them among the dead. Hear me my chiefs. I am tired; my heart is sick and sad. From where the sun now stands, I will fight no more forever."[11]

In the power of these utterances, you hear not just the bitterness of defeat and the outrage at loss, but a revelatory communication that strives with great dignity to offer the world a healing balm from the soul of suffering itself.

## 295

I believe that the fullest suffering has the clearest moral urgency with the least ambiguity. Those among the poor who understand this by having witnessed it close up have the moral lead. Suffering plus insight trumps insight plus empathy. As James Baldwin wrote:

"I do not mean to be sentimental about suffering – enough is certainly as good as a feast – but people who cannot suffer can never grow up, can never discover who they are. That man who is forced each day to snatch his manhood, his identity, out of the fire of human cruelty that rages to destroy it knows, if he survives his effort, and even if he does not survive it, something about himself and human life that no school on earth – and, indeed, no church – can teach..."[12]

Defeat rightly apprehended nourishes the imagination, not only the compensatory but also the creative and the exact imagination by means of which the sciences of the future will emerge. There's an incalculable richness of imagination waiting to be tapped in the souls of oppressed people, a wealth that must not be wasted.

W. E. B. Dubois called the ties between failure and moral leadership the progressive hope for American culture.

"Herein the longing of black men must have respect," he wrote, "the rich and bitter depth of their experience, the unknown treasures of their inner life, the strange renderings of nature they have seen, may give the world new points of view and make their loving, living and doing precious to all human hearts."[13]

And again:

"...there are today no truer exponents of the pure human spirit of the Declaration of Independence than the American Negroes," he insisted. "There is no true American music but the wild sweet melodies of the Negro slave; the American fairy tales and folklore are Indian and African; and, all in all, we black men seem the sole oasis of simple faith and reverence in a dusty desert of dollars and smartness."[14]

Baldwin pursued the same vision:

"This past, the Negro's past, of rope, fire, torture, castration, infanticide, rape; death and humiliation...yet contains, for all its horror, something very beautiful... one eventually ceases to be controlled by a fear of what life can bring; whatever it brings must be borne...It demands great spiritual resilience not to hate the hater whose foot is on your neck, and an even greater miracle of perception and clarity not to teach your child to hate. The Negro boys and girls who are facing mobs today come out of a long line of improbable aristocrats – the only genuine aristocrats this country has produced...I am proud of these people not because of their color but because of their intelligence and beauty. The country should be proud of them too, but, alas, not many people in this country even know of their existence."[15]

When they recognize the readiness of the moment, the oppressed become a force for change. The moral lead that is due them they take. But not always. They can only exert the advantage of creatively wrought low-mediated intentions in full turnings that bind their aggression to love and wisdom.

## 296

Suffering inspires us to draw the remedies for our pains from our pains. The heat draws up elemental forces from the treasury of the neutral traits.

From suffering, we discover new adaptive physiological responses. This deep tissue effort exerts direct pressure from human biology on history. I think of it as the muscle of justice working through time. However, in hard times the muscle of justice brings opposition down on those who use it.

There are no guarantees of success, though eventually suffering and failure reconfigure the content of the turning points of love and wisdom that play out in the signal events. From these changes, new expressions in human nature do make their way into the world.

## 297

Because of their suffering, the creative energy it has generated, and the moral lead it has bestowed, the American underclasses, nourished by an African-American *avant-garde*, have become the creative minority with the best chance of transforming Western culture. Certainly, they have a long history of service to the lands of the Western Hemisphere, and the most just claims.

Because they are less enthralled by the dominant cultures, they live less in thrall to them. Their cries are more substantial, more authentic. Because they nurse the fewest illusions they have the soundest intellectual and emotional foundation to achieve the breakthroughs to the next epoch. As they add alloys to the melting pot in mixed ethnic marriages in the heart of Western nations, they have the best chance to bring us healing.

The Black, Native American and Hispanic peoples in the Western hemisphere carry the healing energy. For good reasons they hold the moral lead:

o   They know both full and empty suffering. The rich only know empty suffering.

- o They are less likely to be deterred by fear of the fragmentation of the self in turning points.
- o Most are not shackled to the conceptual apparatus of the "winners". They know what has disappointed and disillusioned them. They know what needs to be fixed. Losing is their starting place. It's their route to inwardness. They can get stronger from it.

## 298

The black experience in America is greatly shaping global culture through the arts. Its contributions ring with political relevance because the content of their full suffering and the insights and values that come from it *are most in touch with the needs of oppressed people everywhere.* The African-American population knows the secret crimes in the American heart better than anyone else.

My historical prediction is that the influence of Black culture in America, now spreading worldwide, by affiliating with underclass movements everywhere will finally close a circuit of tremendous power. It is most likely to happen on the world stage when Black American soldiers sent to bases in Africa to fight the War on Terror see the suffering of native Africans from close up, and, at home, when urban uprisings reveal the true demographics of life "inside the Beltway."[16]

## 299
### *Why Me?*

Both empty and full suffering ask the "Why me?" question.

"Why me?" fundamentally distinguishes all suffering from pain. Job sitting in the shavings of his boils asked "why me?"

But the "why me" of empty and full suffering have different answers, for though the rich and the poor are broken alike in love and wisdom, their wounds knit and heal differently.

Only those who endure full suffering can answer the "why me?" question rationally, appealing to justice. They know that their oppressors cause the main part of their suffering. In empty suffering the "Why me?" question has no good answer because there is no appeal to justice.

This removes moral parity from the forms of suffering. Between full and empty suffering, only full suffering has unambiguous moral urgency.

## 300

In its secret dimension, empty suffering promotes the failure of empathy. It makes the sufferers cruel to each other.

Full suffering has a secret dimension too. The secret dimension to full suffering is that it *includes* empty suffering. Full sufferers have both.

Only those who have endured full suffering with open awareness, and have not been destroyed by it but have been able to preserve authentic changes in love and wisdom, can recognize and surmount their empty suffering too. And by an internal alchemy, this reaction of full on empty suffering pushes them past the "why me?" question to the "what's next?" question. They become forward thinking in their suffering.

Full suffering takes precedence over empty suffering, therefore, not as a matter of right or because the prosperous grant the right to the oppressed, but because the moral lead belongs to those whose changes have given them the most relevant creative insights and redemptive visions into the healing process.

To the extent the victims of empty suffering have been misguided by successomania, however, they too may need to experience and acknowledge the failure coming from emptiness before they can change. Here, in this failure, we may find our real transformational power. And here the creative minority spanning the dominant and underclasses, may find the way back to full engagement.

## 301

In his marvelous poem *September, 1939,* W. H. Auden calls these people "ironic points of light." He writes

*Defenseless under the night*
*Our world in stupor lies;*
*Yet, dotted everywhere,*
*Ironic points of light*
*Flash out wherever the Just*
*Exchange their messages...*

The points of light, though few and scattered, carry culture. The illuminated people love and grow wise and reach out to each other. Auden's wish is that

*May I, composed like them*
*Of Eros and dust,*
*Beleaguered by the same*
*Negation and despair,*
*Show an affirming flame.*

## 302

Many times before, natural or man-made catastrophes have sown seeds of renewal. Sometimes over-expansion or imperial overreach has provoked economic collapse, as in the Wallbank thesis on the decline of Rome. Sometimes resource depletion has done it, as in the over-grazing and desertification of North Africa following the Agricultural Revolution. The Collapse that Jared Diamond warns about may open the way to renewal. The Black Death immediately preceded the Renaissance in Europe. The period of the Warring States set the stage for classical Chinese culture. The French Revolution and the politics of modernity rose on the back of a corrupt and bankrupt Old Regime.

## 303

To make use of failure for personal renewal, you have to know that you have failed, that you have not held onto the most important things in life, that your turnings have been weak, that your generous impulses have been too sparing, that giving has not been receiving for you, that you have eaten the bread of shame. To fail in the midst of prosperity is to see yourself naked, without cosmetic remediation. Only the people who can see themselves this way while remaining sane possess the power to grow from suffering. Only persons who get access to themselves by rejecting successomania, consumerism, addiction and juvenile arrest can sense how much of their promise they have lost. Only when they experience their present losses with humility will the ethical mind open up in simple neediness and when it does, they will find their way to right giving and receiving.

Giving and receiving (which really turn out to be expressions of love and wisdom) are the rhythms by which goods are exchanged. The ways the means of livelihood that truly undergird "making a living" are distributed along the long legs leading to our turning points. Achieving balance in the rhythms of giving and receiving during approach-separation and withdrawal-return will both require and inspire powers of judgment, courage, kindheartedness and temperance hardly accessible to those engrossed in the four pathologies and afflicted with empty suffering.

We have our best chances for making progress at the fault line where full and empty suffering meet and touch. In the vast variety, even cacophony, of communications along the creases between communities, the creative ferment is growing. There, carried in the gestural substructure of memes, fan-shaped destiny is opening routes to the new human nature.

### End Notes

[1] The plot content of almost every TV series shows our yearning for community. First, they focus on natural sized sodalities; the crew of Star Trek Voyager or Enterprise on adventure shows, the tavern in "Cheers" in TV sitcoms, ER, Seinfld, NYPD Blue, House, Friends, etc., all have this in common: inside them we find real dominance quilts. Sex and violence are less the core than the outer fringe of their content. The group is the important thing. Their real appeal is that they gratify our yearning for a natural social life. Programs like Seinfeld dispense almost entirely with sex and violence, with significant doings altogether, in order to focus exclusively on the feelings of belonging, despite the characters' idiosyncrasies. Nothing happens in the show, but the premise touches on the signal events. They involve the post break-up relationship of Jerry and Elaine, his struggles with his parents, Costanza's inability to find love, Kramer's goofy alienation that he proudly makes into his way of being in the world.

As in real life, the little natural sit-com community is often besieged by the city, by unscrupulous strangers, by occupational and money problems, and in the crime shows, by murder and mayhem. But next week all the principals are back together for another go-round. The mythos of belonging requires that the main characters never die and the groups rarely split up. We need them too much. Captain Kirk is still ready to gather the crew of the Enterprise for one last mission.

[2] Ibid p.67)

[3] John Stuart Mill. On Liberty. Dutton. P. 133

[4] Emerson. *Self Reliance.* p.155

[5] Emerson. *The writings of Ralph Waldo Emerson.* Modern Library 1940. Self R. p. p. 158. He insisted that the recognition of this unique inward capacity "corrects the capital mistake of the infant man, who seeks to be great by following the great, and hopes to derive advantages from another – by showing the fountain of all good to be in himself, and that he, equally with every man, is an inlet into the deeps of Reason." Emerson. An Address. p.70

[6] (ibid p. 146)

[7] Jacob Needleman. *Money and the Meaning of Life.* Doubleday. 1991. p. 41-2

[8] Jacob Needleman. Money and the Meaning of Life. P.__

[9] Jeremy Campbell, *Winston Churchill's Afternoon Nap.* Simon and Schuster. 1986. p.128

[10] James Baldwin. The Fire Next Time. The Dial Press. 1963. p.67

[11] U. S. Secretary of War report 1877, p. 630

[12] The Fire Next Time p.112-115

[13] W. E. B. DuBois. *The Souls of Black Folk.* New American Library 1969. p.138

[14] ibid p. 52

[15] The Fire next Time p.112-115

# POSSIBLE FUTURES

# 18. NEW SCIENCE

My youthful insights on the Golem were true: we *have* external-ized our capacities in technology in order to effect interior changes in ourselves (warmth, comfort, stimulation.) We have designed a culture for ourselves that intentionally seeks to externalize our internal capabilities through technology. Juggernaut's cart gets us where we're going. But past a certain threshold of complexity, our journeys take more not less time and we waste more resources.

First technology connected us to nature. Now it connects us to the connection. As a result, we're facing three unintended long range consequences.

o Technological civilization pushes us away from the use of our own body systems for internal regulation. We are increasingly alienated from ourselves.

o Technological externalizations have built up a momentum of their own. They cannot change without calling out new externalizations. The economy relies on increasing layers of connections.

o The thickening membrane, while producing markets for goods and services, destroys natural habitats, including the internal human habitat of the mind's relationship to the body. This breach keeps us from experiencing our own real needs or recognizing the encroaching pathologies that makes us ill or stupid.

Global warming, resource depletion, species extinctions, mass migration, cultural anomie, enchantment to media, worship of celebrity, propaganda, disinformation, truthiness, ongoing regional wars, the collapse of economies and ecological webs are signs and symptoms of it. They show that we've reached a limit, that the current norms of expansion and acceleration are neither sustainable, desirable nor inevitable. We don't see this last most obvious factor because we've been habituated to mania. We need to sober up.

### 305

Who needs to sober up? The usual experts, economists and politicians. But more importantly, life scientists and physical scientists. It is their research that will have the most impact on the future. Working scientists are best situated now to bridge the gap between the ethical and technical minds.

Any historically powerful reconciliation between ethics and technics would have to change the way we do science and use its findings. There is no stronger route for change. Science is the central enterprise of the Fifth Epoch. It upholds our standards of verification. Its directions for research shape culture and industry at the root. Our historical momentum is closely linked to the progress of the sciences and its technical applications.

A reconfigured science, with scientists less under the thrall of successomania, consumerism, addiction and juvenile arrest, could free us to explore our creative resources in powerful new ways.

We have no better intervention point from which to incite cultural change. Science is the fulcrum. Our hands are on the lever.

### 306

But how can scientists rely on ethical insights in their work when to all appearances they are utterly reliant on the technical mind and its links to quantification and repeatability? The technical mind sits at the core of the scientific method. Some critics argue that the ethical grounding would have to come from outside, that we have to impose an ethic on science to keep it safe.

In *Enough!*, Bill McKibbin argues for self-restraint.

"If germline genetic engineering ever starts,: he tells us, "it will accelerate endlessly and unstoppably into the future, as individuals make their calculations that they have no choice but to equip their kids for the world that's being made. Once the game is under way, in other words, there won't be moral decisions, only strategic ones."[1]

But the self-restraint that scientists need is not part of the doing of science. It's outside of science. It's human restraint. "What makes us unique is that we can restrain ourselves. We can decide not to do something we are able to do. We can set limits on our desires. We can say "Enough!"

He argues that "poverty and illness may not require the highest tech; maybe they can be dealt with in the world we currently inhabit, with the kind of small and steady scientific and cultural progress we're used to."[2]

How does he expect to keep things "small and steady" when the culture shows such a strong desire to push on farther, faster and bigger? Where does he get the notion that we're used to small and steady scientific progress? How does he expect to protect us from the juggernaut? He's not that confident. "As we've seen, meaning turns out to be fragile – we can either pile sandbags around it to keep it safe, or watch it wash away."[3]

This argument from the loss of meaning suffers enormously because McKibbin does not show what meaning consists in. Without dealing with the contents of human nature, with love, wisdom and aggression, his solutions seem unworkably thin.

He may call his nostrum self-restraint, but in practice, some of us would certainly have to restrain others. We spend enormous energy restraining each other already. McKibbin's political solutions would require new agencies, new laws and regulations and a new cadre of regulators, hopefully not dominated by lobbyists. This approach presupposes a functioning democracy with a well-informed and caring citizenry. We don't have that now and we're not likely to get it soon. Lacking a creative ruling class, lacking the sensitivity to recognize the moral lead of the underclasses, and unable to conceive of non-establishment types as a revolutionary vanguard, the greatest likelihood is that McKibbin would end up turning over the management of the *Enough!* agenda to another elite of mediocre experts, and the successomaniacs among them would quickly rise to the top.

But why force controls from outside and disguise them as self-restraint when modern science has very sturdy moral foundations of its own? Where are they, you may well ask? They've been there all along in its commitments to openness and freedom of inquiry. Freeman Dyson described it as being "...based on a fundamental open-mindedness, a willingness to subject every belief and every theory to analytical scrutiny and experimental test... We scientists are by training and temperament jealous of our freedom. We do not in principle allow any statement whatever to be immune to doubt."

In support of which he quotes the saying carved over the door of the Royal Society of London, "*Nullis in Verbia*," and translates it as "No man's word shall be final."[4]

## 307

This commitment to truth, freedom and openness hasn't kept us from resource depletion, pollution, climate change, weapons proliferation, etc. Why not? Because socio-economic pressures, when they drive the technical mind, erode these commitments and replace them with different goals centering on wealth, success and esteem. As Pope John Paul II observed in an apostolic letter, "The pre-eminence of the profit motive in conducting scientific research ultimately means that science is deprived of its epistemological character, according to which its primary goal is discovery of the truth."[5]

In its quest for contact with reality, the technical mind can serve science well only when it relies on the desire for truth. That desire comes from the ethical mind *in science*.

## 308

Richard Feynman knew this from long experience. "People are not honest," he pointed out. "Scientists are not honest at all, either. It's useless. Nobody's honest. Scientist's are not honest. And people usually believe that they are. That makes it worse. By honest I don't mean that you only tell what's true. But you make clear the entire situation. You make clear all the information that is required for somebody else who is intelligent to make up their mind."[6]

Corporate sponsored science obstructs the search for truth by reneging on openness, which makes the problem worse. Good scientists are forced into competitive isolation at a time when they need maximum conviviality. Competitive isolation draws one toward the four pathologies.

**309**

In *Personal Knowledge*, Michael Polanyi, a mid-twentieth century chemist and philosopher of science, saw this happening. He made efforts to restore the root of the ethical mind in science by insisting that aesthetic feelings were indispensable to its search for truth. He considered "intellectual beauty as a guide to discovery and a mark of truth."[7] Beauty and truth were tied together. Beauty was a way to truth.

"Three things have been established beyond reasonable doubt," he argued, "the power of intellectual beauty to reveal truth about nature; the vital importance of distinguishing this beauty from merely formal attractiveness; and the delicacy of the test between them, so difficult that it may baffle the most penetrating scientific minds."[8]

That suggestive phrase "intellectual beauty" harks back to Shelley's *Hymn to Intellectual Beauty,* a poem that affirmed his commitment to the "awful Loveliness" of the truths of nature learned by "studious zeal or Love's delight."[9]

*I vowed that I would dedicate my powers*
*To thee and thine – have I not kept the vow?*
*With beating heart and streaming eyes, even now*
*I call the phantoms of a thousand hours*
*Each from his voiceless grave…*

I take this zeal to reflect the devotion, constancy and enthusiasm scientists have for discovery. Only the pursuit of beauty joined to the search for truth can guide the technical mind into the heart of nature. The best young researchers share this passion. But strictures on openness and loss of conviviality frustrate it.

Polanyi believed that

"… a philosophic movement guided by aspirations of scientific severity has come to threaten the position of science itself. This self-contradiction stems from a misguided intellectual passion – a passion for achieving absolutely impersonal knowledge which, being unable to recognize any persons, presents us with a picture of the universe in which we ourselves are absent." As a result: "…it has now turned out that modern scientism fetters thought as cruelly as ever the churches had done. It offers no scope for our most vital beliefs and it forces us to disguise them in farcically inadequate terms."

The misuse of the technical mind put him in "opposition to a universal mechanical interpretation of things [that} impairs man's moral consciousness."[10]

## 310

Polanyi was caught in a bind. How could scientists allow "vital beliefs" resting on subjective ethical and aesthetic considerations into their experimental work without threatening the very basis of the scientific method? To do so, some philosophers have argued, is to commit the *Naturalistic Fallacy.* A. C. Ewing commented "...there seems to be no possibility of validly deducing ethical propositions by some sort of logical argument from the nature of reality without first assuming some ethical propositions to be true; or at least if there is, the way to do so has not yet been discovered by anybody." That is, one cannot seek an "ought" from the "is" of nature. That effort is absolutely vain, misguided and fatal to scientific progress.[11]

Turning the tables on the naturalistic fallacy, Father Copelston, in his history of Western philosophy, wrote "the only reason for so describing it would obviously be the belief that goodness is a 'non-natural' quality."[12]

## 311

The issue goes to the fit between us and nature, a fit that in all its rhythms, patterns, parts and processes evolved over billions of years. Our senses (senses evolved by nature too) register it as beauty.

To our evolved eye, evolution seems to select for beautiful patterns even where they have no survival benefit. The patterning is intrinsic in nature. It exists on every scale.

Even mollusks buried in the mud have beautiful shells.

And we, since we are the stuff of nature, and account ourselves beautiful too, respond to beauty by opening ourselves to it wherever we encounter it. That is to say, we look for beauty, we want it

to be there; we try to create and preserve it; we search for the link between beauty and truth; we accept it as part of nature.

The working scientist assumes that the aesthetic sensibility that makes a flower beautiful, that makes the blending of colors in the evening sky sublime, that makes an apple delicious, that makes the electromagnetic spectrum a delight to consider, that finds beauty in the Fibonnaci series and revelation in the Phi of a seashell's curve, is in fact the way beauty expresses truth and truth manifests as beauty.

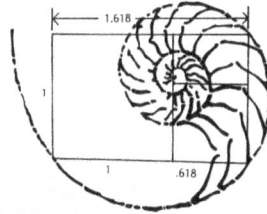

We have established that the standard of beauty comes from the standard of truth which comes, in turn, from the standards of cognition, which come from the evolved nature of our sense organs and processing equipment, which for adaptive reasons suffuse our organism as tangibly as the whorls on a seashell. It is how we order experience.

We find beauty not only spatially but also temporally, in the flow of events. They can be beautiful too. They have their music; they dance. The beauty present in the moment unfolds in time. The great dramatists show this in their work.

But where does goodness come in?

## 312

Polanyi thought to bring goodness into science via the sense of personal responsibility. He wrote that "...no sincere assertion of fact is essentially unaccompanied by feelings of intellectual satisfaction or of a persuasive desire and a sense of personal responsibility"[13]

The link goes deeper than that I think. If openness bears on the search for truth by upholding the ethical mind, as Feynman argued, and if the ethical mind has physiological links to empathy and sympathy, as I have shown, and if beauty and truth are linked, as Polanyi and many other scientists have insisted, it may be that goodness comes into the mix deeper down. Socrates tried to prove that truth, beauty and goodness were one.

## 313

Perhaps the ravishment by beauty in the sciences that Polanyi celebrated connects directly to the natural root of Agape. Socrates pointed out that we naturally love what is beautiful and want it near us, and we do not want to destroy what we love. If so, there must be a love that joins beauty to goodness. Having already admitted the aesthetic energy in truth itself, we now discover, at least through caring not to destroy, that it inclines us to goodness. In beauty, we experience the evolved power of an addictive yearning for contact with reality. We seek it because it "tastes good". It's a dependency rooted as firmly in our natures as the suckling infant's connection to the mother's breast. From that pinnacle of connectedness flows the sugary sweet milk whose flavor is both desirable in itself, necessary for survival and the root of sugar addictions.

The broader truth here is that the beauty in a theory, at the same time that it is a mark of the correspondence between the theory and nature, signifies a correspondence between the inner and outer world of the person conceiving of the theory. That person who happens to be so lucky as to inhabit the part of the cosmos capable of reflecting on itself, makes choices between alternative understandings, feels Spinozan emotions and does something in the world.

Our flow of passions in science, seen from this perspective, is part of a wholeness, an affirmation of full human presence in the joint field of nature-and-self.

## 314

Goethe's science followed this line. It was his response to the mathematical abstractions of the Newtonian worldview that he thought were taking us away from reality. The greatest disaster of modern physics was its "insulating the experiment from man, and attempting to get to know nature merely through artifices and in-struments." He argued, like Kant, that the only reality we could know lay at the interface between man and nature. No matter how many instruments or layers of abstraction intervened between the object and the person, the last step of the transmission would pass through the human sensory sheath, inevitably bringing the science to the flesh, though it might come back drained and desiccated in meaning and relevance.

Goethe believed that beauty could reveal laws of nature. He was the one major poet who lived a life of scientific research, he spent decades experimenting with optics, anatomy, geology, bot-any and the morphology of plants.

As Erich Heller describes it in *The Disinherited Mind,* Goethe's scientific work sprang from a faith in the connectedness of man with nature and that "Goethe's ethical and scientific con-victions are mere aspects of that faith... in a perfect correspon-dence between the inner nature of man and the structure of external reality, between the soul and the world."[14]

## 315

Goethe's approach has had a curious afterlife in Wittgenstein. In collaboration with Frederich Weisman, one of logical positivists in the Vienna Circle of the late twenties the Vienna Circle, Witt-genstein wrote in 1929:

"Our thought here marches with certain views of Goethe's which he expressed in the Metamorphosis of Plants. We are in the habit, whenever we perceive similarities, of seeking some common origin for them. The urge to follow such phenomena back to their origin in the past expresses itself in a certain style of thinking...namely the arrangement as a series in time. (And that is presumably bound up with the uniqueness of the causal schema). But Goethe's view shows that this is not the only possible form of conception."

Ray Monk, in his biography of Wittgenstein, explained the Goethe connection this way: "Goethe's morphology... sought to recognize living forms *as such*, to see in context their visible and tangible parts, to perceive them as *manifestations* of something within. Wittgenstein's philosophical method, which replaces theory with 'the synopsis of trivialities', is in this same tradition." [15]

Wittgenstein's later work reminds one of Goethe's focus on morphology applied to language usage. Wittgenstein searched for "family resemblances" in order to describe common meanings without going beyond experience.

## 316

As I understand Wittgenstein's position, when a family resemblance was pointed out, you either saw it or you didn't. When you did, the family resemblances gelled into a "gestalt," a term that Wittgenstein adapted from Wolfgang Kohler, who himself took it from Goethe. In a gestalt you get the image to hold fast, like those optical illusions that flip-flop between an old crone and a young woman – you get it to hold as a young woman, for example, and the crone goes away. When the gestalt changes and the other face comes back, you know you are slipping from one family of usage to another.

**Beauty-Crone illusion**

With trained sensory and cognitive perspectives I believe a person can see both the crone and the beauty at once, and watch them flip-flop, and experience the changes as part of a larger pattern, a "field" including both. The field itself, with proper training, can lay open before the senses as a gestalt. It congeals when the brain learns to relax into the field of endeavor, when a person achieves what the Taoists call *wu wei,* spontaneous action. The trick is to have the presence of mind to stay there, make choices, do work, conduct experiments, explore, remember and describe.

By persisting in a durable gestalt, we will be able to overcome the either/or character of experience without falling into a baffled chaos of perceptions or an obscuring metaphysical belief system. But persist in a bright, busy actuality.

Whether Goethe found an alternative to grouping events along the causal time series, and whether those gestalts could produce a better way of doing science remains to be seen. I follow a different path; I want to plunge deeper into the time series itself.

## 317

With the inwardness of scientists strengthened by personal training in meditation and self-examination, and with their careers guided by saner markers for success, the choice of projects pursued and experiments performed would certainly change. The changes might bring the ethical and technical minds into better balance.

Experiments don't design themselves; observations don't record themselves. We know what we know because of the choices we made. Had we undertaken other projects, we would have made other discoveries. Some might have taken us down different routes of inquiry.

Encourage us differently and we will discover differently.

In areas hardly touched by science, much simple observation and experimentation remain to be done.

Mendel founded the science of genetics by growing pea plants in his kitchen garden. Darwin observed, collected thought, wrote, and consulted with local farmers, and amateur naturalists, using hardly any big-budget scientific equipment. Galileo's budgets were minimal. Einstein did his work with pad and pencil.

## 318

With different discoveries in pursuit of different goals, we will in future times develop new sciences with new kinds of applications. If the technical and ethical minds rise together to the historical challenges to protect biodiversity and restore the balance in nature, the balance *in us* will be improved. We will cooperate better. We will care differently about the things we value. With the technical mind strengthened, we will invent tools that tune us to the rhythms of life across scales from cells to societies. With the ethical mind vitalized, we will harmoniously integrate our discoveries with basic ecological and human needs.

Along the way, we will uncover the fundamental approach-separation and withdrawal-return pathways in nature. We will use them to configure love and wisdom in wider circles and deeper interactions.

Scientists working freely in a spirit of conviviality, I am confident, will generate strong forces for moral progress.

## 319

But how would different goals and better communications change the scientific method? Would it give it sharper powers of apprehension? Would openness to intellectual beauty give scientists a way to make moral discoveries from within the sciences themselves? Would scientists see themselves as activists on the leading edge of the search for goodness, as do doctors who save lives? Could scientists revere the scientific method as their finest tool, and also see it as their ethical contribution to human progress?

## 320

Francis Bacon in the *New Atlantis* envisioned a utopian brotherhood of scientists dedicated to human advancement through an ethically elevated science. They lived and worked together in the House of Solomon. They observed three maxims.

"The first, that we do not so place our felicity in our knowledge, as we forget our mortality: the second, that we make application of our knowledge to give ourselves repose and contentment, and not distaste or repining: the third, that we do not presume by the contemplation of nature to attain to the mysteries of God."[16]

In his biographical study of Bacon, *The Man Who Saw Through Time*, Loren Eisley reported that he "warned that knowledge without charity could be as dangerous as the modern world has finally discovered it to be."[17]

"Mere power and mere knowledge exalt human nature but do not bless it," Bacon argued, "We must gather from the whole store of things such as make most for the uses of life."

Eisley traced out the consequences of our not following this route: "... the worlds drawn out of nature are human worlds, and their imperfections stem essentially from human inability to choose intelligently...Instead of regarding man as a corresponding problem, as Bacon's insight suggested, we chose, instead, to concentrate on that natural world which he truthfully held to be protean, malleable and capable of human guidance. Although worlds can be drawn out of that maelstrom, they do not always serve the individual imprisoned within the substance of things."[18]

## 321

A revision of science that regarded man "as a corresponding problem," and drew out of nature "human worlds," would take us to the mind/body edge. To get there we would have to develop not only new instrumentation and experiments, but also deeper turnings that transfigured the psychic organization of scientists through modification of the passing rules.

It implies that as the core concerns of the people doing science develop, the way they do science, as well as what they do science about, changes.

Our core concerns do change. They are changing now as we fight to keep love and wisdom intact and convergent against mounting cultural and environmental injuries.

Our core concerns are time-bound and come into consciousness through a creative, morally plural, neurally competitive effort. Scientists who choose what to work on, who decide what to deal with first, how to observe and measure it, how to present the findings and where to go with them from there, will surely play a central role in that effort.

The environmental crisis challenges the scientific enterprise to play new heroic guiding roles. If we rise to the occasion, we will come out the other end as fuller human beings with a better understanding of the world.

If we succumb to the pressures, we will become drones in a new repressive social order.

Which way will we go depends on whether we find the "effective rival" to moral indifference that eluded Polanyi. We can do it by looking in places he neglected.

**322**

F.C. S. Northrup found such a place. He considered immersion in the "undifferentiated aesthetic continuum" of the Eastern mindset the real jointure between the scientist and nature. He believed scientists could use this "effective rival" to build a bridge connecting nature, science and morality. He believed this project would firm up the world's foundations. It would solve the "basic problem underlying the ideological issues of these times..."[19]

In the aesthetic continuum, we meet the world completely. Nothing of us is excluded from it. Our material and mental interests come together in the aesthetic continuum. We do science, make art, discover ethical truth and live full-bodied lives in it. It is our field of endeavor.

**Notice Man in Boat**

You and I are immersed in the aesthetic continuum now. The words in this book are my embracement of you and your reading is your embracement of me at this moment. But the moment has per-

378

ished and past, and something else has replaced it. Now it too has passed. All organic nature flows and perishes as Whitehead understood.

From the Eastern perspective, a person immersed in the aesthetic continuum with the skills to keep balanced there can learn to do the least disturbing thing in the turning dramas of life consistent with realizing an intention, and do it from the smallest and most provisional sense of self he or she can hold.

But this spontaneous action cannot be insouciantly or randomly applied. It must beautifully flow with the inflections of the aesthetic field that makes *us* part of the truth.

In Northrup's view, this gift of the East to the West "makes the aesthetic immediacy, the communal, compassionate, emotional quale of all immediately apprehended things, whether they be human or non-human natural objects, something ultimate and primary in their own right." [20]

The rhythms in the aesthetic continuum endlessly pulse. They cohere or dissolve in us as we form and dissolve in the field that creates us and itself forms and dissolves. Jack Kerouac put it a simple way in his novel *Desolation Angels:* we are "passing through the that which passes through". The *I Ching* describes the way becoming goes through changes as it moves in the world. The sages who compiled the ancient Chinese classic elaborated a set of 64 patterns, built on a math to the base two, that express the basic changes in images.

Right now we are immersed in different sets of changes as they rise and fall. By knowing our place we can guide ourselves or at least understand our own changes. But we must perform that act of knowing in a state of immersion ourselves. Tossing coins or counting yarrow stalks puts you there in the moment.

## 323

Western scientists, Northrup continues as he develops his detailed argument, "fall into the error of supposing that the flashes, i.e. the aesthetic objects, are the mere appearances of unseen and unaesthetic molecules and electrons..." [Here he is discussing the Wilson Cloud Chamber apparatus. When you do this, you] "forthwith require an observer... to clarify the relations between the aesthetic appearances and the supposedly non-aesthetic molecules".

You're not just you any longer, you're "you" the observer. Something has split off to watch. You've got 1) a postulated reality of molecules and electrons banging into each other, 2) an observer who holds an explanation for this atomic and electromagnetic realm in mind (and his mind is also a postulated reality) and 3) a world of colors, shapes and sensory appearances that we live in, and though it means everything to us, it has no ontological status.[21] Northrup wanted to replace the Western "three-termed relation of appearance" with the Eastern "two-termed relation of epistemic correlation".

## 324

What would it actually take to do science in the aesthetic continuum? Can we really apply the field approach to nature in all its sensory subtlety? Can scientists rigorously use the aesthetic presence and the *wu wei* attitude it requires, in controlled experimental settings? Can such experiments be repeated? Quantified? Observed from outside?

In other words, how can an experiment command the full presence of the scientist when the conditions of repeatability and verifiability demand on the contrary that anyone doing the same experiment under the same conditions gets the same results? This seems to require the absence rather than the presence of the scientist in the aesthetic field.

## 325

One approach would be to study the scientific possibilities of immersion much more closely. To do that we would have to find ways to dwell in the aesthetic field for longer stretches of time (spans matching the contours of the withdrawal-return pattern as a whole.) If we could do that, we could deploy attention from one kind of awareness to another from within the continuum without jarring ourselves out; we could see the crone and the beauty simultaneously and *know ourselves as knowers* far better than we do now. That would mean developing self-examination skills to study gestalts from within. It would give interoceptive, exteroceptive and proprioceptive sensations equal status and equally good access to information about the world.

## 326

In fact, a science based on immersion flourished in the European Renaissance only to be stifled during the Reformation.

Giorgio De Santillana, one of the important recent historians of science, treats the influence of the Protestant Reformation on modern science in an interesting, non-obvious way. In his view, the religious wars demonized the Renaissance love of the natural world, the female body, sexual expression and magic. By doing so, it suppressed the powers of personal presence and aesthetic immersion. The scientific worldview was drained of color, literally. The scientists focused their attention on primary qualities. With mass, position, and velocity their data points, they hoped to comprehend everything. In the effort, they developed the standards of induction, experiment and quantitative analysis we work with today.

The seventeenth-century scientists from Galileo through Newton opened amazingly productive paths that the Renaissance magi, because they insisted on personal presence, could not follow. But their unprecedented predictive powers over nature came at a cost: they broke the bonds between the technical and ethical mindsets. They had to absent themselves.

## 327
### *Natural Magic and Science*

The Renaissance magicians tried to harness personal turning points to the turning points in nature through timing. The science in this, such as it was, rested on a belief in the power of imagination to repattern nature through low-mediated intentions sent into the world at the right moment, and in those days, the timing was determined astrologically.

Marsilio Ficino, Lorenzo Medici's tutor, the translator of Plato into Latin, consultant to artists on the thematic content of their paintings, and a central Renaissance thinker, authored numerous books on magic. Of his approach, de Santillana, wrote

"Ficino's theory is currently described as mystical Neoplatonism; yet contemplative withdrawal looks more like the literary side of it, while its intellectual ambition is strongly anchored to reality. It strives, like Pico's, [Pico della Mirandola, Ficino's protégée, polymath, bringer of Cabala to Renaissance magic] toward the conquest of 'natural magic,' the capacity for command that comes to man's soul from standing thus in the cockpit of the universe... It believes that we can reach out for as yet unknown harmonies and powers... The true distinction of man [is that he has] the power to share in the properties of all other beings, according to his own free choice. He is a universal and protean agent of transformation, hence it behooves him to orient his soul properly towards the good, so as not to use his powers wrongly."[22]

Here we come upon a clear description of the fusion between the ethical and technical mind mediated by free choice and immersion in nature ("standing thus in the cockpit of the universe.")

Frances A. Yates, the indispensable historian of Renaissance magic, treats the technique with its personal moral character in more detail. "Internally in the soul or the imagination," Ficino tried to create "a 'figure of the world' and to keep the inner attention concentrated on its images." By this effort the magus attained "through the magical organisation of the imagination a magically powerful personality, tuned in, as it were, to the powers of the cosmos." [23]

The repatterning of nature could only proceed in the presence of the magician.

### 328

Did the Renaissance magi advance the scientific enterprise? Yes. Did they do magic? No. Did their work guide the scientific revolution of the 17th century? Hardly. And yet the European Renaissance has to be considered a high point in Western culture, its artistry perhaps unsurpassed. How could its explosion of creativity fail to contribute to the sciences?

William H. McNeil, the world historian, argued that you have to look to the art to evaluate the science in the Renaissance; you have to appreciate the efforts of those great practitioners who, in making beauty, created a revolutionary new mathematics of space and light. McNeil points out that their work caught the central themes of Western science.

"This truly remarkable definition of a new and distinctively Western style of painting involved a sophisticated mathematization of space and an intellectual reorganization of intuitive optical experience.

Italian painting thus presaged the mathematical development of natural science that came to full expression only in the seventeenth century."[24]

**329**

And let's not forget Leonardo. Santillana considered Leonardo da Vinci the crucial figure linking art and science. He insisted that Leonardo's renderings of human anatomy, birds in flight and the flowing forms of water took empirical observation to degrees of penetration it had never achieved before (or possibly since.)

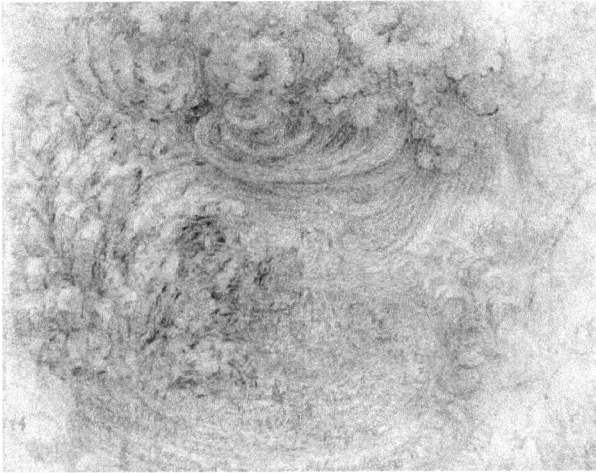

**The Deluge, both aesthetically whole and scientifically exact**

"He was the most original natural philosopher of his own time. That time did not possess what we call science, but it possessed art in a sense that is lost to us. Leonardo's guiding idea was not that the eye alone is able to see reality; but that the trained intent eye, the eye 'knowing how to see,' which controls the skilled hand, can come as close to the hidden structure of reality as it is possible for man to read – insofar as he has redesigned it himself."[25]

Leonardo's "trained intent eye" which, like Goethe's "exact imagination," sought to discover laws of nature revealed through beauty, may have made discoveries that we still haven't been able to express in the language of science. Santillana quotes an astonishing passage from Leonardo's notebooks to illustrate this:

"Write the tongue of the woodpecker and the jaw of the crocodile. Write the flight of the fourth kind of chewing butterflies and of the flying ants, and the three chief positions of the wings of birds in descent... Write of the regions of the air and the formation of clouds, and the cause of snow and hail, and of the new shapes that snow forms in the air, and of the trees in cold countries with the new shapes of the leaves... Write whether the percussion made by water upon its object is equal in power to the whole mass of water supposed suspended in the air, or no."

He comments on this amazing train of associations:

"His indefatigable endeavor surveys the whole terrain of experience in search of the outline of a science as yet dimly seen but which he thinks can be eventually grasped only from the whole. It is a science which is expected, on the first level, to yield the laws of shock and fall, and also of dynamic equilibrium through the principle of virtual velocities (on this Leonardo had more penetrating insights than most of his successors) but should then proceed to levels not simply reducible to mechanics, nor dominated by the mechanical model."[26]

Fritjof Capra, in *The Science of Leonardo,* draws the many areas of Da Vinci's scientific interests together in an accessible and insightful study. Freeman Dyson, who regularly reports on efforts to do science "eventually grasped only from the whole", urges us in that direction too. But so far the adherents of the holistic approach have neglected the power of "the trained intent eye" and what I have described as the "exact imagination" as tools for investigating the subtle patterns informing the whole. Without the trained intent eye and the exact imagination, however, personal presence, phenomenological analysis and self-examination lose their essential underpinnings.

### 330

Can one make a reasonable argument supporting "the trained intent eye" in science? Can one claim that the Renaissance magi, though on the right track, *didn't go far enough into presence* to reach the phenomenological/physical frontier where caring and seeing could fuse? Can we do better?

One approach is to consider the possibility that certain information only comes to us through the secondary qualities of sensation – through the beauty that Goethe experienced. Perhaps synergistic effects occur when certain primary qualities interact neurologically in certain states of consciousness, and may only be perceived, (fused with colors shapes and sounds) as secondary qualities. Perhaps traditional Chinese medicine follows this path.

Further, we transduce the secondary qualities to primary qualities as they travel through the central nervous system. Electro-chemical impulses from the retina, cochlea, nasal receptors, etc. become frequencies, wave trains, moving electrical charges. Voltage potential differences and electromagnetic field effects carry information. The information gets in on bursts of pulses. The body transmits it by pulse train modulation using nature's own versions of wavelet, amplitude and frequency modulation.

Though in all cognition reality makes its ultimate impression on consciousness through mass and energy, the Leonardo seen by Santillana took on the subtleties of sensation directly and penetrated its gestalts. Through skilful means, combining art and science, he may have succeeded in discerning the primary qualities as they streamed from the secondary qualities at the transduction points in his nervous system. We still have a great deal to learn about attending to our own states of consciousness, a project the Dalai Lama has repeatedly recommended in his conferences with scientists. He discussed these scientific possibilities in his recent book, *The Universe in a Single Atom,* where he described

"a vast body of practices that involve the use and enhancement of visualization and imagination, and various techniques for manipulating the vital energies in the body to induce progressively deeper and subtler states of mind... they may suggest unexpected capacities and potentials within the human mind..."

"...if the scientific study of consciousness is ever to grow to full maturity – given that subjectivity is a primary element of consciousness – it will have to incorporate a fully developed and rigorous methodology of first-person empiricism. It is in this area that I feel there is a tremendous potential for established contemplative traditions, such as Buddhism, to make a substantive contribution to the enrichment of science and its methods."[27]

Not least, it opens possibilities for being present at the transduction point from primary to secondary qualities.

### 331

I am suggesting we put more, not less, rigor into our experimental designs:

1) That we focus more narrowly on the unexplored venues where mind and body meet by analyzing our personal qualities of comportment, bearing, quietude, sensory acuity and meditation as the "personal knowledge" Polanyi sought,
2) That by including these experiences we apply the "observer effect" to the sciences of mind too,
3) That we use this data to re-evaluate which are and which are not the nuisance variables in an experiment.
4) That by doing this, we stop using personal presence as a nuisance variable,
5) That through aesthetic immersion of the sort Northrup recommended we will receive hints of higher correlations that appear at the interface of biophysics and phenomenology and fit the standards Goethe proposed when he described "the manifestation of secret laws of nature, which, were it not for their being revealed through beauty, would have remained unknown for ever,"
6) That we apply Clarkes law, described below, to cognitive science with the hope and expectation of doing work here "indistinguishable from magic" in experiments on dream and trance states, precognition, telekinesis, remote viewing, etc.
7) That our accomplishments will shift the borders between scientific and magical thinking with great benefit to psychic health and moral judgment.

### 332

As we attain better voluntary control over autonomic functions, we will become capable of engaging nature with fewer, saner ego investments, less delusion and less dependence on cultural conditioning. With personal presence restored, perhaps nature will open to us in ways she has never shown before.

Focusing scientific research on the mind/body frontier will give us the creative zest to make discoveries that nurture our own human-heartedness from within the flow of aesthetic immersion. Perhaps we will find the place where beauty, truth and goodness meet. As love and wisdom converge, we may get the strength, courage and smarts to deal with radically expanded possibilities only accessible to deep caring. New discoveries merging phenomenology with psychophysiology would give us access to unexplored regions of the mind.

Exciting times, but not lighthearted ones, because we will be relying on the strength of the ethical mind to create a viable world under the Damocles swords of nuclear proliferation, economic oppression, cultural collapse, climate change, overpopulation, racism and aggression.

## 333

All told, the effort to restore the rudiments of a science of personal presence in a time of depersonalization and powerlessness like ours could help reconcile the conflicts between the ethical and technical mindsets. As observer-participants in the new science of presence, we could study our own resonant relationships with nature. Using advanced brain imaging systems, we might find correlations between the frequencies and amplitudes of thoughts and images and the natural processes entering through the senses that underlie them. We could analyze them musically.

Since love and wisdom, as we have reconceived them here, are the primary carriers of natural rhythms in human life, we can anticipate their showing up in neuroscience in the frequency characteristics of approach/separation and withdrawal/return oscillations in the nervous system.

This knowledge, made certain by our personal presence at its source, would help widen the range of volitional over autonomic responses. Would this not open new opportunities of advancement for our creative powers? And if our creative powers can be augmented by changes in passing rules bringing new inputs from the neutral traits, might this not favor a reconciliation of our technical and ethical interests at a time when we need it badly?

Do we have examples of it? The poets got there first.

## 334
### The Magic in The Tempest

Prospero, Shakespeare's great magician, was perhaps the last and best representation of the Renaissance magi. And he did all of his magic for wisdom and love.

In briefest outline, the magical repatternings converge on combined turning points in love and wisdom. At the turning point Prospero restores his dukedom, ends his island exile, liberates his familiar spirit Ariel, sees his daughter Miranda betrothed to the shipwrecked Ferdinand.

When Prospero accomplishes these tasks, he breaks his staff and buries his magic book in the sea.

Prospero's magic requires his full participation in the workings of nature. Mind and body, memory, intentions, historical knowledge and his gorgeous words all play parts, words particularly, because they live on the mind/body interface.

Stephen Orgel in the *Oxford Shakespeare* edition of *The Tempest* describes Prospero's magic. "From one aspect, Prospero's art is Baconian science and Neoplatonic philosophy, the empirical study of nature leading to the understanding and control of all its forces."[28]

Harold Bloom emphasizes the non-theurgic nature of the magic. "Evidently, Prospero is a true scholar, pursuing wisdom for its own sake...His quest is intellectual, we might even say scientific, though his science is as personal and idiosyncratic as Dr. Freud's."[29] Commenting on his speech *"We are such stuff/ As dreams are made on; and our little life/ Is rounded with a sleep,"* Bloom tells us: "Prospero's great declaration confirms the audience's sense that this is a magus without transcendental beliefs, whether Christian or Hermetic-Neo-Platonic."[30]

Here wisdom reaches toward magic by treating reality as a dreamlike field of endeavor in which the mage can seek and find the points of contact to induce turning point dramas.

The binding force in the aesthetic continuum is time itself, time expressed in the periodicities and opportunities rising and falling in nature, in language rhythms, in the breath of the body, in heartbeat, and nerve traffic. "Prospero's awareness of the drama of time, his ability to seize the instant," Orgel argues, "in large measure constitutes the source of his power."[31] He never loses track of time.

This immersion in time rests on the steadiness and power of his personal presence, on his rhythmical entwinement in the moments that spiral out from his sensory awareness, intention and imagination. As Bloom asserts, "Prospero's mastery depends on a strictly trained consciousness, which must be unrelenting."[32]

Prospero's tuned consciousness apparently can perceive the flow of time directly. He streams along in time independently of the outer material changes we now take as markers of its passage.

*Now does my project gather to a head:*
*My charms crack not; my spirits obey; and time*
*Goes upright with his carriage. How's the day?*
(V.i.1)

## 335

The world conjured by Prospero has mass and matter, but it is vibratory matter, shimmering, made of finer stuff, an abode for sprites and spirits too, something like the Buddhist universe that flips in and out of existence millions of times a second, a luminous matter of the photonic sort we posit in quantum electrodynamics. It is time-centered, floating on temporality. That which passes is time. Thus, Prospero tells Miranda and Ferdinand

*Our revels now are ended. These our actors,*
*As I foretold you, were all spirits, and*
*Are melted into air, into thin air:*
*And like the baseless fabric of this vision,*
*the cloud capped towers, the gorgeous palaces,*
*The solemn temples, the great globe itself,*
*Yea, all which it inherit, shall dissolve,*
*And, like this insubstantial pageant faded,*
*Leave not a rack behind.*
(IV.i. 148)

He does not say the world is unreal, but that the "baseless fabric" *is* the space-time continuum. The fine matter/energy of reality, which flows as a process, is mutable; it is a *perishing* that, with timing, presence and insight, you can change in low-mediated ways. The spirit helpers Prospero directs are real too, real in the same way the clouds are real, real as cloud castles, real as the photonic cosmos. Ariel travels on the mind/body interface where words themselves manifest as deeds, and then perish in the instant. Like Shakespeare's language, Ariel operates as an engine of change. When Prospero releases him/her/it, Ariel vanishes into larger nature.

The language of magic tries to seize the potentialities in the moment impeccably well, without waste or excess. The spells move in the undifferentiated aesthetic continuum. However, they have no magic to them. The spiritual-mechanical advantage comes from the wonderful expression of perfectly timed approach-separation and withdrawal-return patterns. Every spell makes a change in love or wisdom that favors their convergence. To the outsider they look like Jungian synchronicities.

## 336
### *Magic and Freedom*

In *Irrational Man*, William Barrett wrote, "the figure of the magician is as it were, the primitive image of human freedom." He explained that "to free oneself, to break the chains of a situation, whether inner or outer, that imprisons one is to experience something like the magical power that commands things to do its bidding. The figure of the magician is as it were, the primitive image of human freedom."[33]

But how primitive is it? Jacob Needleman, in *Money and the Meaning of Life*, connects sorcery with the philosophic traditions concerning the "Way in Life."

"Throughout history," Needleman writes, "the idea of the way in life has been spoken of as the 'path of the warrior' or as the 'teaching for kings.' Both the warrior and the king represented, in literal fact and symbolically, the individual engaged in all the forces of life, as opposed to the priestly class or the ascetic removed or protected from many of the influences that permeate the greater world. Often, this idea of the way in life was transmitted as the 'way of the magician,' that is, in the language of sorcery. Again, it is a matter of the individual who confronts and masters all the forces, high and low, that constitute reality..."[34]

The language of sorcery and the *Way in Life* both seek empowerment by low-mediated means and subtle timing. But the borderlands where imagination and intention fuse to empower deeds are hard to describe. The magical aura, while real enough in the experience, later seems hardly communicable to others. The numinous moments pass by too quickly for us to put them into words. Their tracery fades. We doubt ourselves. Maybe we were dreaming, maybe not.

If there's a better language for describing the rare and recondite, we haven't found it yet. What about the promises of religion? How much like magic to turn the wafer and wine into the body and blood of Christ! C. S. Lewis describes religious miracles as "revelations of that total harmony of all that exists. Nothing arbitrary, nothing simply 'stuck on' and left unreconciled with the texture of total reality can be admitted. They will not be like unmetrical lumps of prose breaking the unity of the poem; they will be like that crowning metrical audacity which, though it may be paralleled nowhere else in the poem, yet, coming just where it does, and effecting just what it effects, is (to those who understand) the supreme of the unity of the poem's conception."[35]

Between miracles, magic, and the strictures of science, where does the truth lie?

## 337

Arthur C. Clarke argued that the accomplishments of any sufficiently advanced science were indistinguishable from magic as the primitive mind conceived it.

Clarke was certainly not promoting magic. He was undertaking a phenomenological analysis of how we register certain experiences. By his reckoning, the glamour that marks the boundaries between magic and ordinary experience keeps changing. When science advances, it pushes back the frontiers of magic. However, following an historical catastrophe in which scientific knowledge was lost, particularly if parts of the technology remained behind for the "new primitives" to marvel over, magical understandings would advance.

We must not take a one-sided view. As science advances, for instance, skills once considered ordinary get to seem magical. They languish on the fringes of possibility. Few can now track game from faint signs, or know the medicinal uses of thousands of native plants or predict the weather by the fur of animals or hear earthquakes coming, or commune with the spirits of the ancestors by detecting the tracery of their presence in places they frequented or objects they used. Who can enter animal consciousness, whisper to horses, and shift the energy fields in the body sufficiently to induce mass hallucinations, conduct shamanic cures, detect the meridian lines in Chinese medicine or follow ley lines on the earth? A great deal of knowledge about nature, including verifiable techniques for inducing change, has probably been lost in the advance of science. And should the sciences wane in a new dark age, they won't necessarily come back.

Therefore, if "magic" is a moving locus of the inexplicable, if that is all it is, we can expect there always to be magic because there will always be fuzziness on the borderlands where we lose old things and discover new ones. The magical/technological frontiers will keep shifting as long as there is something new to learn and something old to forget, and we have no idea where the waves of change will take us.

### 338

Some thinkers maintain that belief in God or the Sacred is essential to our happiness, that our nagging meaninglessness is loneliness for God, that we have pushed God out of the modern world by the power of the outer over the inner, and this has led to the eclipse of all genuine religious experience communally experienced. Robert Coles argues this way in *The Secular Mind*.

He tries to cross the phenomenological/ontological frontier by taking a hard look his own inwardness, an effort he considers "the last gasp of the sacred."

"One prays at the very least on behalf of one's own kind, though unsure, in a secular sense, to whom or what such prayer is directed, other than, needless to say, one's own secular mind, ever needy of an otherness' to address..."

C. S. Lewis took a much stronger position. He argued that we could not do without God because God exists. The incarnation was real. Jesus came down to earth in a virgin birth. He lived, died, and rose from the tomb.

Arnold Toynbee, a practicing Christian of the same generation, believed that Western civilization could only solve the world problems through a renewed faith in the Christian God. The Enlightenment failed precisely because it lost this faith. Christian ethics alone won't do. We need to pray.

"And, inasmuch as it cannot be supposed that God's nature is less constant than Man's, we may and must pray that a reprieve that God has granted to our society once will not be refused if we ask for it again in a humble spirit and with a contrite heart."[36]

## 339

My take is that secular humanism has not failed. It's just that the two hundred plus years since the Enlightenment have been too brief a time for us to have developed a firm sense of our own moral convictions. We still do not conceive of ethical choice as central to our experience of life. We haven't found our own bright inwardness. We do not know where to aim the gratitude that naturally wells up in us on many happy occasions. It seems to want to soar up to some being outside ourselves. But look closely at your own experience and you will see that the sudden surge of gratitude is thankfulness *sui generis*, in and of the moment entirely. The assignment of the gratitude to a "more", a "beyond", is a gratitude that seeks the Helping Hand behind the helping hand. It is often a trick of the mind that makes us forget the courage in action we needed to be fully human.

Why does gratitude always need a recipient? Isn't the sudden surging feeling in itself thankfulness for the wholeness and rightness of the present moment? Perhaps we are inclined to thank God because our long infant dependency predisposes us to attach our gratitude to parents. Our infant awe goes to our parents, so we give our adult appreciation to God and from that build a language that refers gratitude to a higher power.

But bad things happen when we give our gratitude away: we lose hold of our peak experiences; the zest goes out of the passing moment. When we rein in the playfulness of life as a grand improvisation, we turn off the switch that alerts us to the perception of eternity *as now*, and forget that it certainly cannot exclude now.

As with "gratitude," so with "faith" "salvation," "aspiration" and "atonement." They have religious connotations. To resume the journey to freedom today requires us to struggle against the shackles of language itself. The language of ethical inwardness, in all its nuances and associations, still belongs to religion by default. We have not yet developed a secular language for ethical choice. The simplest hopes and fears have theological resonances: forgiveness, compassion, justice, forbearance, patience, redemption and sacrifice all trail back to God. We can barely receive our deepest inward *apercus* without entangling them in theocratic or magical language. Which means we cannot conceive of our choices clearly – or our responsibility for them.

**Flammarion Woodcut**

Future psychologists will understand that we experience the phenomenologically bracketed "magical", "mystical" and "providential" in the mundane world whenever the enhanced power of intention, ignited in turning points, thrusts us into the present, awakens us to beauty, and elicits from us surges of energy, alertness and enthusiasm.

What we feel most immediately then is a *frisson*, a thrill that reflects the sudden coming into place of a new set of passing rules and model change. The experience of the passing rules forming on the mind/body edge links our exteroceptive with our interoceptive interests. It hints at possibilities for the reunion of self and world, just as the presence of the magus in the experimental situation seemed to stamp the field of endeavor with a more complete pattern of human nature – but this time for real.

Kepler, on discovering the laws of planetary motion, wrote of his moment: "that for which I have devoted the best part of my life to astronomical contemplation, for which I joined Tycho Brahe... at last I have brought it to light, and recognized its truth beyond all my hopes... So now since eighteen months ago the dawn, three months ago the proper light of day, and indeed a very few days ago the pure Sun itself of the most marvelous contemplation has shone forth... nothing holds me."[37] Michael Polanyi paraphrases this as "Having made a discovery, I shall never see the world again as before. My eyes have become different; I have made myself into a person seeing and thinking differently."

What we needed when we woke up from our long entrancement to religion was a slow and measured inward journey to the guiding forces of ethical discovery, not a simple or dreamy passage contrary to common sense, but a fierce one, clearly and strongly debated. What we got instead was an outward shove into exploration, enterprise, industry and expansion. We never had the time to make sustained contact with our own creative inwardness. Confronted with the amazing vigor of the technical mind in its alliance with aggressive ambition in the Industrial Revolution, we defaulted on our wholeheartedness. As the naïve promises failed and the emptiness overwhelmed us, we retreated to existential grit, to spiritual stoicism, to solipsism, to magical thinking or to faith in divine providence.

## 340

With secular minds, we will treat life less as tragedy than comedy, a deep comedy, the human comedy. People will deal with the flaws in human nature with more humor and less revulsion. Devoted Sancho Panzas will protect their beloved Don Quixotes.

We will accept our turning points as comic resolutions. For humor manifests in all resolutions. We feel it bubble up in the experience of relief itself. To be occupied totally by a laugh in a crucial moment, liberates the energy for creative achievements.

Humor puts us in the present. A good laugh clears the mind. It makes the foibles of our struggles for authenticity tolerable. Kierkegaard, in his early life, before he was distracted by religion, wrote that in the hardest times mastered irony has redeeming power. It "limits, renders finite, defines, and thereby yields truth, actuality, and content... He who does not understand irony... lacks *eo ipso* what might be called the absolute beginning of the personal life."[58]

With ironic detachment, we can laugh at the power of our emotions to tie us in knots. The laugh, while it lasts, or when we recall it later, teaches us not to grieve over our deficiencies and inadequacies. It sees the grief as another trick of the mind to keep us out of action. With irony, we cut ourselves down to size. Kierkegaard concluded many years later

"In order not to be distracted by the finite, by all the relativities of the world, the ethicist places the comical between himself and the world...The ethicist is... ironical enough to perceive that what interests him absolutely does not interest the others absolutely; this discrepancy he apprehends, and sets the comical between himself and them, in order to be able to hold fast to the ethical in himself with still greater inwardness."[39]

## 341

Happier, more self-deprecating, in closer touch with the engines of history, and with a better understanding of our own temperaments and motivations, we may yet develop new kinds of "spirituality". We'll withdraw the projection of gratitude from an unfathomable Beyond. We'll start to grow up. We'll understand why the Buddha launched a spiritual movement that treated God as a mental projection and taught his students that deeds undertaken for a mental projection were actually deeds done for oneself. I am not espousing an atheistic or agnostic position here. God keeps his own mysteries. If the Creator wants to speak to me, I'm ready to listen, even if it kills me. But doesn't there come a time when the baby must learn to walk on his or her own? Mightn't even a good God, with a providential interest in humanity, withdraw His hands from supporting us so that we can know what it is like to stand or fall on our own? With our sense of meaningfulness in our own hands we would have to use our evolved equipment to lead us not back to God's cradle but ahead to life in community. For loving community is our soundest spiritual home. Love and wisdom are designed by nature to hold us together in it. In the arms of community genuine happiness, meaning, creativity, moral choice and full belonging can thrive. As Einstein wrote,

"In their struggles for the ethical good, teachers of religion must have the stature to give up the doctrine of a personal God, that is, give up that source of fear and hope which in the past placed such vast power in the hands of priests. In their labors they will have to avail themselves of those forces which are capable of cultivating the Good, the True and the Beautiful in humanity itself."

(Albert Einstein Science and Religion. 1941.)

Ursula Le Guin has her main character wrestling with this problem of terminology in *The Telling.* Travelling as an observer among the suppressed indigenous people of the planet Aka, she notes:

"There are no Akan words for God. Gods, the divine," she told her noter…"On Aka, *god* is a word without referent. No capital letters. No creator, only creation. No eternal father to ward and punish, justify injustice, ordain cruelty, offer salvation. Eternity not an endpoint but a continuity… No afterlife, no rebirth, no immortal disembodied or reincarnated soul. No heavens, no hells. The Akan system is a spiritual discipline with spiritual goals, but they're exactly the same goals it seeks for bodily and ethical well-being. Right action is its own end. Dharma without karma.

"She had long debates with her noter about whether any word in Dozvan or in the older and partly non-Dozvan vocabulary used by 'educated people' could be said to mean sacred or holy. There were words she translated as power, mystery, not-controlled-by-people, part-of-harmony. These terms were never reserved for a certain place or type of action."[40]

As with the Akans, our spirituality will be secular and non-theistic someday. We will build it on our direct experience of the transience of the self, an illusion-free spirituality. We will get to it not by recovering truths from ancient times but in a series of scientific discoveries. Our demeanor will not have the 'late imperial' tone of Roman Stoicism. It will not be skeptical, cynical, ascetic or epicurean.

We will have epiphanies and ecstasies and we will trust their wholesomeness. Our spiritual system will affirm the power of turning points, personally and historically. We will accept their cosmic role in restoring wholeness. We will take on the Cabalistic principle that the turning from below evokes a turning from above. A turning from *here, now,* a turning that turns the whole from within.

### 342

With secular spirituality stabilized by a firmer hold on inwardness, the ethical mind will flourish. We will live more modern and venturesome lives than we do now. We will experience the convergence of love and wisdom.

In practice, these changes will expand the 'aggredi' of good aggression and prepare us to enter new fields of action safely.

Should this happen, we will have changed the human nature in ourselves *by ourselves*, changed it as we did before in the Agricultural and Industrial Revolutions. And new deep currents will stream through us into the world. We will be a little older, just out of childhood, adolescents perhaps, but no longer infantile. (*Childhood's End* has been a recurrent theme in science fiction.) Our 6,000 years of recorded history will turn out to be an early stage in human history. We will have taken ourselves to the beginning of our young maturity.

We will still be wild, untamed, restive and bold, barely out of our teenage years, but our creative juices will be flowing. With nationhood and religiosity behind us, we will build a new planetary culture with vigor and hope, precisely the spirit befitting a young world civilization attaining its first maturity.

Then the global civilization, conscious of the approach/-separation, withdrawal/return and dispersal/aggregation rhythms that move us through the natural world, will reach the threshold of its next stage. We will leap off the planet. The age of space colonization will begin wholeheartedly. We will find our way to the moons and planets of the solar system, and then beyond, in the great pulse of dispersal, roughly analogous to the Renaissance mariners' discovery of the New World that led to the modern world. The great dynamic of human civilization will continue to unfold: migration followed by settling, followed by migration, followed by settling. As Freeman Dyson put it,

"The destiny which I am preaching is not the expansion of a single nation or of a single species, but the spreading out of life in all its multifarious forms from its confinement on the surface of our small planet to the freedom of a boundless universe. This unimaginably great and diverse universe, in which we occupy one fragile bubble of air, is not destined to remain forever silent. It will one day be buzzing with the murmur of innumerable bees, rustling with the flurry of feathered wings, throbbing with the patter of little human feet."[41]

On this pathway, our main achievements will not be in bio-technology, nanotechnology, robotics, artificial intelligence or any of the other amazing technologies. Those advancements may come, they almost certainly will, but we will take them in stride without celebrating them as the apex of human accomplishment. They will give us new products, treatments, processes, vehicles that take us where we want to go; they will be means to an end and not ends in themselves. The important thing will be where we want to go and why. The power sources for it may be solar arrays or nuclear fusion plants. But the call to adventure will be the real energy and the choices we make with it will come from the creative power of the ethical mind expressing its good judgment in community life - which means that our real breakthroughs will rest on new family, marriage and community relations. And the outflow of our new emergent skills will be nurtured in those venues.

In those days, the wider expressions of human nature will come into the world with less dependency on mechanical devices or techniques, because love and wisdom will be their sources and we will gather them as essentially moral energies in the time sensitive freedom of our turning points. We will feel more comfortable living a full community life. We will trust its benefits and let ourselves enjoy real safety.

On this cultural avenue, we will live compliant to the times. With fuller consciousness, we will learn to recognize and connect with our real opportunities and to seize them when they come. And we will find these close to hand, because doing what is close to hand is our perpetual starting place and there is always something to do close to hand.

Our transformational changes inevitably enter the world where our feet touch the ground, where we wake from sleep, where we first make eye contact in the circles of affiliation around us. However, not everything close to hand ought to be undertaken because not everything is relevant or ready.

We will develop skills for knowing the right moments and the right combinations. We will know both restraint and license. We will become proficient users of low-mediated intentions. The baby will learn to walk.

**343**

A Native American saying sums it all up. "The white man builds a big fire and sits far away, the Indian builds a small fire and sits close up."

The thinking behind this suggests that there is a path on which we can use self-regulation strategies (some based on internal mind/body techniques) to readjust our relationship to technology. We don't always need a big fire. We can stay warm by sitting close up to a small one or by self-regulating our thermal physiology. To an extent based on our mind/body skills, we have the means to acquire what external technology supplies on our own, with a much lower expenditure of energy and materials. Thoreau used a similar formulation concerning warmth in *Walden*.

The point I am making here is that action-oriented low-mediated intentions we discussed earlier follow the path of the small fire. And on certain fannings of possibility, I foresee this route of development bringing with it not only new human capabilities but *new sorts of techn*ology that will reduce energy use, lead us to discover new renewable resources, including some internal to ourselves, and give us a deeper reverence for and closer contact with nature. Under certain fannings of possibility I can foresee us changing the thrust of technology altogether. On these radials of destiny, we would continue to invent more daringly than ever, but our devices, culled from deeper human-heartedness, would serve new purposes. They would become instruments that teach us how to free ourselves from them, akin to biofeedback monitoring devices. In medical technology, for instance, we would make heart pacemakers that eliminate the need for pacing by retraining the heart, electronic implants that grow new tissue and replace themselves, etc.

We'd develop a wider learning model. Our focus would shift from therapy to education. Education in inwardness would thrive, deepening our connection with subtle energy sources inside us and with the rhythms in nature. Moreover, these would open new theoretical approaches to inducing change through time by efficient means.

In certain scenarios I can even envision us developing new sensory, cognitive and imaginative faculties: electromagnetic senses beyond the visual spectrum (already operating in the electrochemical gradients on cell membranes), radio frequency senses, magnetic senses (that some animals use for location), senses for fabulous agility, dexterity, speed, stamina and focus, combined senses, synesthesias suited for perceiving gestalts.

Our better understanding of timing will give us a modest edge in predicting likelihoods in personal and historical venues. And with that edge, we will develop better judgment and more peaceful approaches to life than we have now. And this will not only make us more effective citizens, it will gentle us down, because it will eliminate action in excess of what we need to accomplish a goal. And these advancements would make us more punctual and calmer too. We'd be less arrogant and selfish because we'd recognize that much of what we've seen as strength of character is really an illusion. We'd be happier and more willing to work together too. Seeing through the veils of illusion with comic immediacy, we might even discover new ways to empower low-mediated intention. We might learn how to bring ourselves into lucid dream states reliably and with full consciousness, to improve insight and foresight through conscious recall, and to facilitate healing.

**344**

The four ethical discoveries themselves travel the path of the small fire. By managing to ease the aggressive overload on the stress response they stimulate new self-regulation capabilities. With better voluntary control, we may be able to moderate the by-play between stress and aggression and aid the re-fusion of aggression with love and wisdom, restoring its old conservative function. In times to come these capabilities could work as catalytic agents to spark transformations that bring the ethical and technical mind-sets closer together.

And in small venues, when people are functioning well, their wisdom journeys, tempered by realigned aggression, will produce innovations that favor freedom and justice.

At significant moments, love and wisdom, with modest pressure, will be able to soften and shift dominance and territorial relationships. In the aftermath, entrepreneurial energy will not be so intractably associated with financial profit but will rise to broader adventures, including service and amity and the improvement of life in mixed enterprises, some set up as non-profits, others employee owned, others established as the foundation arms of profit-making corporations.

As the aggressive components of human nature fall increasingly under the influence of love and wisdom (using the dynamic definitions we have explained here,) our technical and ethical sensibilities will converge too, merge at times, and then pull apart to restore their primordial reciprocal relationship.

As the new elements in human nature surface from the treasury of the neutral traits, we will reframe many of our old duties and privileges. We will have to reset the stages of life to conform with lengthened life expectancy in the prosperous world, and then worldwide, perhaps extending the period of youth into the fourth decade and middle age to the ninth - after that sagacity.

### 345

Many science fiction novels show how these things might happen. They depict future human societies whose strengths are not material. Some portray successful human societies enjoying every resource of the arts and sciences, but living in decentralized small communities in balance with nature, many with fully developed non-theistic spiritual practices. The science in these utopian novels is of mind. The plot lines trace the advancement of cognitive abilities. They explore new powers of communication, of joined consciousnesses, of shared dreams, of telepathy and magical repatternings of nature. High technology, space flight, virtual reality, robots and extended life spans play parts in these utopian science fiction novels, but the real excitement in the stories comes from the expansion of human capabilities and the deepening of social ties that go with it.

Ursula LeGuin in her *Earthsea* novels, *The Dispossessed* and *The Telling,* deals with these possibilities. Dorothy Bryant covers similar ground in *The Kin of Atta are Waiting for You*, as does Aldous Huxley in *Island,* Norman Spinrad in *Songs from the Stars*, Isaac Asimov in the *Foundation* series (for example, the Second Foundation's mastery of inner personal change,) Arthur C. Clarke in *The City and the Stars* and many others. Modern science fiction explores utopian possibilities in ways unequaled since the Renaissance.

I particularly liked *The City and the Stars* when I read it long ago. Alvin, its young protagonist, finds his way out of a labyrinthine self-sustaining, million-year-old world city equipped with every imaginable means to support and entertain its vast population, (for the most part happily, peacefully and with respect for personal freedom.) Once outside, Alvin discovers a low-tech, decentralized telepathic civilization living in close touch with nature. He makes friends with the youth there. Together the young people uncover the lost history of their common origins. The mind-centered and technology-centered civilizations join their destiny lines, and that begins a process of reconciliation between the ethical and technical minds.

## 346
### *Convergence on the Path of the Small Fire*

In our present mindset, we deplete the energy for convergent turnings by surrendering to the conflicts between the technical and ethical minds as if they were intractable (see # **184**). We are stuck here for three reasons. Our thinking is too linear. We refuse reversal. We cannot move smoothly between inwardness and outwardness. Consequently, we cannot align the turning points in love and wisdom. We pull them apart before they can energize each other. Our loves are blind, our wisdoms cold. This makes us opaque to ourselves and more dangerous to each other.

Continuing failures of convergence degrade the quality of leadership. The failures spread widely through the population, blighting the growing tip of cultural evolution. When love is absent from wisdom and wisdom from love, parents fail their children and children dishonor their parents.

But human nature changes and so do its competencies and interests (see # **173**.) A shift toward low mediated intentions would stimulate inwardness and support the ethical mind. With better balance between inwardness and outwardness, we will make saner decisions. We will handle perturbation more robustly. We will hold more firmly to the resonances between personal and environmental rhythms. We will return to homeostasis sooner. In this ambiance, love and wisdom will choose (and be able) to reach for each other in crucial, conscious turning points.

\* \* \*

### *End Notes*

[1] Bill McKibbin *Enough.* Times Books. 2003. p.35

[2] Ibid p. 143

[3] Ibid p.198

[4] Freeman Dyson. Infinite in All Directions. p. 11

[5] Quoted in "The Dawn of McScience." Richard Horton. *NY Rev of Books.*

3/11/0

[6] Richard Feynman. *The Meaning of it All.* Perseus books. 1998. p. 106

[7] Ibid p.300

[8] ibid p.149

[9] *I vowed that I would dedicate my powers*
   *To thee and thine – have I not kept the vow?*
   *With beating heart and streaming eyes, even now*
   *I call the phantoms of a thousand hours*
   *Each from his voiceless grave: they have in visioned bowers*
   *Of studious zeal or love's delight*
   *Outwatched with me the envious night –*
   *They know that never joy illumined my brow*
   *Unlinked with hope that thou wouldst free*
   *This world from its dark slavery,*
   *That thou – O awful Loveliness,*
   Would give whatever these words cannot express.[9]

[10] (ibid p. 153). ibid p.142

[11] A. C. Ewing. *Ethics.* The Free Press. 1953. p.9

Bertrand Russell, at around the same time, wrote: "I do not think there is, strictly speaking, such a thing as ethical knowledge... Certain ends are *desired*, and right conduct is what conduces to them." Bertrand Russell. Principles of Social Reconstruction. p. 37

To this Father Copleston,, added sardonically: "'would that everyone had an aversion from cruelty' is no more describable as true or false than 'would that everyone appreciated good claret.' Hence there can be no question of proving that the judgment 'cruelty is bad' is true or false." Copleston. *A History of Philosophy.* Vol 8. Doubleday Image. 1966. p.237

[12] ibid p. 169

[13] Michael Polyani. Personal Knowledge. U. Chicago Press, 1962 p. 27

[14] All quoted in Erich Heller, *The Disenchanted Mind*, Farrar Strauss, 1957.

[15] Ray Monk. Ludwig Wittgenstein. The Duty of Genius. Penguin 1991. p.303

[16] Sir Francis Bacon, Advancement of learning and New Atlantis. p.297

[17] Loren Eisley, *The Man Who Saw Through Time*. Charles Scribner &sons. p.39

[18] ibid. p.99

[19] Northrop. *The Meeting of East and West*. Macmillan. 1950. p.442

[20] (ibid p.396) Further, "the aesthetic continuum within the essential nature of all things is, to use the language of Shakespeare, 'Such stuff as dreams are made on.' But there are other differentiations in this ineffable, all-embracing aesthetic component than the introspected images which constitute dreams. There are also the immediately inspected images which constitute the colors, fragrances, and flavors of the sky, the earth, the flowers, the sea, and other natural objects. The aesthetic component is therefore also the stuff that these are made of." (ibid p.462)

[21] The East, by contrast, takes the aesthetic appearances as real. Their explanations are "built out of concepts by intuition, whereas Western doctrine has tended to be constructed out of concepts by postulation. 'Blue' in the sense of the sensed color is an example of concept by intuition. 'Blue' in the sense of the number of the wave-length... is... a concept by postulation... In the Cartesian and Lockean foundations of traditional modern Western culture the material substances and the mental substances are theoretically inferred rather than immediately apprehended factors." (ibid p.448-449)

Accept blue to be blue, at least start with that, allow the aesthetic continuum to be real in and of itself and "the last ground is removed for regarding the aesthetic component as a mere appearance of the theoretic factor, or as the principle of evil. Both components are equally real and primary, and hence good, the one being the complement of the other." (ibid p. 450)

[22] (Giorgio de Santillana, ed. *The Age of Adventure*. Mentor. 1956. p. 14-15)

[23] Frances A. Yates. *Giordano Bruno and the Hermetic Tradition*. 1964. Vintage Books. P.77) (ibid. p.192)

[24] William McNeill. *The Rise of the West*. p.555

[25] ibid p. 69

[26] ibid p.71

[27] Dalai Lama. *The Universe in a Single Atom*. Morgan Road Books. 2005. P. 156, 160.

[28] Oxford *Shakespeare*. 1987 *The Tempest*. p.20

[29] Harold Bloom. *Shakespeare – The Invention of the Human.* Penguin. 1998. p.670

[30] ibid p. 681

[31] ibid p.50

[32] ibid P.680

[33] William Barrett. *Irrational Man* Anchor Books, 1958, p.129

[34] Jacob Needleman, Money and the Meaning of Life, p.82

[35] C. S. Lewis. Miracles. 1947. Collins Fount Paperbacks. P. 65

[36] Arnold Toynbee. *A Study of History,* two volume Abridgement. Oxford, 1946. Dell Edition Vol I p.628

[37] Kepler quoted in Polyani. *Personal Knowledge.* U. Chicago Press. 1958 p. 143

[38] Kierkegaard. *The Concept of Irony.* 1841. Indiana U press. 1965. #339. In ironic detachment we suspend belief in the power of emotions to validate us in our hang-ups by laughing at them. This comical aspect of irony teaches us not to grieve over our deficiencies and inadequacies as another trick of the mind to keep us out of action.

[96] *Concluding Unscientific Postscript.* p.450-1

[40] Ursula Le Guin. *The Telling.* Ace Books. 2001. p.95-6.

[41] Freeman Dyson. *Infinite in all Directions.* Harper and Row, 1985. p. 134

# 19. ACCEPTANCE

## 347

Love and wisdom lack constancy. They keep changing on us. We live with the uncertainty that at the highest, furthest, deepest, or strongest moment we will find ourselves shifting down to a lower or lesser state. The physiology of desire and satiation work that way. The time of high achievement is the instant when diminishment and loss begin. The moment we fully realize the loss, the gain begins. Every meeting is a parting, every learning a forgetting. We endure calamitous disruptions and suffer unhappy endings. We can die too soon or too late or stupidly or in terror. Nothing lasts. Everything ends. As the Buddha said, "Whatever is of the nature of arising, all that is of the nature of cessation."

## 348

Our contradictions make us generous and stingy, isolated and social, credulous and skeptical, self-seeking and altruistic, giving and taking, ambitious and apathetic. Fear and hope endlessly wrestle inside us. We function that way because,

- o The suppressive neural overlays in the triune brain pull us in different directions
- o The conflicts between our modular personalities tie us to separate streams of memory and meaning
- o The rapid fusing and decoupling of aggression with love undermines our trust in ourselves and each other
- o Our sense of uniqueness keeps us in a perpetual shoving match with the commonalities of our lot.

## 349
### *Residual Suffering*

Many wisdom traditions trace the sources of suffering in our nature to change. Change brings impermanence, fragmentation. We get no rest. We suffer when we don't rise to the needs of the occasion. We suffer when we try to cling to what passes. We endure halts, blocks and sudden alterations.

The Buddha called this *Dukkha.*

"In the happiest moments there is Dukkha, because they end. In the brightest insights is Dukkha because we want to believe in them. The Noble Truth of Dukkha is this: Birth is suffering; aging is suffering; sickness is suffering; death is suffering; sorrow and lamentation, pain, grief and despair are suffering; association with the unpleasant is suffering; dissociation from the pleasant is suffering; not to get what one wants is suffering – in brief, the five aggregates of attachment are suffering."[1]

The cravings and aversions we experience along the legs of love and wisdom fit the Dukkha pattern. There's nothing to cling to and no stable self to do the clinging. Nobody lurks behind the agglomeration of sensations, thoughts, perceptions and feelings even in our most memorable experiences. Nobody directs the passing show.

The Buddha proposed a radical solution to suffering: Let go, stop clinging. He taught a method leading to the cessation of attachment. "These two entanglements – belief in an ego-personality and the conception of personal attainment – must be utterly destroyed and never again permitted to rise to define the true Essential Mind."[2]

The renunciation of attachment, however, does not bring with it a refusal to act. On the contrary, it opens the Buddhist to action in the untroubled, almost carefree way that Gandhi described as "renounce and enjoy."

He wrote of Hinduism, which shares the Buddhist analysis of the ego, that "He who gives up action falls. He who gives up only the reward rises. But renunciation of fruit in no way means indifference to the result. In regard to every action one must know. He, who, being thus equipped, is without desire for the result, and yet is wholly engrossed in the due fulfillment of the task before him, is said to have renounced the fruit of his action."[3]

Gandhi's twentieth-century version of spiritual immersion makes one big mistake. Neither Buddhists nor Hindus consider a single lifetime sufficient to accomplish it. You cannot let go of attachment so readily. Past actions deeply etch the "stains" of craving, aversion and delusion in our being.

Religious minded people solve this problem by maintaining that it is the illusion of a constant self that makes us mistakenly believe that our existence ends at death. It doesn't. The self that is continuously being annihilated and recreated, many times a second in Buddhist psychology, keeps right on pulsing. Death counts as just another interval in the pulse/interval rhythm; at most, there is a brief hiatus. "When the Aggregates arise, decay and die, O Bhikkhu, every moment you are born, decay and die."[4]

The Buddhist treats death as another rearrangement of energies, a throwing off certain components of the aggregates of the self in order to take on others. We slough our skins. It is a stage in the karmic drama that continues over many lifetimes. We build up karma, and then work it off. Even the great meditation masters had to endure numerous incarnations to achieve release. The Buddha himself had to be reincarnated 500 times before he attained perfect enlightenment.

### 350

So much depends on rebirth. Without reincarnation, the karmic process cannot accomplish release. Nobody wants to believe his or her suffering is meaningless. It seems to me that the Eastern world needs the karmic system, and the rebirths that drive it, more for emotional than metaphysical reasons. They need it to validate personal suffering, to console themselves that our misery is not a superfluous spawn of nature but part of a redemptive design that will lead to its eventual cessation.

I don't buy that. The evidence overwhelmingly suggests that we live only once, in this present life, the one that is fleeting by. Nothing comes after.

How Buddhists, who expend such great efforts to overcome false hopes, can allow themselves to cling to the wildest hope of all, that they have many lifetimes to reach their goals, I cannot understand. Or rather, I see it coming from the same fallible human-heartedness that makes us all want to believe there is more to life than this.

## 351
### *King Solomon*

King Solomon rejected belief in an afterlife.

"For that which befalleth the sons of men befalleth beasts; even one thing befalleth them: as the one dieth, so dieth the other; yea, they have all one breath; so that a man hath no preeminence above a beast: for all is vanity. All go unto one place; all are of the dust, and all turn to dust again."
(Ecclesiastes 3:19-20)

He recognized no karmic system. There was nothing you could do to achieve permanent release from suffering. Death is the release. Life is full of difficulties.

Solomon's understanding of suffering, like the Buddha's, came from a long meditation on the human condition. However, Solomon conducted his whole search from within the world of attachments. He didn't meditate in the forest as the Buddha did. He was a lover, a father, a merchant, a rich man, a builder, a ruler, a judge, a court poet and cultural leader. He never relinquished his roles.

Though he sought wisdom and prayed to God for it – and God answered his prayers; so the Bible says – he never found a path to the cessation of vanity and vexation. He insisted you deceive yourself if you think you have found one. Moreover, you end up living a smaller, more blinkered, less worthy existence by nurturing false hopes. You drain your vitality in masturbatory self-satisfaction or self-disgust.

In the contorted world of broken vessels and scattered sparks, nobody attains full self-realization or true wisdom. Nobody transcends the ordinary strains of life.

"Then I beheld all the work of God, that a man cannot find out the work that is done under the sun: because though a man labour to seek it out, yet he shall not find it; yea further: though a wise man think to know it, yet shall he not be able to find it."
(*Ecclesiastes* 8:17)

## 352

As Solomon never found a solution to vanity and vexation in God's providence, Kafka's hero, K., never finds a way to reach the castle. Neither do we. The human body/mind is too weak and vagrant to get that far. We're too flawed to grasp hold of reality for long. Our minds work on dualisms. We're too subject to internally generated pleasures, pains and anxiety states to become whole or fully conscious or "illuminated." It's an illusion to think we can fully control the mind, that its stains can be scoured clean. The spirit is unsteady *sui generis*; we cannot train it to steadiness.

For brief joyful periods, the Dukkha may lift. When it does you sing and dance and compose a new song to God.

Then suffering returns.

Solomon taught that knowing our limits was better than not knowing them, though wisdom confers few benefits beyond the solace of knowing itself. Maybe knowing gives you a slight edge. It helps you cope. It brightens the playing field, but it does not help you win. Spiritual quests of the sort that Hindus and Buddhists pursued do not release you from the vanity of life in the one lifetime you have to do it. A Saddhu just transfers his vanity to his loincloth.

## 353

Solomon encouraged us to accept our fallibility and live fully in the face of incompletion:

"Let thy garments be always white; and let thy head lack no ointment. Live joyfully with the wife whom thou lovest all the days of the life of thy vanity, which he has given thee under the sun, all the days of thy vanity: for that is thy portion, and in thy labor which thou takest under the sun. Whatsoever thy hand find to do, do it with thy might; for there is no work, nor device, nor knowledge, nor wisdom, in the grave, whither thou goest."
(Ecclesiastes 9: 7-10)

Full engagement, moderated expectations. Courage. Creative enthusiasm. I call this living on the once-only plan. Do it with thy might. Indeed!

## 354
### *The Once-only Plan*

On the once-only plan, you accept that there is no radical solution to suffering. Dukkha can never be eliminated, excised, or extirpated. Basing a life on the hope that you can get a second chance in a subsequent incarnation diminishes the soul. We do better coping with the Dukkha inextricably woven into the *oneness* of life.

On the once-only plan, you accept that the self is a natural illusion supported by your physiology in the same way that a stage magician supports his illusions with apparatus. Like the audience at a magic show, we may yearn to be taken in, but we don't have to be duped. For the sake of our freedom, it is better not to be fooled. By learning to see through the smoke and mirrors and by resisting the diversion of attention from the sources of the illusion, and the sleight of hand, we can perceive and engage our modular parts. We can penetrate the veil of the I-sense. From behind the scenes, in the penumbra of consciousness, we can watch the brain's illusions piecing together the world out of best-guess approximations and then filling in the gaps with more guesses in an ongoing act of creative imagination that gives reality its seeming continuity.

Life is better on the once-only plan. Without rebirth, without sure knowledge of results, the Buddhist virtues of altruism, generosity, service, compassion, the prospect of work without concern for reward become really tremendous accomplishments because without the karmic working off, without a final reckoning or release, this lifetime becomes our only chance, our precious "once in a lifetime" opportunity.

In oneness, we raise the stakes for freedom because we recognize that the mounting risks reach their peaks in our turning points.

## 355

The oneness that is part of the once-only plan brings *nowness* with it. The virtue in action that brings meaning and purpose to my life only occurs in the present, and I have my present and you have yours and they are never the same. This gives our developmental processes and the signal events that mark them a paradoxical quality. They are both universal and unique. As lived, nobody has encountered love or loss before you, yet everyone has.

In practicing virtue in action, the needful thing is to have the courage to endure the chaos when the turning dramas come. That is when the opportunities to see and understand vastly expand. In those moments, we come closest to learning how life really works. In those moments, we feel the pulse of the rhythms of nature beating in us, distinguishable in their separate frequency bands. We experience the fullness as we would hear with startling clarity the play of instruments in an orchestra.

## 356

If you "live joyfully with the wife of your youth" (or husband or lover, or with whatever love and wisdom you find,) and have the luck, help and common sense to keep from getting too terribly lost, you will, by Solomon's counsel, live life fully. In doing so your "I-ness" will slip into the background. You will still have to deal with your part-personalities. Your field-selves will continue to wrangle and succeed each other. But you will play your roles with more gusto by knowing they are roles. And self-consciousness will slip into the background. You will still prize your uniqueness, but as an instrumentality for doing, not a state of being, not an image seen in a mirror, but a living presence leaping in through a window.

We are built like finely tuned violins, made for playing, not for looking at.

As we become less attached to the ego, we become more spontaneous. Moreover, this spontaneity does not lead to helplessness or passivity, but to eager activity, filled with choices, lit by heightened awareness of possibilities.

Eventually even the Watcher recedes. One finds oneself dancing on the free edge of the moment, careless of heights. One identifies less with each dance step and more with the music as a whole. One takes on the transformative moments with less personal investment *and* more fully. Easily the deeds fall clear of the doer to make their own way in the world. And the Prime Doer recedes.

# 357
## *Courage*

It takes courage to live by the once-only plan.

We usually have our courage backward. On the long reaches of love and wisdom, when it is useful to be steady, we're impatient with our shortcomings, flaunt our willingness to change and yearn for decisive action. But in the turning hour when the crisis comes and every tremor is evocative of change, we hesitate and resist. When steadiness is called for, we're ready to change but when change impends, we want to hang tough.

Three kinds of courage help us through our turning dramas. And these apply equally to our personal and historical turning dramas.

The first lets us be brave during the long reaches of the voyage. This stalwartness, this steadfastness, is courage not to shorten the leg going toward the turning point, to stick with the consequences of our actions, to endure danger and reverses, yet, like Odysseus, swept beyond the known world, to be mindful of the journey home. This first courage helps bring us to our turning points with strength and presence.

When we engage the turning process itself, we need a second kind of courage, to submit to chaos, to stay present with eyes wide open while the world falls apart. It means letting our old understandings go, including our sense of ourselves. Odysseus, his final ship gone, his raft destroyed, clinging to a spar, is washed ashore. He sleeps. Awakened by the sound of teenage girls at play, he gathers his wits, pulls a branch from a bush to hide his nakedness, then stands before the girls and politely asks for help. This courage in sheer naked awareness, bare and open in the face of total loss, has at its heart not excess but reserve.

We need a third courage too: to seize the time when conditions change and we fly out of the turning to a new radial on the next long leg. The situation here is turbulent, intense, delicately poised between possibilities. Forward movement requires the courage to affirm your intention with an implementing gesture. We cannot leave it for later; we have to do it now.

In *Shambala, The Sacred Path of the Warrior*, Chogyam Trungpa Rimpoche relates this courage to gentleness, and describes it as "energy beyond aggression." He explains that

"The fundamental aspect of bravery is being without deception... Usually if we say someone is brave, we mean that he is not afraid of any enemy or he is willing to die for a cause or he is never intimidated. The Shambala understanding of bravery is quite different. Here bravery is the courage to be – to live in the world without any deception and with tremendous kindness and caring for others... When you develop bravery, you make a connection with the elemental quality of existence."[5]

This quality, this *Jen*, human-heartedness, he applies not to life in the cloister but to living "in the world without any deception." Moreover, that includes the business world, most importantly, in its next entrepreneurial outpouring under new rules of corporate governance.

## 358

The wise people of the East generally don't emphasize the moments of choice as dramatic turning points. But there are exceptions. In the Baghavad Gita, the warrior hero Arjuna confronts just such a choice. He is out reviewing his troops on the eve of battle. His charioteer drives him down the line between the opposing armies. Instead of seeing the "us" against the "them", Arjuna sees the "us" on both sides – friends, relatives, teachers, beloved colleagues. He falls to the floor of the chariot overwhelmed with grief. He prays to Krishna for guidance. Immediately, the charioteer turns into Krishna, who addresses Arjuna as a friend. First, Krishna tells him to stand up like a man. "This despair and self-pity in a time of crisis is mean and unworthy of you, Arjuna. How can you have fallen into a state so far from the path to liberation?[6]

Once Arjuna stands up he overcomes the grosser part of his "merely personal" anguish. By that act, he breaks the inertial hold of character on action. This prepares him to make genuine choices.

Krishna advises him to "Arise with a brave heart and destroy the enemy." But Arjuna does not *want to* fight. "How can I ever bring myself to fight against Bishma and Drona," he says, "who are worthy of reverence...I don't even know which would be better, for us to conquer them or for them to conquer us... I will not fight".

Krishna proceeds to give him many reasons why he should fight. He argues from reincarnation to show that nobody ever really dies, not friend nor enemy. "As the same person inhabits the body through childhood, youth and old age, so too at the time of death he attains another body. The man of wisdom is not deluded by these changes." (2.13)

He argues that Arjuna's choice to fight, if he truly immerses himself in the flow of action, will generate no new karma.

*The seer says truly*
*That he is wise*
*Who acts without lust or scheming*
*For the fruit of the act:*
*His act falls from him,*
*Its chain is broken,*
*Melted in the flame of my knowledge.*
*Turning his face from the fruit,*
*He needs nothing:*
*The Atman is enough.*
*He acts, and is beyond action.*[7]

This desire to "need nothing" worked for Arjuna. *The Atman is enough.* He is a warrior by caste and experience. He leads his army into battle.

That kind of submission to a social role worked well enough in classical India because personal action fit into an unchanging traditional order of life. You were a soldier, you fought. You were a priest, you officiated. You were a garbage hauler, you hauled garbage. But it doesn't won't work in the West where the culture deeply depends on personal distinction and self-actualization. According to Joseph Campbell,

"In the Indian tradition all has been perfectly arranged from all eternity. There can be nothing new, nothing to be learned but what the sages taught from of yore. And, finally, when the boredom of this nursery horizon of 'I want' against 'thou shalt' has become insufferable, the fourth and final aim is all that is offered – of an extinction of the infantile ego altogether: disengagement or release (moksa) from both 'I' and 'thou.'"

"In the European West, on the other hand, where the fundamental doctrine of the freedom of the will essentially dissociates each individual from every other, as well as from the will in nature and the will of God, there is placed upon each the responsibility of coming intelligently, out of his own experience and volition, to some sort of relationship with – not identity with or extinction in – the all, the void, the suchness, the absolute, or whatever the proper term may be for that which is beyond terms."[8]

As the new human nature emerges in the coming sixth age, we will have both. We will establish a unique relationship with the "suchness" *by becoming our individual selves.* At the same time, our actions will fall cleanly from us, their "chain... broken", because we will yearn more to contribute to the world than to aggrandize ourselves.

That's the Western contribution to world culture. Learn to trust your uniqueness. It is never too late to start. When you establish the focal distance that lets you become the exact lens you are, you enter the flow of life with a keener sensory presence and a more daring address to uncertainty. Even in a turbulent, changing world in crisis your transformational powers will count for something: they will help you get your balance; they will create parity between inwardness and outwardness; they will help strengthen the weak leg (see # **97-99**.) So endowed, you will experience the signal events authentically. Your choices will make you canny and clever without turning you sour. You will find responses to hate and folly. You will counter malicious interference before the ruthless ones take you down.

## 359
### *Odysseus*

Odysseus had been a childhood hero of mine. He united primordial cunning with penetrating judgment. He was brave, willing to go where life took him. I admired his focus on the action at hand; it impressed me how he could use action as a springboard for knowledge. I admired how his power grew, and how he made his grief part of his power. Odyssean character, I believed, made us less self-involved, less self-interested, larger, more primal, and more open to a flowing interplay with events, more able to sail into our turning points in full career. I loved Tennyson's poem on Ulysses:

*I am a part of all that I have met;*
*Yet all experience is an arch wherethrough*

*Gleams that untravelled world, whose margin fades*
*Forever and forever when I move.*[9]

What a shock I got when I learned later that Dante cast Odysseus into Hell. He was thrown there not because of the violent and treacherous parts of his character, but because he gave false counsel, because he didn't stay home after returning to his wife and son, and he persuaded his crewmembers to sail away with him. In the circle of Hell reserved for false counselors, Odysseus tells Dante, how he failed his family duty.

"Not fondness for a son, nor duty to an aged father, nor the love I owed Penelope which should have gladdened her, could conquer within me the passion I had to gain experience of the world and of the vices and worth of men..."[10]

After the Trojan War and his ten-year battle to return home and his campaign to restore his kingdom, he set off again. All along, a deep selfishness dominated him – a need for distinction, excellence, for the accumulation of experience. It made him blind to love, heedless of the continuities in the rhythms of relationship, oblivious to the needs of others and irresponsible toward his subordinates. Why did his wisdom fail in the crucial moments? Because he kept it apart. It did not converge with love.

## 360
### *Love and Wisdom Fighting Back*

I have learned from life that love and wisdom can fight back. They can resist assault. Particularly during turning points in the signal events, they can bring aggression back to its primordial role as a preserver of rhythm. On the strength of aggression alloyed to affection, many lovers, fighting off terrible reverses, surmounting every obstacle, have thrived in nearly impossible conditions. And people have grown wise in the most inhospitable settings too, because the fierceness of their real needs for meaning required satisfaction regardless of conditions, and with fierce intent they commanded the *aggredi* to rise up against injustice. With brilliant means, they confronted opposition.

The aggressive energy you gather up to surmount assaults on rhythmic love and wisdom not only keeps you engaged with the essential meaning-bestowing virtues of life, it helps keep you apt to your times.

The recruitability of aggression, its readiness, its *arousability,* its quick bifurcation from the stress response, thrusts you into the moment (see # **133, 135, 154, 155**.) To be apt in love is to be truly there for another person *now*, concretely, not in the sweet by and by. To love, care, and never give up demonstrates the true Spinozan impulse to self-preservation; it asserts our need for others (see # **234**.) Aggression works in the service of wisdom too. It's there when you resist the urge to throw aside a hard project. It gives you staying power. It keeps you from following false hopes. It rejects mediocrity.

To stand always tentative and open on the path of possibility, and to be willing to countenance aggression in the oneness of your life while risking its abuses in an imperfect world imperfectly understood - that takes something more than courage, nerve, judgment and aptness. *It takes the desire to win.*

### 361

Fighting back is one thing; winning is another. To win the historical struggles in a Dukkha laden world where nobody wins, your deeds of love and wisdom have to do more than express themselves authentically in small settings: they have to enter the wider struggle and transform territorial and dominance relationships. This they can only do when many people who have been driven into their depths or intimacies by related psychological and socio-economic problems return to social life bearing common solutions based on shared views of their human potentials and powers. In the return stage of wisdom, they find each other. In the approach stage of love they discover mutual commitments. Many important changes fly in under the radar when cultures falter and fail. Sometimes in perilous circumstances the resonant turnings of many people have synergistic effects. When anonymity is joined to anonymity moral currents are generated that change history.

But how do you get there from here?

## 362
### *Keep On Keeping On*

Look closely and you will see that in all of its present inadequacy love and wisdom are still flowing in your life. Somewhere you are attempting to love and trying to get wise. Wherever love is flowing, keep it flowing. Wherever wisdom seeks for meaning, keep it searching. Know that you are already on the paths, and because you are, even if you are floundering and flailing there, even if you do not know what you are seeking, or where you are going, keep on. Keep on keeping on. You are heading toward transformational opportunities.

By shedding egotism in favor of deeds, and for letting those deeds cut the path to individuation itself, you get the power to act in the moment according to its needs. With aptness, deeds find their own way. However, you can only live a life of full engagement and follow this strategy if you have developed the skills to recognize a "me" who is there when the "I" has blown apart, a "me" of small moves and little hopes and finite interests who has shed the illusions of reference to a permanent self. To find this "me" takes secular spiritual attainment.

Living in the moment without an overweening self, we become clearer valves for giving and receiving. Deeds performed this way require great finesse, and precise timing, since the assertion of a clear intention timed to the turning moment must be delicate and strong and apt enough to meet and join the world's own energies. When they do, it feels as if volition is moving from the world to the world, finding its own way through the open valve of you.

Only when we have this strong delicacy, this resonant rhythmic precision, do we achieve poise in the heart of time. Our deeds then fall clear of the doer.

## 363

Nature ordains the value of life to be the full play of our propensities. The primordial center of each living being is uniquely tuned and differently sensitized to input. We are peerless transmitting/receiving devices. You will find your unique ways and I will find mine and we will help each other.

Only this courageous openness leads to the thorough exploration of our genome and to the fullest possible expression of our potentialities. Our real task, then, is to struggle imperfectly with love and wisdom and to individuate, not discouraged by our temporal limits.

It is good to touch the primordial core of our being. The deeds of love and wisdom that start there awaken to the needs of others. More beings surround us. More beings bring more life.

Doing the best we can from the center of our uniqueness, our journeys of wisdom will produce innovations that will move us toward freedom and justice. And from our loves will come the strength to care. And our caring will stir caring in others. And caring establishes a deep communion with nature.

When the rhythms of love and wisdom converge in our own lives, winding about each other durably, with aggression the third in the threefold cord, we begin to access the natural sources of meaning through which rectification of social injustices occur. With this impetus, we can organize our aspirations into something truly great.

Stephen Spender wrote:

*I think continually of those who were truly great*
*Who, from the womb, remembered the soul's history*
*Through corridors of light where the hours are suns,*
*Endless and singing. Whose lovely ambition*
*Was that their lips, still touched with fire,*
*Should tell of the spirit clothed from head to foot in song.*
*And who hoarded from the spring branches*
*The desires falling across their bodies like blossoms.*

## 364

The deed carries the mind into the world better than a thousand volumes of philosophical reflection. By living in rhythmic congruity with nature, we gain the strength to be ourselves, and when we are ourselves we see that what really matters beyond the achievements and disappointments of life are the deeper rhythms of love and wisdom. We weave them as they weave us. We are our own flying shuttles. The patterns that take hold in the weft and weave are those we have earned or spoiled by living. You only get the gifts of self-realization in the moments you give them away. That's the nature of true bestowal. The spoiled efforts you cannot give away. They go with you.

Reality offers startlingly more than we know or can anticipate. And sometimes it delivers crushingly less than we merit. As lived, however, it is perpetually new. This newness brings with it an uncertainty that, if we are brave, is enlivening rather than debilitating, amusing rather than terrifying, inspiring rather than deflating.

Of course, we can make a mess of it. We are only human. We are never fully formed. We don't really know how to handle ourselves because we don't quite know what is happening or where we are going. Fortune can come bounding in with hard surprises. We can get sick or die too soon. We can lose our bets. Or we can make fine choices and life can still turn out badly. There are no guarantees, not for outcomes or intentions.

Yet we must live in the moment before judging it, because doing precedes knowing. To live well we must say Yes before we say No. We need to welcome the moment - this one, the next one, the last. We learn by living, and since we have hardly engaged the fullness of our capacities yet, we do not know how much there is yet to learn, or whether what we learn will hold up or whether we have gone too far or not far enough. No wisdom holds forever. Our understandings and our motivations vary, develop, evolve, and pass away. I wrote this in a poem when I was thirty:

*We change and our changings change*
*And we leave behind us a refuse of small wisdoms*
*As the price we pay*
*To be present at our turnings.*

That is my best understanding of the way to freedom! Would that I could practice it better myself! The risks get higher as life goes on, perhaps because there is less time left to rectify mistakes, and more regrets, and one's energy diminishes. But the greatest risk and the major drain on our energy is not taking the risk.

### 365

Our saving grace is that transformation is always possible. Even up to the last breath in the final moment giving can become receiving. One kind of love can turn into another. Wisdom can refresh itself in other wisdoms. Love and wisdom can converge on each other.

Once you start making the changes, you engage the deep turnings. Then the past comes back to you in a new way, it make sense in ways you never suspected. The mistakes you made were not for nothing. The past can be redeemed. Doors open. You can call forth more life.

### *The end*

### *End Notes*

[1] Dhammacakkappavattanutta
[2] Surangama Sutra in Buddhist Bible p. 220
[3] Quoted in Eswaren. Gandhi the Man. Nigiri Press. 1978. p.110
[4] Prmj 1 p.78
[5] Chogyam Trungpa. *Shambala.* Shambala books. 1984. p.106-109
[6] Baghavad Gita I.2, Easwaren translation, p. 49 vol. 1 *The End of Sorrow*
[7] Baghavad Ghita Chap IV. Translated by Swami Pravananda and Christopher Isherwood
[8] Joseph Campbell. The Masks of God: Oriental Mythology. p.22
[9] Tennyson. *Ulysses*
[10] Dante. *Inferno.* Canto 26. John D. Sinclair Translation

# List of Illustrations

P. 129 Yin/Yang with Trigrams

P. 133 Blake Jerusalem etching

P. 137 Ixion. Ixion's Torture, Amphora c. 330 BC Staatliche Museen, Berlin, Germany

P. 150 Photo, 1970

P. 162 Graylag Goose display. Courtesy Konrad Lorenz. On Aggression.

P. 207 Lascaux cave painting

P. 208 Lascaux Bird Shaman

P. 209 Laura Rose drawing

P. 211 Abu Hureya. www.edunetconnect.com

P. 232 Minoan Women. Fresco. Palace at Knossos.

P. 267 Triune brain. Courtesy, P. D. Maclean, 1990

P. 294-297 My drawings

P. 298 Bifurcation tree. Computer generation by Conrad Schneiker. In Hameroff. quantumconsciousness.org

P. 298 Japanese Cherry Blossoms Koganei. Hiroshige Utagawa. 1843-1847

P. 301 Acceleration graphic representing Kurzweil's Law of Accelerated Returns

P. 313 Spinoza's microscope lens design. Courtesy Kvond.wordpress.com

P. 320, 321 My drawings

P. 345 Mesmer's tub. Courtesy www.deeptrancenow.com/gallery.htm

P. 393 Hubble photograph. 2005. Starv838 Monocerotis dust clouds

P. 393 Phi of seashell

P. 394 Alchemy. Etching. From Michael Maier's Atalanta Fugiens

P. 395 Japanese nursing mother

P. 397 Beauty/Crone illusion

P. 401 Chinese landscape painting.

P. 407 Da Vinci Deluge. Notebooks.

P. 419 Flammarian Woodcut

Back Cover portrait by Laura Rose

# INDEX

# A

Abandonment, 261
Abu Hureya, 227
acceleration, 216, 232, 241, 288
acceptance, **407**
ADDICTION, **240**
aesthetic continuum, 380
aesthetic sensibility, 371
Agape, 61, 372
aggredi, 187, 299, 398
aggregation and dispersal, 39, 82
aggression, 233, 254, 293, 418
agon, 166
Agricultural Revolution, 201, 234
ah, hah! experience, 321
altruism, 64, 136, 137, 154, 185, 207, 213, 242, 244, 268, 412
amino acids, 36, 37, 39
amygdala, 258
animal masters, 200
anxiety, 77, 109
appetitive aggression, 154
approach and separation, 6, 41, 146
Arendt, Hannah, 167
Aristotle, 129
Aryans, 248
Atlantic conveyor, 202
Atlantis, 248
atomic individualism, 285
atttunment, 8
Auden, W. H., 359
Auschwitz, 185
authenticity, 80
autism, 236
autistic disorders, 257
autocatalytic reactions, 30
autonomic nervous system, 68, 78

# B

Bacon, Francis, 376
Baghavad Gita, 415
Baldwin, James, 354
band life, 48, 82, 196, 204
Bar Yosef, Ofer, 201
Barrett, William, 390
base two mathematics, 286
beat frequencies, 14
beauty, 301, 369
Begley, Sharon, 330
belonging, 54, 165, 167, 254, 279, **341**
Bentov, Itzhak, 37
Berry, Wendell, 238
bifurcation, 127, 144
biodiversity, 375
biofeedback, 301, 333, 335, 400
bio-oscillating systems, 286
Biophysics, 10
biotechnology, 399
Birth, 82
birth trauma, 51
black experience, 358
Blake, William, 159, 231
blood pressure., 336
Bloom, Harold, 388
Bloom, Paul, 185
*Book of Balance and Harmony*, 50, 308
boom times, 202
brain waves, 20, 23
brain., 205
Braudel, Fernand, 281
Brave New World, 245
breath work, 333
Buddha, 407, 408

# C

Cabalism, 87, 351
calcium oscillations, 19
Calcium oscillations, 18

www.ingramcontent.com/pod-product-compliance
Lightning Source LLC
Chambersburg PA
CBHW031805190326
41518CB00006B/201